Lecture Notes in Physics

Managing Editor

W. Beiglböck
Assisted by Mrs. Sabine Landgraf
c/o Springer-Verlag, Physics Editorial Department II
Tiergartenstrasse 17, D-69121 Heidelberg, Germany

The Editorial Policy for Proceedings

The series Lecture Notes in Physics reports new developments in physical research and teaching – quickly, informally, and at a high level. The proceedings to be considered for publication in this series should be limited to only a few areas of research, and these should be closely related to each other. The contributions should be of a high standard and should avoid lengthy redraftings of papers already published or about to be published elsewhere. As a whole, the proceedings should aim for a balanced presentation of the theme of the conference including a description of the techniques used and enough motivation for a broad readership. It should not be assumed that the published proceedings must reflect the conference in its entirety. (A listing or abstracts of papers presented at the meeting but not included in the proceedings could be added as an appendix.)

When applying for publication in the series Lecture Notes in Physics the volume's editor(s) should submit sufficient material to enable the series editors and their referees to make a fairly accurate evaluation (e.g. a complete list of speakers and titles of papers to be presented and abstracts). If, based on this information, the proceedings are (tentatively) accepted, the volume's editor(s), whose name(s) will appear on the title pages, should select the papers suitable for publication and have them refereed (as for a journal) when appropriate. As a rule discussions will not be accepted. The series editors and Springer-Verlag will normally not interfere with the detailed editing except in fairly obvious cases or on technical matters.

Final acceptance is expressed by the series editor in charge, in consultation with Springer-Verlag only after receiving the complete manuscript. It might help to send a copy of the authors' manuscripts in advance to the editor in charge to discuss possible revisions with him. As a general rule, the series editor will confirm his tentative acceptance if the final manuscript corresponds to the original concept discussed, if the quality of the contribution meets the requirements of the series, and if the final size of the manuscript does not greatly exceed the number of pages originally agreed upon. The manuscript should be forwarded to Springer-Verlag shortly after the meeting. In cases of extreme delay (more than six months after the conference) the series editors will check once more the timeliness of the papers. Therefore, the volume's editor(s) should establish strict deadlines, or collect the articles during the conference and have them revised on the spot. If a delay is unavoidable, one should encourage the authors to update their contributions if appropriate. The editors of proceedings are strongly advised to inform contributors about these points at an early stage.

The final manuscript should contain a table of contents and an informative introduction accessible also to readers not particularly familiar with the topic of the conference. The contributions should be in English. The volume's editor(s) should check the contributions for the correct use of language. At Springer-Verlag only the prefaces will be checked by a copy-editor for language and style. Grave linguistic or technical shortcomings may lead to the rejection of contributions by the series editors. A conference report should not exceed a total of 500 pages. Keeping the size within this bound should be achieved by a stricter selection of articles and not by imposing an upper limit to the length of the individual papers. Editors receive jointly 30 complimentary copies of their book. They are entitled to purchase further copies of their book at a reduced rate. As a rule no reprints of individual contributions can be supplied. No royalty is paid on Lecture Notes in Physics volumes. Commitment to publish is made by letter of interest rather than by signing a formal contract. Springer-Verlag secures the copyright for each volume.

The Production Process

The books are hardbound, and the publisher will select quality paper appropriate to the needs of the author(s). Publication time is about ten weeks. More than twenty years of experience guarantee authors the best possible service. To reach the goal of rapid publication at a low price the technique of photographic reproduction from a camera-ready manuscript was chosen. This process shifts the main responsibility for the technical quality considerably from the publisher to the authors. We therefore urge all authors and editors of proceedings to observe very carefully the essentials for the preparation of camera-ready manuscripts, which we will supply on request. This applies especially to the quality of figures and halftones submitted for publication. In addition, it might be useful to look at some of the volumes already published. As a special service, we offer free of charge LaTeX and TeX macro packages to format the text according to Springer-Verlag's quality requirements. We strongly recommend that you make use of this offer, since the result will be a book of considerably improved technical quality. To avoid mistakes and time-consuming correspondence during the production period the conference editors should request special instructions from the publisher well before the beginning of the conference. Manuscripts not meeting the technical standard of the series will have to be returned for improvement.

For further information please contact Springer-Verlag, Physics Editorial Department II, Tiergartenstrasse 17, D-69121 Heidelberg, Germany

J. M. Arias   M. I. Gallardo   M. Lozano   (Eds.)

Response of the Nuclear System to External Forces

Proceedings of the V La Rábida International
Summer School on Nuclear Physics
Held at La Rábida, Huelva, Spain
19 June - 1 July 1994

 Springer

Editors

José Miguel Arias
María Isabel Gallardo
Manuel Lozano
Departamento FAMN, Facultad de Física
Universidad de Sevilla, Aptdo 1065
E-41080 Sevilla, Spain

ISBN 3-540-59007-2 Springer-Verlag Berlin Heidelberg New York

CIP data applied for

Typesetting: Camera-ready by the editors
SPIN: 10127105   55/3142-543210 - Printed on acid-free paper

Preface

The V La Rábida International Summer School, whose proceedings are contained in this book, was entitled "Response of the Nuclear System to External Forces".

Subjects at the forefront of nuclear research, bordering other areas of many–particle physics, such as electron scattering at different energy scales, new physics with radioactive beams, multifragmentation, relativistic nuclear physics, high spin nuclear problems, chaos, the role of the continuum in nuclear physics, or recent calculations with the shell model were presented to a dynamic group of young nuclear physicists. The meeting provided the audience with an opportunity to discuss and assess their feelings about the main future lines of development of nuclear physics.

We would like to express here our deep gratitude to all the professors who kindly accepted our invitation to lecture at this summer school. Without their friendly and continuous collaboration the school and this book would have been impossible. They made an important effort in presenting their talks in a pedagogical way not only at the school but also in their written version presented in this book. Their accesibility and disponibility during the school facilitated informal discussions with students in the rest hours in extra "beach" and "bar" sessions. Special thanks are given to the students. They, together with the lecturers, contributed to give the school an excellent, from the academic point of view, and a warm, from the personal relationship side, atmosphere.

From the less "poetic", but very important, financial point of view, this school would not have been possible without substancial support from several Spanish organizations. First, the School was held at the Universidad Internacional de Andalucía, Sede Iberoamericana de Santa María de La Rábida. We are really indebted to the heads of this University, Prof. J. Marchena and Prof. E. Garzón, for their economical help as well as for their cooperation in organizative aspects of the school. Our gratitude goes to them and the whole staff of this University. Second, the national Spanish government, through the Dirección General de Investigación Científica y Técnica (DGICyT), and the regional government, through the Consejería de Educación y Ciencia de la Junta de Andalucía, gave the school financial support. We would like to acknowledge them for their attention even in these economicaly bad times. Special thanks are given to the Dean of the Physics Faculty of the University of Sevilla, Prof. Antonio Córdoba, for his kind help when it was needed. Finally we would like to thank the University of Sevilla, through its Vicerrectorado de Extensión Universitaria, and the bank Caja de Ahorros y Monte de Piedad de Huelva y Sevilla, for their financial assistance.

This edition of the school, as all the previous ones, was benefited by the enthusiastic cooperation of our collaborators from the Departamento de Física Atómica, Molecular y Nuclear of the University of Sevilla before, during and after the school. Our gratitude to the administrative personal of our Department José Díaz, Chari González and Charo Cadierno for their continuous assistance in the organization of the School.

Sevilla, Spain   J.M. Arias
October 1994   M.I. Gallardo
                                                                                      M. Lozano

Contents

● Electron Scattering 1
  J.D. Walecka

● Elementary Nuclear Excitations Studied with Electromagnetic
  and Hadronic Probes 13
  A. Richter

● Probing Nucleon and Nuclear Structure with High-Energy
  Electrons 61
  B. Frois

● Relativistic Theory of the Structure of Finite Nuclei 95
  P. Ring

● Semiclassical Description of the Relativistic Nuclear Mean
  Field Theory 115
  X. Viñas

● Photonuclear Reactions 131
  E. Oset

● Notes on Scaling and Critical Behaviour in Nuclear
  Fragmentation 153
  X. Campi and H. Krivine

● The Continuum in Nuclei 181
  R.J. Liotta

● Spherical Shell Model, a Renewed View 195
  A. Poves

● High Spins and Exotic Shapes 211
  S. Åberg

● Heavy Ion Scattering Problems; Regular and Chaotic Regimes 231
  C.H. Dasso, M. Gallardo and M. Saraceno

- Deterministic Chaos in Heavy-Ion Reactions 251
  M. Baldo, E.G. Lanza and A. Rapisarda

- Nuclear Level Repulsion; Order vs. Chaos and Conserved
  Quantum Numbers 263
  J.D. Garrett, J.R. German and J.M. Espino

- Nuclear Physics and Nuclear Astrophysics with Radioactive
  Nuclear Beams 273
  J. Vervier

- List of Participants 289

Electron Scattering

John Dirk Walecka

College of William and Mary
Williamsburg, Virginia, 23187
CEBAF
Newport News, Virginia, 23606

These lectures are divided into two parts. First, I will give an elementary introduction to electron scattering. This material is based on lectures I gave last summer at the 6th Annual Summer School in Nuclear Physics Research held at North Carolina State University in Raleigh, North Carolina. I have written up the first part for the proceedings of this school. Then I will give an overview of the present status of electron scattering, including a description of CEBAF. The second part is based on two talks I gave recently at conferences. Since the material in the second part appears in the published literature in Refs. [1, 2], I will simply refer students to that published material.

1 Electromagnetic Interactions

Non-Relativistic Scattering of a Charged Lepton - Born Approximation. Suppose one scatters a non-relativistic lepton of charge ze_p with $z = \pm1$ from the nucleus. The interaction takes place through the Coulomb potential

$$V(x) = \frac{ze^2}{4\pi} \int \frac{1}{|\mathbf{x} - \mathbf{x}'|} \rho_N(\mathbf{x}') d^3 x' \tag{1}$$

The scattering amplitude is given in first Born Approximation by

$$f_{\text{B.A.}}(\mathbf{k}', \mathbf{k}) = -\frac{2\mu}{4\pi\hbar^2} \int e^{-i\mathbf{q}\cdot\mathbf{x}} V(x) d^3 x \tag{2}$$

Here μ is the reduced mass, $\hbar\mathbf{k}$ is the inital momentum, $\hbar\mathbf{k}'$ is the final momentum, and $\hbar\mathbf{q}$ with $\mathbf{q} = \mathbf{k}' - \mathbf{k}$ is the three-momentum transfer whose magnitude is given for elastic scattering by $q^2 = 2k^2(1 - \cos\theta) = 4k^2 \sin^2\theta/2$. For a spherically symmetric nuclear charge density $\rho_N(x)$, the Fourier transform of the potential in Eq. (1) yields[1]

$$\int e^{-i\mathbf{q}\cdot\mathbf{x}}\frac{1}{|\mathbf{x}-\mathbf{x}'|}\rho_N(\mathbf{x}')d^3x d^3x' = \frac{4\pi}{\mathbf{q}^2}\int e^{-i\mathbf{q}\cdot\mathbf{y}}\rho_N(y)d^3y$$

$$= \frac{4\pi}{\mathbf{q}^2}F(\mathbf{q}^2) \qquad (3)$$

Here $F(\mathbf{q}^2)$ is the nuclear "form factor". Now use $e^2/\hbar c = 4\pi\alpha$ where $\alpha \approx 1/137.0$ is the fine-structure constant. The differential cross section then follows from the square of the modulus of the scattering amplitude as

$$\frac{d\sigma}{d\Omega} = \frac{4\mu^2}{\hbar^4}(\hbar c\alpha)^2\frac{1}{\mathbf{q}^4}|F(\mathbf{q}^2)|^2$$

$$= \frac{(\hbar c\alpha)^2}{16E_0^2\sin^4\theta/2}|F(\mathbf{q}^2)|^2$$

$$= \sigma_{\text{Rutherford}}|F(\mathbf{q}^2)|^2 \qquad (4)$$

Here $E_0 = \hbar^2 k^2/2\mu$ is the incident energy and $\sigma_{\text{Rutherford}}$ is the familiar cross section for scattering from a point charge. Experimental measurement of this cross section evidently determines the Fourier transform of the nuclear charge density[2]

$$F(\mathbf{q}^2) = \int e^{-i\mathbf{q}\cdot\mathbf{y}}\rho_N(y)d^3y$$

$$= \int \frac{\sin qy}{qy}\rho_N(y)d^3y \qquad (5)$$

Note that $F(0) = Z$, the total nuclear charge.

Nuclear Physics . Suppose now that one extends the analysis to deal with the internal quantum dynamics of the nuclear target. The nuclear charge density then becomes an operator in the nuclear Hilbert space

$$\rho_N(\mathbf{x}) \rightarrow \hat{\rho}_N(\mathbf{x}) ; \qquad \text{Nuclear Density Operator} \qquad (6)$$

In first quantization, for example, with a collection of structureless nucleons, the nuclear density operator takes the form

$$\hat{\rho}_N(\mathbf{x}) = \sum_{j=1}^{Z}\delta^{(3)}(\mathbf{x}-\mathbf{x}_j) \qquad (7)$$

The analysis of the scattering amplitude in Eq. (3) indicates that one now requires the nuclear transition matrix elements of the operator

$$\hat{F}(\mathbf{q}) = \int e^{-i\mathbf{q}\cdot\mathbf{y}}\hat{\rho}_N(y)d^3y \qquad (8)$$

---

[1] Use $\int e^{-i\mathbf{q}\cdot\mathbf{x}}(e^{-\lambda x}/x)d^3x = 4\pi/(\mathbf{q}^2 + \lambda^2)$; now let $\lambda \rightarrow 0$.

[2] One actually measures the square of the modulus of the form factor, but since it is real here and $F(0) = Z$, one can track through the zeros and determine both the sign and magnitude.

Take the momentum transfer **q** to define the z-axis and expand the plane wave appearing in this expression according to

$$e^{-i\mathbf{q}\cdot\mathbf{x}} = \sum_{J=0}^{\infty} \sqrt{4\pi(2J+1)}(-i)^J j_J(qx)Y_{J0}(\Omega_x) \qquad (9)$$

This gives

$$\hat{F}(\mathbf{q}) = \sum_{J=0}^{\infty} \sqrt{4\pi(2J+1)}(-i)^J \hat{M}_{J0}(q)$$

$$\hat{M}_{JM}(q) \equiv \int j_J(qx)Y_{JM}(\Omega_x)\hat{\rho}_N(\mathbf{x})d^3x \qquad (10)$$

The quantities $\hat{M}_{JM}(q)$ are now *irreducible tensor operators* (ITO) in the nuclear Hilbert space. The general proof depends on the fact that the nuclear density is a scalar under rotations; in the case where Eq. (7) holds, these multipoles consist of a sum of single particle radial functions multiplied by spherical harmonics and the result is evident. The great advantage of identifying an ITO is that one can now use the general theory of angular momentum (Ref. [3]), in particular the Wigner-Eckart theorem states that the matrix element of an ITO taken between nuclear eigenstates of angular momentum results in[3]

$$\langle J_f M_f|\hat{M}_{JM}|J_i M_i\rangle = (-1)^{J_f - M_f} \begin{pmatrix} J_f & J & J_i \\ -M_f & M & M_i \end{pmatrix} \langle J_f||\hat{M}_J||J_i\rangle \qquad (11)$$

This result has two invaluable features: it gives the explicit dependence on the nuclear orientation (all M's), and it contains the angular momentum selection rules (the J's must satisfy the triangle inequality). The average over initial states and sum over final states $\overline{\sum_i}\sum_f$ for a nuclear transition to a discrete state (this can include elastic scattering) is then immediately performed using the orthogonality property of the 3-j symbols

$$\frac{1}{2J_i+1}\sum_{M_i}\sum_{M_f} \begin{pmatrix} J_f & J & J_i \\ -M_f & M & M_i \end{pmatrix} \begin{pmatrix} J_f & J' & J_i \\ -M_f & M' & M_i \end{pmatrix}$$

$$= \frac{1}{2J_i+1}\frac{1}{2J+1}\delta_{JJ'}\delta_{MM'} \qquad (12)$$

Hence the nuclear physics is now contained in the following expression

$$\overline{\sum_i}\sum_f |\langle J_f M_f| \int e^{-i\mathbf{q}\cdot\mathbf{y}}\hat{\rho}_N(\mathbf{y})d^3y|J_i M_i\rangle|^2 = \frac{4\pi}{2J_i+1}\sum_{J=0}^{\infty} |\langle J_f||\hat{M}_J(q)||J_i\rangle|^2 \qquad (13)$$

---

[3]We assume here that the nuclear target is heavy and localized and that the nuclear eigenstates can be characterized by their angular momentum.

This sum is actually finite since the nuclear matrix elements vanish unless the selection rules are satisfied.

Relativistic (Massless) Electrons . The cross section for the scattering of relativistic (massless) electrons through the Coulomb interaction can now be obtained from the previous results through the following modifications:

1) Replace the transition matrix element for the projectile $e^{-i\mathbf{q}\cdot\mathbf{x}}$ by $e^{-i\mathbf{q}\cdot\mathbf{x}}$ $u^\dagger(\mathbf{k}')u(\mathbf{k})$ which includes the overlap of the Dirac spinors for the electron.

A simple calculation with the Dirac wave functions then gives (Ref. [4])[4]

$$\frac{1}{2}\sum_{s_1}\sum_{s_2}|u^\dagger(\mathbf{k}')u(\mathbf{k})|^2 \;=\; \frac{1}{2}(1+\cos\theta)=\cos^2\frac{\theta}{2} \tag{14}$$

2) Replace μc^2 in the numerator of the scattering amplitude by the full final electron energy $\hbar k'c$; this factor arises from the appropriate incident flux and density of final states in Fermi's Golden Rule.

3) Make use of the four-momentum transfer $q_\mu^2 = \mathbf{q}^2 - q_0^2$ where $q_0 = k' - k$ to write the point cross section. This quantity satisfies

$$q_\mu^2 = 4kk'\sin^2\frac{\theta}{2} \tag{15}$$

for both elastic and inelastic transitions.

The resulting differential cross section then takes the form

$$\frac{d\sigma}{d\Omega} \;=\; \sigma_{\text{Mott}}\frac{q_\mu^4}{q^4}\frac{4\pi}{2J_i+1}\sum_{J=0}^{\infty}|\langle J_f||\hat{M}_J(q)||J_i\rangle|^2$$

$$\sigma_{\text{Mott}} \;\equiv\; \frac{\alpha^2\cos^2\theta/2}{4k^2\sin^4\theta/2} \tag{16}$$

Here σ_{Mott} is the cross section for scattering a Dirac electron from a fixed, point charge. Note that this quantity can also be written

$$\sigma_{\text{Mott}} = \frac{4\alpha^2 k'^2\cos^2\theta/2}{q_\mu^4} \tag{17}$$

---

[4] Use

$$\frac{1}{2}\sum_{s_1}\sum_{s_2}|\bar{u}(\mathbf{k}')\gamma_4 u(\mathbf{k})|^2 \;=\; \frac{1}{2}\frac{1}{4kk'}Tr\gamma_4(-i\gamma_\mu k_\mu)\gamma_4(-i\gamma_\nu k'_\nu)$$

$$=\; -\frac{1}{2kk'}(2k_4 k'_4 - k_\mu k'_\mu)$$

$$=\; \frac{1}{2kk'}(\mathbf{k}\cdot\mathbf{k}' + kk') = \frac{1}{2}(1+\cos\theta)$$

Long-Wavelength Limit. In the limit that the momentum transfer goes to zero, an expansion of the spherical Bessel functions reduces the multipole operators in Eq. (10) to the form

$$\hat{M}_{JM}(q) \rightarrow \frac{q^J}{(2J+1)!!} \int x^J Y_{JM}(\Omega_x)\hat{\rho}_N(\mathbf{x})d^3x \tag{18}$$

If R characterizes the size of the nuclear target, then the multipole operators go as $(qR)^J$ and the lowest allowed multipole dominates in the limit $qR \rightarrow 0$. Recall it is a property of Fourier transforms that the equivalent wavelength at which the system is examined bears an inverse relation to the momentum transfer

$$|\mathbf{q}| \equiv \frac{2\pi}{\lambda} \tag{19}$$

In electron scattering, the product qR can be made arbitrarily large by going first to larger scattering angles at fixed incident energy, and then by going to higher and higher energy electrons.

Gamma Decay. Consider a nuclear transition $|J_iM_i\rangle \rightarrow |J_fM_f\rangle$ with the emission of a photon. The hamiltonian governing this electromagnetic process is

$$H' = -\frac{e_p}{c} \int \hat{\mathbf{J}}_N(\mathbf{x}) \cdot \mathbf{A}(\mathbf{x})d^3x$$

$$\mathbf{A}(\mathbf{x}) = \sum_{\mathbf{k}} \sum_{\lambda=1,2} \left(\frac{\hbar c^2}{2\omega_k \Omega}\right)^{1/2} (a_{\mathbf{k}\lambda}\mathbf{e}_{\mathbf{k}\lambda}e^{i\mathbf{k}\cdot\mathbf{x}} + \text{h.c.}) \tag{20}$$

Here $\mathbf{A}$ is the vector potential for the quantized radiation field and the hamiltonian is written in the Schrödinger picture. In this expression $\mathbf{e}_{\mathbf{k}1,2}$ are a set of unit vectors orthogonal to $\mathbf{k}$, $\omega_k = kc$, $a^\dagger$ (a) are the creation (destruction) operators for the photons, and we use periodic boundary conditions in a big box of volume Ω.

It is convenient to first make a canonical transformation to photon states with circular polarization. This leads to an expression for the vector potential where one now replaces $\sum_{\lambda=1,2} \rightarrow \sum_{\lambda=\pm1}$ with $\mathbf{e}_{\mathbf{k},\pm1} \equiv \mp(\mathbf{e}_{\mathbf{k}1} \pm i\mathbf{e}_{\mathbf{k}2})/\sqrt{2}$. The nuclear matrix element for photoemission then takes the form

$$H'_{fi} = -\frac{e_p}{c}\left(\frac{\hbar c^2}{2\omega_k \Omega}\right)^{1/2} \langle f| \int \mathbf{e}^\dagger_{\mathbf{k}\lambda}e^{-i\mathbf{k}\cdot\mathbf{x}} \cdot \hat{\mathbf{J}}_N(\mathbf{x})d^3x|i\rangle \tag{21}$$

Now introduce the following expansion for the plane wave times the unit vector (Ref. [3, 5])

$$\mathbf{e}^\dagger_{\mathbf{k}\lambda}e^{-i\mathbf{k}\cdot\mathbf{x}} =$$

$$-\sum_{J\geq1}\sqrt{2\pi(2J+1)}(-i)^J\left\{\frac{1}{k}\nabla \times [j_J(kx)\mathbf{\mathcal{Y}}^{-\lambda}_{JJ}] + \lambda j_J(kx)\mathbf{\mathcal{Y}}^{-\lambda}_{JJ}\right\} \tag{22}$$

Here the vector spherical harmonics are defined by ($e_{k0} \equiv k/|k|$)

$$\mathcal{Y}_{lJ}^{M} \equiv \sum_{m_l m_s} \langle l m_l 1 m_s | l 1 J M \rangle Y_{l m_l}(\Omega_x) e_{m_s} \tag{23}$$

Equation (22) is simply an algebraic identity. Its great utitlity lies in the fact that it allows one to again make an expansion of the required nuclear transition operator in ITO

$$\int e_{k\lambda}^{\dagger} e^{-i k \cdot x} \cdot \hat{\mathbf{J}}_N(x) d^3 x =$$
$$- \sum_{J \geq 1} \sqrt{2\pi(2J+1)}(-i)^J [\hat{T}_{J,-\lambda}^{\text{el}}(k) + \lambda \hat{T}_{J,-\lambda}^{\text{mag}}(k)] \tag{24}$$

The electric and magnetic multipole operators are defined by

$$\hat{T}_{JM}^{\text{el}}(k) = \frac{1}{k} \int \{\nabla \times [j_J(kx)\mathcal{Y}_{JJ}^{M}(\Omega_x)]\} \cdot \hat{\mathbf{J}}_N(x) d^3 x$$
$$\hat{T}_{JM}^{\text{mag}}(k) = \int [j_J(kx)\mathcal{Y}_{JJ}^{M}(\Omega_x)] \cdot \hat{\mathbf{J}}_N(x) d^3 x \tag{25}$$

The decay rate now follows from Fermi's Golden Rule

$$d\omega_{fi} = \frac{2\pi}{\hbar}|H'_{fi}|^2 \delta(E_f + \hbar\omega_k - E_i)\frac{\Omega d^3 k}{(2\pi)^3} \tag{26}$$

Since the electromagnetic multipoles have opposite parity, it follows that the good parity of the nuclear states implies

$$|\langle J_f||\hat{T}_J^{\text{el}} + \lambda\hat{T}_J^{\text{mag}}||J_i\rangle|^2 = |\langle J_f||\hat{T}_J^{\text{el}}||J_i\rangle|^2 + |\langle J_f||\hat{T}_J^{\text{mag}}||J_i\rangle|^2 \tag{27}$$

We leave it as an exercise for the reader to show that a combination of the above results leads to the following expression for the decay rate for photon emission

$$\omega_{fi} = 8\pi\alpha k c \frac{1}{2J_i + 1} \sum_{J=1}^{\infty} (|\langle J_f||\hat{T}_J^{\text{el}}(k)||J_i\rangle|^2 + |\langle J_f||\hat{T}_J^{\text{mag}}(k)||J_i\rangle|^2) \tag{28}$$

In fact, this is a general expression for the decay rate for photon emission for any heavy, localized quantum mechanical system; it is exact to order α. The multipole operators appearing in this expression now contain a factor c^{-1} and are dimensionless.

2   Electron Scattering

The amplitude for the scattering of a relativistic electron from a nuclear target can be calculated to order α in time-independent perturbation theory by combining the first order Coulomb amplitude arising from Eq. (1) with the second order amplitude for the exchange of a transverse photon of momentum $\hbar q$ coming from Eq. (20) (and its analog for the electron). Since the Coulomb and transverse multipoles carry different amounts of angular momentum along the q axis, they do not interfere after the sum and average over nuclear orientations. It should therefore not be too surprising that the differential cross section can be written in the following form (see e.g. Ref. [6])

$$\frac{d\sigma}{d\Omega} = 4\pi\sigma_{\text{Mott}}\frac{1}{2J_i + 1}\left\{\frac{q_\mu^4}{\mathbf{q}^4}\sum_{J=0}^{\infty}|\langle J_f||\hat{M}_J(q)||J_i\rangle|^2\right.$$

$$\left. + \left(\frac{q_\mu^2}{2\mathbf{q}^2} + \tan^2\frac{\theta}{2}\right)\sum_{J=1}^{\infty}(|\langle J_f||\hat{T}_J^{\text{el}}(q)||J_i\rangle|^2 + |\langle J_f||\hat{T}_J^{\text{mag}}(q)||J_i\rangle|^2)\right\} \quad (29)$$

Several features of this result are of interest:

- The nuclear matrix elements obey all the selection rules discussed above; in particular they vanish unless $J_f + J_i \geq J \geq |J_f - J_i|$.

- Because of the unit helicity of the photon, the sum over the transverse multipoles starts with $J = 1$; in contrast, there is a $J = 0$ Coulomb monopole.

- The momentum transfer $\hbar|\mathbf{q}|$ can take any value in electron scattering.

- There are 3 lepton variables in electron scattering (k, k', θ) or equivalently (q^2, ω, θ) where the energy transfer $\hbar\omega$ is given by $\omega/c \equiv k - k'$. The Coulomb contribution and that arising from transverse photon exchange can be separated by keeping the first two variables (q^2, ω) fixed and varying the electron scattering angle θ, or by working at $\theta = 180^o$ where only the transverse term contributes.

- It has been assumed here that the nucleus is heavy and this is the laboratory cross section. If nuclear recoil is included in the density of final states, the result is to multiply this expression for the cross section by a factor r where $r^{-1} = 1 + (2\hbar k/M_T c)\sin^2\theta/2$. We leave the demonstration of this result to the reader.

Construction of the nuclear current at various levels of the description of the nucleus and calculation of nuclear matrix elements is described in Ref. [7].

Covariant Analysis. Let us revisit our analysis of electron scattering and start from the beginning in an explicitly covariant manner. The S-matrix with one-photon exchange can be written in the form[5]

$$S_{fi} = -\frac{ee_p}{\Omega}\bar{u}(k')\gamma_\mu u(k)\frac{1}{q^2}\int e^{-iq\cdot x}\langle p'|J_\mu(x)|p\rangle d^4x \quad (30)$$

---

[5] We now revert to units where $\hbar = c = 1$. We use a metric with $x_\mu = (\mathbf{x}, it)$. Our gamma matrices are hermitian and satisfy $\gamma_\mu\gamma_\nu + \gamma_\nu\gamma_\mu = 2\delta_{\mu\nu}$. Also $\gamma_5 = \gamma_1\gamma_2\gamma_3\gamma_4$.

The momenta appearing in this expression are now all four-vectors and the four-momentum transfer satisfies the relation $q = k' - k$.[6] One can use translational invariance on the nuclear matrix element to write in the continuum limit

$$\int e^{-iq\cdot x}\langle p'|J_\mu(x)|p\rangle d^4x = (2\pi)^4\delta^{(4)}(p' + q - p)\langle p'|J_\mu(0)|p\rangle \tag{31}$$

The T-matrix is then identified from the expression

$$S_{fi} = -\frac{(2\pi)^4}{\Omega}i\delta^{(4)}(p' + q - p)\overline{T}_{fi} \tag{32}$$

The cross section follows in the standard manner

$$d\sigma = \overline{\sum_i}\sum_f 2\pi|\overline{T}_{fi}|^2\delta(W_f - W_i)\frac{\Omega d^3k'}{(2\pi)^3}\frac{1}{I_{\rm inc}}\left[\frac{(2\pi)^3}{\Omega}\delta^{(3)}(\Delta\mathbf{p})\right] \tag{33}$$

The last factor takes into account the fact that up to the final step, one is really working in a big box with periodic boundary conditions so that

$$(2\pi)^3\delta^{(3)}(\Delta\mathbf{p}) = \int_{\rm box} e^{i\Delta\mathbf{p}\cdot\mathbf{x}} \equiv \Omega\delta_{\mathbf{p}_f,\mathbf{p}_i} \tag{34}$$

where the last term is a Kronecker delta satisfying

$$[\delta_{\mathbf{p}_f,\mathbf{p}_i}]^2 = \delta_{\mathbf{p}_f,\mathbf{p}_i} \tag{35}$$

We leave it as an exercise for the reader to show that the incident flux in any frame where $\mathbf{k}\|\mathbf{p}$ can be written for a massless electron as

$$I_{\rm inc} = \frac{1}{\Omega}\frac{\sqrt{(k\cdot p)^2}}{\varepsilon E_p} \tag{36}$$

This relation is immediately verified in the laboratory frame where $E_p = M_T$ and $k\cdot p = -\varepsilon M_T$.

The square of the T-matrix then leads to the cross section in the form

$$d\sigma = \frac{1}{\sqrt{(k\cdot p)^2}}\frac{4\alpha^2}{q^4}\eta_{\mu\nu}W_{\mu\nu}\frac{d^3k'}{2\varepsilon'} \tag{37}$$

As an element of transverse area, this cross section must take the same value in any frame where $\mathbf{k}\|\mathbf{p}$, and indeed, it has now been written in an explicitly Lorentz invariant form.

The lepton tensor appearing in this expression is defined by

$$\begin{aligned}
\eta_{\mu\nu} &= -\frac{1}{2}2\varepsilon\varepsilon'\sum_{s_1}\sum_{s_2}\bar{u}(k)\gamma_\nu u(k')\bar{u}(k')\gamma_\mu u(k) \\
&= k_\mu k'_\nu + k_\nu k'_\mu - k\cdot k'\delta_{\mu\nu}
\end{aligned} \tag{38}$$

---

[6]The quantity q^2 now denotes the four-momentum transfer; the three-momentum transfer will henceforth be explicity denoted by $\mathbf{q}^2$.

The hadronic target contribution is similarly summarized in a tensor of the form

$$W_{\mu\nu} = (2\pi)^3 \overline{\sum_i} \sum_f \delta^{(4)}(q + p' - p) \langle p|J_\nu(0)|p'\rangle \langle p'|J_\mu(0)|p\rangle (\Omega E_p) \quad (39)$$

This Lorentz tensor can be analyzed through the following observations:

- Conservation of the electromagnetic current implies $q_\mu W_{\mu\nu} = W_{\mu\nu} q_\nu = 0$

- The only remaining four-vectors with which to construct this tensor are p_μ and q_μ

- The only remaining Lorentz invariant variables are q^2 and $p \cdot q$

As a result, the target response tensor must take the form

$$W_{\mu\nu} = W_1(q^2, q \cdot p) \left(\delta_{\mu\nu} - \frac{q_\mu q_\nu}{q^2}\right) +$$
$$W_2(q^2, q \cdot p) \frac{1}{M_T^2} \left(p_\mu - \frac{p \cdot q}{q^2} q_\mu\right) \left(p_\nu - \frac{p \cdot q}{q^2} q_\nu\right) \quad (40)$$

Note that in the laboratory frame the Lorentz invariants take the form

$$q^2 = 4\varepsilon\varepsilon' \sin^2 \frac{\theta}{2}$$
$$\frac{p \cdot q}{M_T} = \varepsilon - \varepsilon' \quad (41)$$

We again leave it to the reader to show that a combination of these results results in a laboratory cross section of the form

$$\frac{d^2\sigma}{d\Omega' d\varepsilon'} = \sigma_{\text{Mott}} \frac{1}{M_T} \left[W_2(q^2, q \cdot p) + 2W_1(q^2, q \cdot p) \tan^2 \frac{\theta}{2}\right] \quad (42)$$

An Example. As an example, consider elastic scattering from a spin zero nucleus. In this case, Lorentz invariance and current conservation imply that the nuclear matrix element must have the form

$$\langle p - q, 0^+|J_\mu(0)|p, 0^+\rangle = \frac{1}{M_T}\left(p_\mu - \frac{p \cdot q}{q^2} q_\mu\right) F_0(q^2) \left(\frac{M_T^2}{E_p E_{p'} \Omega^2}\right)^{1/2} \quad (43)$$

Hermiticity of the electromagnetic current implies the form factor $F_0(q^2)$ is real. The response functions are immediately evaluated in this case to give

$$W_1 = 0$$
$$W_2 = |F_0(q^2)|^2 \frac{M_T^2}{E_{p'}} \delta(W_f - W_i) \quad (44)$$

The cross section then takes the form

$$\frac{d\sigma}{d\Omega} = \sigma_{\text{Mott}} |F_0(q^2)|^2 r \tag{45}$$

Here r is the previously discussed recoil factor.

Parity Violation. Consider now a longitudinally polarized electron beam. If one does nothing more than reverse the electron helicity, then the parity violating asymmetry

$$\mathcal{A} = \frac{d\sigma_\uparrow - d\sigma_\downarrow}{d\sigma_\uparrow + d\sigma_\downarrow} \tag{46}$$

must vanish since the electromagnetic interaction conserves parity to all orders. Parity violation is present in electron scattering to a small extent due to interference with the weak amplitude arising from the exchange of the Z^0 — the heavy, neutral, weak vector boson.[7] If Z^0 exchange is added to γ exchange, the above S-matrix is extended to

$$
\begin{aligned}
S_{fi} =\ & -\frac{ee_p}{\Omega} \bar{u}(k')\gamma_\mu u(k)\frac{1}{q^2}\int e^{-iq\cdot x}\langle p'|J_\mu^{(\gamma)}(x)|p\rangle d^4x \\
& -\frac{G}{\Omega\sqrt{2}}\bar{u}(k')(a\gamma_\mu + b\gamma_\mu\gamma_5)u(k)\int e^{-iq\cdot x}\langle p'|\mathcal{J}_\mu^{(0)}(x)|p\rangle d^4x
\end{aligned} \tag{47}
$$

Here G is the Fermi constant and the weak neutral current is assumed to have the familiar V-A form

$$\mathcal{J}_\mu^{(0)} = J_\mu^{(0)} + J_{\mu 5}^{(0)} \tag{48}$$

In the standard model the electron weak neutral current is given by

$$a = -(1 - 4\sin^2\theta_W) \qquad\qquad b = -1 \tag{49}$$

The use of helicity projection operators for massless electrons

$$P_\uparrow = \frac{1 - \gamma_5}{2} \qquad\qquad P_\downarrow = \frac{1 + \gamma_5}{2} \tag{50}$$

allows one to calculate the asymmetry in a manner directly analogous to that described above for the cross section. The result is (Ref. [6])

$$
\begin{aligned}
\mathcal{A}\left[\cos^2\frac{\theta}{2}W_2^\gamma + 2\sin^2\frac{\theta}{2}W_1^\gamma\right] = & \frac{Gq^2}{4\pi\alpha\sqrt{2}}\left\{ b\left[\cos^2\frac{\theta}{2}W_2^{\text{int}} + 2\sin^2\frac{\theta}{2}W_1^{\text{int}}\right]\right. \\
& \left. -a\left(\frac{2W_8}{M_T}\right)\sin\frac{\theta}{2}\left(q^2\cos^2\frac{\theta}{2} + \mathbf{q}^2\sin^2\frac{\theta}{2}\right)^{1/2}\right\}
\end{aligned} \tag{51}
$$

---

[7] We are here discussing electron scattering from nuclei up to several GeV.

Here the nuclear target response tensors are defined in a fashion similar to that in Eq. (39). The response tensor arising from the interference of the electromagnetic and vector part of the weak neutral current is written as

$$W_{\mu\nu}^{\text{int}} = (2\pi)^3 \overline{\sum_i} \sum_f \delta^{(4)}(q + p' - p) \left[\langle p|J_\nu^{(0)}(0)|p'\rangle\langle p'|J_\mu^{(\gamma)}(0)|p\rangle + \right.$$
$$\left. \langle p|J_\nu^{(\gamma)}(0)|p'\rangle\langle p'|J_\mu^{(0)}(0)|p\rangle \right] (\Omega E_p) \tag{52}$$

We assume the weak vector current is conserved, and thus this tensor must again have the covariant form

$$W_{\mu\nu}^{\text{int}} = W_1^{\text{int}}(q^2, q\cdot p)\left(\delta_{\mu\nu} - \frac{q_\mu q_\nu}{q^2}\right) +$$
$$W_2^{\text{int}}(q^2, q\cdot p)\frac{1}{M_T^2}\left(p_\mu - \frac{p\cdot q}{q^2}q_\mu\right)\left(p_\nu - \frac{p\cdot q}{q^2}q_\nu\right) \tag{53}$$

The tensor arising from the interference of the electomagnetic current and axial vector part of the weak neutral current is defined by

$$W_{\mu\nu}^{\text{V}-\text{A}} = (2\pi)^3 \overline{\sum_i} \sum_f \delta^{(4)}(q + p' - p) \left[\langle p|J_{\nu 5}^{(0)}(0)|p'\rangle\langle p'|J_\mu^{(\gamma)}(0)|p\rangle + \right.$$
$$\left. \langle p|J_\nu^{(\gamma)}(0)|p'\rangle\langle p'|J_{\mu 5}^{(0)}(0)|p\rangle \right] (\Omega E_p) \tag{54}$$

It must be a pseudotensor, and the only one we can make from p and q is

$$W_{\mu\nu}^{\text{V}-\text{A}} = W_8(q^2, q\cdot p)\frac{1}{M_T}\epsilon_{\mu\nu\rho\sigma}P_\rho q_\sigma \tag{55}$$

An Example. Consider again the example of elastic scattering from a spin zero nucleus. In this case the matrix element of the axial vector current must vanish since one cannot make an axial vector from p and q

$$\langle p - q, 0^+|J_{\mu 5}^{(0)}(0)|p, 0^+\rangle = 0 \tag{56}$$

The response tensors are then evaluated as above to give

$$W_1^{\text{int}} = W_8 = 0$$
$$W_2^{\text{int}} = 2F_0^{(\gamma)}(q^2)F_0^{(0)}(q^2)\frac{M_T^2}{E_{p'}}\delta(W_f - W_i) \tag{57}$$

As above, the form factors must be real. The asymmetry in this case thus takes the form

$$\mathcal{A} = \frac{Gq^2b}{2\pi\alpha\sqrt{2}}\frac{F_0^{(0)}(q^2)}{F_0^{(\gamma)}(q^2)} \tag{58}$$

Measurement of this asymmetry at all q^2 thus completely determines the distrubution of weak neutral current in this nuclear system.

3 Overview of Current Status of Electron Scattering

The final talk in this lecture series is an overview of the current status of electron scattering, including a description of CEBAF. This material appears in the published literature in Ref. [1].

References

[1]   J. D. Walecka, *Electron Scattering*, Conference on Perspectives in Nuclear Structure, the Niels Bohr Institute, Copenhagen, June 13-18, 1993 (to be published in Nuclear Physics **A**)

[2]   J. D. Walecka, *Overview of the CEBAF Scientific Program*, A.I.P. Conference Proceedings, **269**, eds. F. Gross and R. Holt, A.I.P., New York (1993) p. 97

[3]   A. R. Edmonds, *Angular Momentum in Quantum Mechanics*, 3rd printing, Princeton University Press, Princeton, New Jersey (1974)

[4]   L. I. Schiff, *Quantum Mechanics*, 3rd ed., McGraw-Hill, New York (1968)

[5]   J. M. Blatt and V. F. Weisskopf, *Theoretical Nuclear Physics*, John Wiley and Sons, New York (1952)

[6]   J. D. Walecka, *Lectures on Electron Scattering*, ANL-83-50, Argonne National Laboratory, Argonne, Illinois (1984); CEBAF (1987)

[7]   J. D. Walecka, *Theoretical Nuclear and Subnuclear Physics*, Oxford University Press, New York (to be published)

Elementary Nuclear Excitations Studied with Electromagnetic and Hadronic Probes [1]

A. Richter

Institut für Kernphysik, Technische Hochschule Darmstadt
D-64289 Darmstadt, Germany

Abstract: In three lectures recent developments in the field of elementary electric and magnetic nuclear excitations are discussed by using electromagnetic probes of different selectivity. The first lecture, after a brief reminder of the features of electron and photon scattering at low energy, deals with two selected examples of electric transitions studied in inclusive and exclusive electron and proton scattering in ^{12}C and ^{40}Ca respectively, i.e. the features of an isospin forbidden electric dipole transition in (e,e′) and the structure of electric giant resonances in (e,e′x) and (p,p′x) reactions are studied. In the second lecture the nature of the orbital and spin magnetic dipole strength with particular emphasis on the physics of the so–called *scissors mode* and the spin magnetic dipole giant resonance in heavy deformed nuclei are discussed. The third lecture is concerned with three examples, illuminating how subnuclear effects might be detected in the nuclear electromagnetic response at low energy. For this, the prime case for quenching of magnetic dipole spin–flip strength in ^{48}Ca and the evidence for meson exchange current enhancement of isovector magnetic dipole strength in ^{24}Mg are revisited, and finally as an interesting outlook of the field the role of Δ–isobars in the nuclear wave function is illustrated by looking at the idea and at first results from a recent electroproduction experiment of positively and negatively charged pions on ^{3}He at medium energies.

1 Introduction

In order to set the scope of these three lectures in a school under the title ”Response of the nuclear system to external forces” let us inspect Fig. 1 in which the general electromagnetic response of the nucleus is illustrated. This response is studied in terms of a double differential cross section $d^2\sigma/d\Omega dE_x$ (which divided by the Mott–cross section is called form factor) as a function of momentum trans-

---

[1] Work supported by the German Federal Minister for Research and Technology (BMFT) under contract number 06DA641I.

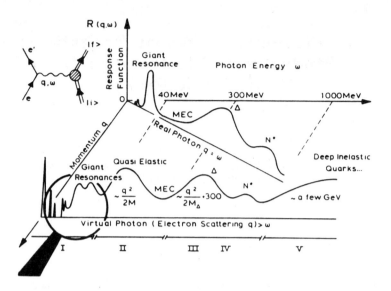

Fig. 1. Response function of a nucleus in an external electromagnetic field generated by real and virtual photons at low energies and the different regimes of nuclear excitations treated in these lectures by way of examples. It is shown schematically how discrete states and unbound giant resonances might be explained in the independent particle shell model by particle–hole excitations. It is furthermore indicated that spin– and isospin–flip $p - h$ excitations on the nuclear level can be influenced by such excitations whereby the nucleon is moved into the $\Delta(1232)$ region about 300 MeV above the nuclear ground state. On the quark level the process is explained by a quark spin–flip.

fer q and energy transfer E_x imported to the nucleus by either real photons or virtual photons in electron scattering.

The discrete states below the threshold energy E_B for nucleon emission can be associated with simple particle–hole $(p - h)$ excitations between (or within) bound orbits in the independent particle shell modell (IPM). The so–called unbound giant resonances correspond to $p-h$ excitations into the continuum. They therefore aquire broader widths. The nuclear excitations – we will discuss this fact in these lectures in particular in connection with spin–isospin–flip M1 excitations – might also be accompanied by excitations whereby the nucleon is moved into the $\Delta(1232)$ isobar region about 300 MeV above the nuclear ground state. Such Δ–hole excitation therefore probe the pion field inside the nucleus.

The consequences of the excitation energy mechanisms depicted in Fig. 2 are the following:

(i) We expect some *clustering* of $p - h$ strength at excitation energies $E_x = 0\hbar\omega, 1\hbar\omega, 2\hbar\omega, 3\hbar\omega, \cdots$ and the corresponding excited states are of alternating parity $\Delta\pi = +, -, +, -, \cdots$. In the IPM one finds $\hbar\omega \approx 41A^{-1/3}$ MeV. This clustering of strength is then a general feature of nuclei and a smooth function of mass number A.

(ii) The nuclear residual interaction causes specific intraband effects, i.e. the excitation strength is further concentrated at a certain excitation energy E_x. In the case of electric giant resonances e.g., the Brown–Bolsterli effect results in a constructive superposition of $p-h$ states and leads to rather narrowly localized strength around E_x. One speaks of collective excitations or loosely of giant resonances. Such a giant resonance usually exhausts a large fraction of an energy weighted sum rule (EWSR) which is often employed as a measure of collectivity.

(iii) The coupling of $1p - 1h$ states to many particle–many hole states leads naturally to a fragmentation of excitation strength (or equivalently of the measured cross section), i.e. to sometimes considerable fine structure in the experimental observables (even for continuum states). In order to detect this fine structure, a high experimental resolution is a prerequisite in nowadays photon and electron scattering experiments on nuclei.

For pedagogical reasons I have sometimes used material from previous lectures on the subject. Furthermore, since parts of the present lectures, especially the ones on the magnetic dipole response of nuclei, were central topics in two consecutive talks at international meetings I have been asked to give just before the present school, there is no point reiterating what I tried to formulate as best as I could once and I sincerely hope the reader will understand that each of the three manuscripts on the topic of the magnetic dipole response is only a slightly modified version of the other. Furthermore, considering the wealth of experimental data presented in the actual lectures the rather limited space allowed for their written version in these proceedings forces me to restrict myself essentially only to a summary of those points which I did discuss orally. Also, the list of references given at the end will necessarily not be complete.

As I have always done in the past (see e.g. [1]) one of the things that I shall try to do in these lectures is to show how the same problem, say the structure and strength of a giant resonance, can often be tackled by various experimental methods. In fact – as Sir Denys Wilkinson has pointed out once – when the problem is a general one it is always desirable that all methods should be tried, since "nature might always be cleverer than we and introduce some effect that we have not thought of and that might show up in an apparently–insensitive approach but not in an apparently–sensitive one". It will become clear from the discussion of the various experimental examples that we always learn most about the nucleus if we put the results from experiments using electromagnetically probes together with those from strongly interacting probes – and in the case of magnetic trasitions to be treated later on also with weakly interacting probes – into a common perspective.

The general plan of the three lectures is the following: Firstly, I will discuss in the spirit of Fig. 1 a couple of illustrative examples from the field of electric transitions. Secondly, I will look at some new results from the orbital and spin magnetic dipole response in nuclei and finally I collect three interesting experimental observations that might tell us something on the role of subnuclear degrees of freedom in low and medium energy electromagnetic transitions.

2 Electric transitions

The following lecture presents two selected examples of what can be learned about elementary electric modes in inclusive and exclusive inelastic electron scattering experiments at low momentum transfers. Experiments as described here have only become possible with a new generation of continuous–wave electron accelerators coupled with high–resolution, large solid–angle magnetic spectrometers for which the superconducting Darmstadt electron linear accelerator S–DALINAC stands as an example.

2.1 Investigations of weak transitions in (e,e') reactions: the example of an isospin forbidden transition to the $J^\pi = 1^-, T = 0$ state in ^{12}C at $E_x = 10.84$ MeV

In self–conjugate nuclei a class of E1 transitions exciting $J^\pi = 1^-$, $T = 0$ states has been observed which are forbidden in the long–wavelength limit. Despite their isospin forbideness fairly large $B(E1)$ values up to 10^{-3} W.u. are observed. Electron scattering at low momentum transfer q is particularly sensitive to small $T = 1$ admixtures because of the different q dependence of the isoscalar and the isovector part of the form factor

$$F(q) \sim |A_0\, q^3 + \Phi\, A_1\, q|\,. \tag{1}$$

Here A_0, A_1 denote the $T = 0, 1$ amplitudes and $\Phi = \pm 1$ is a relative phase. Such investigations have e.g. been performed in the nuclei ^{16}O and ^{40}Ca [2, 3].

We have investigated the transition to the $E_x = 10.84$ MeV, $J^\pi = 1^-$ state in ^{12}C. Data were taken in incident energies $E_0 = 60$ MeV and scattering angles $\theta_e = 117°$, 135° and 155°. This corresponds to momentum transfers $q = 0.47 - 0.53$ fm^{-1}. The main experimental problem lies in the large intrinsic width of this level, $\Gamma = 315$ keV, which severely hampers the detectability on top of the radiative tail in the spectra.

A special interest into this transition was born out of a recent study in low–energy inelastic pion scattering [4]. There, approximate charge symmetry was observed for this transition in the ratio of π^+ and π^- cross sections. However, because of the Coulomb interaction this does not imply a purely isoscalar character of the transition. A preliminary analysis [5], where proton and neutron matrix elements were adjusted to describe π^+ and π^- cross sections simultaneously, indicates an isovector/isoscalar matrix element ratio of ≈ 0.06.

In Fig. 2 the longitudinal form factor for the transition to the 10.84 MeV level is shown as a function of the momentum transfer. Data at higher q from Torizuka et al. [6] are also included. The solid line is a shell–model calculation [7] using a new effective interaction derived for the coupled p– and sd–shell regions [8] and a harmonic oscillator potential.

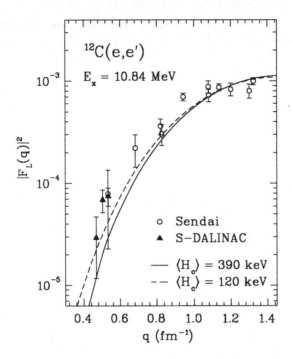

Fig. 2. Experimental values of the longitudinal formfactor of the 10.84 MeV state in ^{12}C. The solid and the dashed lines are calculations with Coulomb matrix elements of 120 and 390 keV, respectively, fitted to the experimental data.

An isospin mixing matrix element of 120 keV is assumed in the calculations. The dashed line shows the result assuming a 390 keV mixing which would correspond to a mixing amplitude of 0.06 similar to the pion scattering results. The calculations had to be normalized by factors 1.504 and 1.435, repectively.

Alternatively, one can extract the relative $T = 0, 1$ contributions from an empirical form factor [2, 3]

$$|F_L(q)| = \left| [A_0 \, q^3 \, b^3 + \Phi \, A_1 \, qb] \, \exp(-q^2 b^2/4) \right| , \qquad (2)$$

where b is the oscillator parameter for ^{12}C. The phase cannot be well determined in the present case, however assuming $\Phi = -1$ as found in the other cases, the $A_0 \, / \, A_1$ ratio is almost identical to the comparable transition in ^{16}O [23]. Since the main mixing $J^\pi = 1^-, T = 1$ level is known at $E_x = 17.23$ MeV one can evaluate the mixing amplitudes in a two-state model

$$|1^-, T = 0\rangle = |\alpha \, \text{IS} - \beta \, \text{IV}\rangle \qquad (3)$$

$$|1^-, T = 1\rangle = |\beta \, \text{IS} + \alpha \, \text{IV}\rangle \qquad (4)$$

where α, β denote the isoscalar and isovector amplitudes and IS, IV stand for the corresponding matrix elements. One obtains $\alpha = 0.999, \beta = 0.042$ corresponding to a 0.2% T = 1 admixture in the wave function. Using the $B(E1)$ strength extracted from the extrapolation of eq. (2) to the photon point and the known partial width Γ_{γ_0} of the 17.23 MeV level, the Coulomb matrix element can be estimated. One finds $\langle H_c \rangle \simeq 140$ keV. This agrees well with the systematics of known isospin mixing matrix elements [9].

2.2 Investigations of giant resonances in (e,e'x) reactions: the example of ^{40}Ca

With the advent of continuous–wave electron accelerators coincidence experiments have become feasible as a powerful tool of giant resonance spectroscopy. These studies are especially attractive, since they allow an efficient suppression of the huge radiative tails which usually limit the extraction of broad structures in inclusive measurements. Some of the new possibilities offered by the coincidence technique are a background–free multipole decomposition through variation of the momentum transfer for constant excitation energy E_x, the use of the decay into specific final states as a "spin filter", or the additional information contained in the angular correlations of emitted particles with respect to momentum transfer q.

On the other hand, one should not conceal that such experiments are difficult and time consuming. The cross sections are very low and experimentally one is plagued by large radiation backgrounds (mainly delta electrons) in the counters. The theoretical description of angular correlations is complicated by ambiguities in the case of nucleon decay, and usable models for α decay are missing at all. Furthermore, the information on the transition strength is usually limited to particular channels. Nevertheless, important new information can be gained in (e,e'x) experiments (see e.g. references in [10]) and results of a ^{40}Ca(e,e'x; x= p,α) study are chosen here to demonstrate this. Because of the limited space available the discussion is confined to the question of multipole decomposition and the fine structure of the decay into the ground states of the daughter nuclei. Further results can be found in [10, 11].

The doubly magic nucleus ^{40}Ca has always been a favoured subject of both experimental and theoretical work. Giant resonances in ^{40}Ca have been extensively studied with electromagnetic and hadronic probes and rather compact giant dipole resonance (GDR) strength centered at 19.5 MeV and a strongly fragmented isoscalar giant quadrupole resonance (GQR) at about 18 MeV have been observed. However, a number of open problems remain. Different results have been reported for the total strength and fine structure of the GQR. A ^{40}Ca(α,α'x) experiment [12] found a roughly equal splitting of the EWSR strength with a second maximum at about 14 MeV in contrast to all microscopic calculations (however, see below). There is also some discussion about the presence and amount of isoscalar giant monopole resonance (GMR) contributions at lower energies.

The relevant informations about excitation and decay in ^{40}Ca are displayed schematically in Fig. 3. The excitation energy region $E_x = 8 - 26$ MeV was studied. Proton and α decay into low–lying states of ^{39}K (p$_0$–p$_3$) and ^{36}Ar (α_0,α_1), respectively, could be resolved. Neutron emission, which was not measured in the present experiment, competes above the threshold energy $E_n = 15.67$ MeV. Photon induced reactions [13] indicate a neutron contribution of about 20% at the GDR maximum excitation energy.

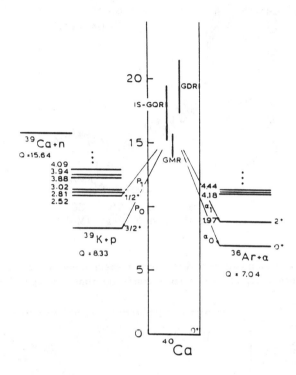

Fig. 3. Scheme of excitation and decay studied in the ^{40}Ca(e,e'x) experiment.

The experiment was performed at the accelerators MAMI A in Mainz and at the S–DALINAC in Darmstadt. In Mainz, 183.5 MeV electrons were used and the scattered electrons were detected with the magnetic spectrometer at angles (momentum transfers) $\theta_e = 22.0°$ ($q = 0.35$ fm^{-1}), $31.4°$ ($q = 0.49$ fm^{-1}) and $43.0°$ ($q = 0.66$ fm^{-1}). In Darmstadt data were taken with a large solid–angle QCLAM spectrometer at $E_0 = 78$ MeV and $\theta_e = 40.0°$ (0.26 fm^{-1}). The decay products were detected relative to the q axis with up to 10 charged particle detector telescopes consisting of $75-100$ μm ΔE and two 1000 μm E–Si counters. The telescopes were placed on a goniometer out–of–plane under an azimuthal angle $\Phi_x = 135°$ in order to cancel the transverse–transverse interference contribution to the reaction cross section.

Branching ratios of the decay into the final channels and total excitation spectra were derived from a 4π integration assuming purely longitudinal excitation. The almost background–free response generated by the coincidence method is beautifully demonstrated in Fig. 4 where an inclusive and an exclusive measurement are compared for identical kinematics.

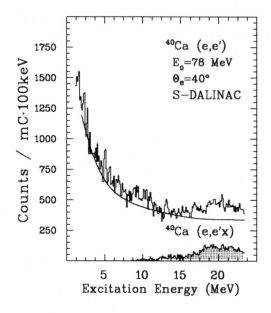

Fig. 4. Comparision of an inclusive and exclusive measurement of the ^{40}Ca(e,e') reaction of identical kinematics. The koincidence condition almost completely suppresses the radiative tail.

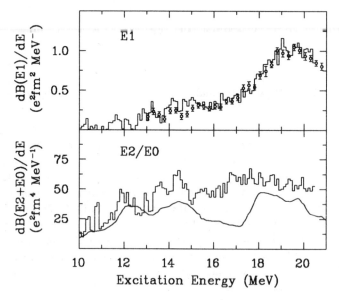

Fig. 5. E1 and E2 (plus E0) strength distributions resulting from the multipole analysis of (e,e'x) cross sections following ref. [14]. The open circles are the difference of total photoabsorption [15] and (γ,n) data [16]. The solid line is a RPA calculation of E2 strength in ^{40}Ca [18].

A multipole strength analysis based on the variation of the momentum transfer was performed following the method described in [14]. It is based on the assumption that the form factors are independent of excitation energy and that the contributing multipoles are limited to $\lambda \leq 2$. The resulting GDR and GQR strength functions are displayed in Fig. 5. One should note that possible GMR contributions cannot be unfolded from the GQR strength with this method because of the similarity of the form factors.

The GDR results are compared to photonuclear cross sections derived from the difference of total photoabsorption and (γ,n) data [15, 16] (open circles). The shape and absolute value of the two curves agree very well. The E2 (plus E0) strength distribution shows maxima around 17, 14 and 12 MeV. The former corresponds to E2 strength observed in numerous previous experiments. The bump around 14 MeV contains some E0 contribution, e.g. a decomposition of the ^{40}Ca(e,e'α_0) channel shows about 50% E0 cross section. The strength around 12 MeV is due to a number of discrete E2 transitions which could be resolved in high resolution (e,e') and (p,p'x) experiments [11, 17]. The solid line presents the result of a state–of–the–art RPA calculation of isoscalar E2 strength including $1p1h\otimes$phonon configurations and continuum coupling [18]. The quality of description of the experimental results is remarkable and, for the first time, a realistic theoretical approach of the strongly fragmented E2 strength with significant parts below $E_x = 15$ MeV is achieved. A detailed analysis reveals that the low–lying strength can be traced back to $2p2h$ ground state correlations which are treated beyond the usual RPA level in the present approach.

The E1 form factor resulting from the analysis is displayed in Fig. 6 and compared to a RPA calculation [19] using the model of separable interactions (MSI). Good correspondence is obtained. Different to the multipole analysis described in [10], where the same model was used for the E2 form factor, in this case the E2 form factors were generated from transition densities given in [18]. The integrated strengths for charged particle decay in the energy intervall $10 - 20.5$ MeV exhaust 58 ± 15 % of the GDR, respectively 80 ± 16 % of the isoscalar GQR energy weighted sum rule. The above value constitutes an upper limit for the GQR charged particle decay strength because of the possible monopole contributions.

Further insight into the contributions of the yet interweaved E2 and E0 strength parts and its fine structure can be gained from comparison to other experiments with different selectivity. We have thus performed a study of the ^{40}Ca(p,p'x) reaction covering the same excitation range [11]. The experiment was executed at the National Accelerator Centre cyclotron at Faure in South Africa with 100 MeV protons using a recently built $K600$ magnetic spectrometer. Scattered protons were measured at $\theta_p = 17°$, $23°$ and $27°$. Charged particle decay was detected with three semiconductor detector telescopes placed on a rotatable table in the reaction plane. A typical energy resolution of $\Delta E = 35$ keV FWHM was achieved in the ^{40}Ca excitation spectra.

Figure 7 presents first results for the α_0 and p_0 channels. The 4π integrated spectra are shown for $\theta_p = 17°$ where E2 strength should be strongly enhanced

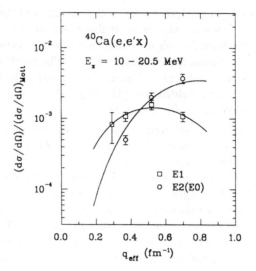

Fig. 6. Form factors resulting from the multipole analysis. The theoretical E1 form factor is a MSI–RPA calculation [19] and the theoretical E2 form factor was generated from the transition densities of the RPA calculation described in Ref. [18].

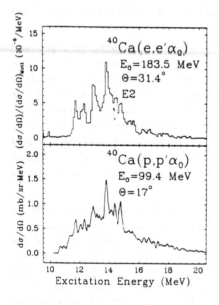

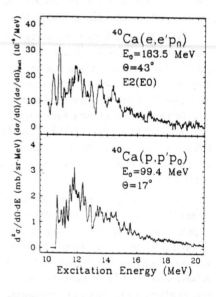

Fig. 7a. E2 spectrum from the multipole analysis of the (e,e'α_0) ACF compared to a 4π integrated spectrum of the ^{40}Ca(p,p'α_0) reaction at an angle where E2 is strongly enhanced.)

Fig. 7b. E2 (plus E0) part of the cross section in the ^{40}Ca(e,e'p_0) channel obtained from the multipole analysis compared to a spectrum of the ^{40}Ca(p,p'p_0) reaction.

compared to other multipolarities. In Fig. 7a the $(p,p'\alpha_0)$ data are compared to the E2 strength obtained from a multipole decomposition of the angular correlation function [20]. All dominant structures are reflected in the (p,p') spectrum with somewhat more fine structure due to the better resolution. Some additional levels are present in the (p,p') data whose multipolarity must be ascertained by a comparison to DWBA calculations.

Figure 7b displays a comparison to the E2(E0) strength distribution resulting from a multipole decomposition of the $(e,e'p_0)$ channel with the method described above. Neglecting the data below $E_x = 11$ MeV, where the (p,p') data are plagued by an efficiency cut–off in the detector, very good correspondence is obtained even on a level–by–level basis. This indicates that little E0 strength is present in the p_0 decay channel.

2.3 Summary

In summary, an electron scattering study of the weak isospin forbidden transition to the $J^{\pi} = 1^-$, $T = 0$ state in ^{12}C at low momentum transfer has been presented. A $T = 1$ admixture of 0.2% and an isospin mixing matrix element $\langle H_C \rangle \simeq 140$ keV can be extracted from the analysis of an empirical form factor. A shell–model calculation with comparable isospin mixing using a harmonic oscillator basis reproduces the shape of the form factor well, but underestimates the strength still by a factor of about 1.5.

Low–multipole ($\lambda \leq 2$) strength in ^{40}Ca was measured in the excitation energy range $E_x = 8-26$ MeV with the $(e,e'x; x=p,\alpha)$ reaction. A decomposition of the E1 and E2 (plus E0) parts utilizing the variation of the momentum transfer reveals very good correspondence of the E1 strength with photonuclear results. The E2 (plus E0) strength is strongly fragmented with a significant part at lower energies ($E_x < 15$ MeV). A recent RPA calculation [18] including $1p1h\otimes$ phonon configurations and coupling to the continuum describes the strength distribution remarkably well. The low–lying part can only be described if $2p2h$ g.s. correlations are taken into account.

Furthermore the quadrupole strength fine structure in the decay to the ^{39}K and ^{36}Ar g.s. was investigated. The shape of the E2 spectrum resulting from a multipole analysis of the $^{40}Ca(e,e'\alpha_0)$ angular correlation functions is in excellent agreement with $(p,p'\alpha_0)$ data taken at a scattering angle where $\lambda = 2$ is strongly enhanced. The same holds for the comparison of $(e,e'p_0)$ and $(p,p'p_0)$ results which indicates that E0 contributions are small in this channel.

Both, the example of a weak nuclear transition in ^{12}C and of strong nuclear transitions in ^{40}Ca thus demonstrate the power of inclusive and exclusive (e,e') and $(e,e'x)$ reactions, respectively, in the study of the electric response in nuclei.

3 Magnetic transitions

In this lecture (and also partly in the next one) I shall deal with a discussion of problems connected with the magnetic dipole response in nuclei. This is a topic of great current interest looked at e.g. from a different viewpoint in low energy electron and photon scattering, in intermediate energy proton scattering and in charge exchange reactions, in which Gamow–Teller giant resonances are excited. Again I shall not attempt to cover the field exhaustively, but cover it partly with some illustrative examples from recent work. I focus the attention of the reader also to a couple of articles which deal with the same topic [21, 22].

Why study the magnetic dipole response? Firstly, the investigation of static nuclear magnetic moments $\vec{\mu}_j$, i.e. of *diagonal* matrix elements of the magnetic dipole operator, and of their spin and orbital parts $\vec{\mu}_s$ and $\vec{\mu}_l$, respectively, has yielded valuable information about nuclear configuration mixing – Arima and Horie showed fourty years ago how the coupling of a valence nucleon (or hole) with $J^\pi = 1^+$ states strongly influences magnetic moments – and mesonic currents. From elastic magnetic electron scattering we even now have information on single particle magnetic density distributions inside the nucleus. Secondly, the measurements of M1 transition strengths, i.e. of the *off–diagonal* matrix elements of the magnetic dipole operator, yield – as we will see in detail – important additional information on the nuclear magnetization. Because of the primarily single particle nature of the M1 operator we can study neutron (proton) moments and transitions as well as the interplay between spin– and orbital magnetization in the case of neutron and proton excitations and we will also be able to shed some light on the question of isoscalar vs. isovector transitions.

The field of the magnetic dipole response has obtained a boost when almost exactly ten years ago an article appeared in the literature [23] with the title "New magnetic dipole excitation mode studied in the heavy deformed nucleus ^{156}Gd by inelastic electron scattering". This article has led truly to a renaissance of high resolution, low energy spectroscopy with electrons, photons and protons, to the development of novel theoretical ideas and to the improvement of already existing nuclear models. The large amount of scientific articles published since the discovery of the mode and the sizable number of articles still appearing (about 20 per year) is remarkable and signals that experimentalists and theoreticians are both fascinated by this elementary nuclear excitation and are driven to understand its very nature.

The experimental search for this orbital mode about a decade ago was driven by a theoretical prediction by Lo Iudice and Palumbo in terms of the so called Two Rotor Model [24] (TRM) and by a fairly cute estimate by Iachello for its expected transition strength in the Interacting Boson Model (IBM), and the latter termed it [25] at first "Nuclear Wobble". Soon after, for the obvious out of phase movement of protons against neutrons, the mode has been called "Scissors Mode". A most natural framework for studying proton and neutron degrees of freedom in collective states of nuclei is the IBM–2 [26]. When including proton and neutron bosons explicitly, besides the symmetric combinations that turn out to be equivalent to the IBM–1 description of nuclear structure, nonsymmetric couplings give rise to a totally new family of states of mixed symmetry [25]. In even–even nuclei the scissors mode leads through small angle vibrations of protons against neutrons to the excitation of $J^\pi = 1^+$ states that are the best examples for mixed symmetry states known so far. The discovery of those states made thus the F–spin concept [26] really meaningful and allowed a unique determination of the strength of the Majorana force which is responsible for the splitting between states of F_{max} and $F_{max} - 1$.

The discovery of the scissors mode and the very question if it is truly of orbital nature has also prompted experimental efforts to investigate the nuclear magnetic dipole spin response in deformed nuclei which hitherto had almost entirely been concentrated on spherical nuclei.

3.1 Qualitative nature of the magnetic dipole response

Both the TRM and the IBM are clearly the simplest macroscopic and microscopic approaches, respectively, towards an understanding of the basic features of the orbital magnetic dipole mode, and a large number of experiments since its discovery have revealed ample information on its excitation energy, its fragmented transition strength, its form factor and on the relative importance of its orbital vs. spin content. For a proper description of all those features more refined theoretical descriptions in terms of the shell model, RPA and QRPA had to be developed (for recent references, see [21, 22, 27 − 31] and references therein).

What is the simplest approach towards the nature of the magnetic dipole response in nuclei? Let us briefly recall the structure of the M1 operator

$$\mathbf{T}(\mathrm{M1}) = \sum_i \left\{ g_l(i)\vec{l}_i + g_s(i)\vec{s}_i \right\} \mu_N \tag{5}$$

with the g's being the usual g–factors for neutrons and protons. After rewriting (5) as a sum of isoscalar and isovector pieces using $t_z(i) = \pm 1/2$ for protons and neutrons, respectively, and neglecting further the small isoscalar piece because the g's of proton and neutron are of about equal magnitude but opposite sign we end up with the following structure of the isovector M1 operator:

$$
\begin{aligned}
\mathbf{T}(\mathrm{M1})_{IV} &= \left\{ \sum_i t_z(i)\vec{l}_i + (g_p - g_n) \sum_i t_z(i)\vec{s}_i \right\} \mu_N \\
&= \left\{ \frac{1}{2}(\vec{L}_p - \vec{L}_n) + 4.71\,\mathbf{T}(\mathrm{M1})_{\Delta T_z = 0} \right\} \mu_N
\end{aligned} \tag{6}
$$

This equation yields already some insight [21]. The isovector strength splits into orbital and spin parts, the first involving $\vec{L}_p - \vec{L}_n$, which, viewed as rotation generator, immediately suggests the scissors notion in a qualitative way. The spin–flip piece is the $\Delta T_z = 0$ component of the Gamow–Teller operator which is enhanced because its coefficient is $(g_p - g_n)$.

Let us look next, very schematically, what happens in an RPA calculation of the excitation strength in a medium heavy or heavy even–even nucleus. The unperturbed particle–hole strengths are scattered from the ground state up to say 10 MeV in excitation energy and the orbital and spin–flip contributions are thoroughly mixed. By turning on the well known particle–hole interaction in the spin–isospin channel, the spin–flip piece of the excitation, carrying the majority of the total strength, is swept up to excitation energies of 10 MeV or higher. The *orbital strength*, however, *hardly moves at all*. It remains low–lying, is *scissors-like* and weakly collective, but its observability is a strong collective effect as a consequence of the fact that the $p - h$ force has moved the competing stronger spin–flip strength up to higher excitation energy.

The weakly collective M1 excitation now becomes an ideal test of microscopic models of nuclear vibrations. Shell models are usually calibrated to reproduce properties of strong collective excitations (lowest $2^+, 3^-$ states, electric giant resonances). Weakly collective phenomena, however, force the models to make real predictions and the fact that the transitions in question are strong on the single-particle scale makes it impossible to dismiss failures as a mere detail. This should be kept in mind in an assessment of the wide variety of models which this new excitation mode has already inspired. The above discussion nowhere mentions deformation which is introduced alongside the discussion of the experimental data.

Returning to the IBM, the M1 operator of eq. (5) in fermion space has its image in boson space

$$\mathbf{T}^B(\mathrm{M1}) = \{g_\pi L_\pi + g_\nu L_\nu\}\, \mu_N \tag{7}$$

with g_π and g_ν being the respective proton and neutron boson pair g–factors and L_π and L_ν the corresponding orbital angular momenta. These pair g–factors can be estimated from an analysis of g–factors of first excited 2^+–states [32]). The expected M1 strength in the SU(3) limit of IBM–2 (most nuclei to be discussed below are good rotors and are sufficiently well described in this limit) is given (in μ_N^2) by

$$B(\mathrm{M1})\!\uparrow = \frac{3}{4\pi}\frac{4N_\pi N_\nu}{N_\pi + N_\nu}(g_\pi - g_\nu)^2. \tag{8}$$

That the simple picture of the nuclear magnetic dipole response is at least approximately correct is shown in Fig. 8 by using the three nuclei ^{56}Fe, ^{156}Gd and ^{238}U as examples. The mean excitation energy of the orbital mode is approximately given by $E_x \simeq 66\delta A^{-1/3}$ MeV with δ being the ground state quadrupole deformation.

The spin magnetic dipole strength recently found in inelastic proton scattering [33] lies at $E_x \simeq 41A^{-1/3}$ MeV and thus exhibits an excitation energy dependence reminiscent of the shell model. As is seen in Fig. 8 the ratio of orbital to spin strength is indeed small, indicating that the spin strength is the really collective part of the total M1 strength.

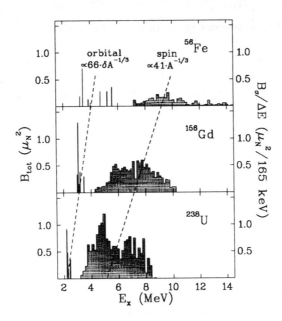

Fig. 8. The nuclear orbital and spin magnetic dipole response in a medium heavy, a heavy and a very heavy nucleus derived from experiments with electromagnetic and hadronic probes, respectively.

3.2 Magnetic dipole response in heavy deformed nuclei
3.2.1 Overview

I will now discuss recent advances in the study of the magnetic dipole response in heavy deformed nuclei. The salient features of the scissors mode unraveled in high-resolution electron and photon scattering experiments are the following:

- The center of gravity of the orbital M1 strength distribution lies in rare earth nuclei at $E_x \simeq 3$MeV.
- The total strength is $\sum B(\mathrm{M1}) \simeq 3\mu_N^2$ and the maximum strength that is carried in the transition to an individual state is roughly $1.5\,\mu_N^2$.
- In the nuclear transition current the orbital part dominates over the spin part and one has typically $B_l(\mathrm{M1})/B_\sigma(\mathrm{M1}) \simeq 10/1$.
- The summed transition strength up to $E_x \simeq 4$ MeV is proportional to the quadrupole ground state deformation.

Extreme forward angle inelastic scattering experiments of protons at medium energy indicate the following with respect to the M1 spin–flip resonance:

- It is located at excitation energies $E_x \simeq 5 - 10$ MeV and is characterized mostly by a double–humped structure.
- The total transition strength is $\sum B(\mathrm{M1}) \simeq 11\mu_N^2$.

3.2.2 Search for the scissors mode in even–odd nuclei

As the situation with even–A nuclei is fairly well understood, the next question concerns odd–A nuclei for which only scarce data exist. The search for the scissors mode has so far been negative [34] in ^{165}Ho but clearly positive in ^{163}Dy as has been demonstrated beautifully by a Giessen/Cologne/Stuttgart collaboration [35] in a nuclear resonance fluorescence experiment. The measured ground state transition strengths with $\Delta K = 1$ in a chain of Dy isotopes is shown in Fig. 9.

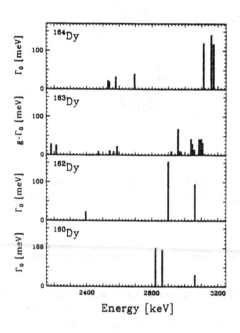

Fig. 9. Dipole strength distributions in four Dy isotopes. The strength is given in units of the g.s. decay width Γ_0 and for ^{163}Dy in form of the spin dependent decay width $g\Gamma_0$ (from [35]).

For ^{163}Dy it has not been possible to determine the spin of the excited state due to nearly identical angular correlation functions for the three possible decay chains. Therefore the ground state decay widths have been multiplied with the appropriate spin weighing factors g. From the figure it is obvious, however, that the detected strength fits both, energetically and in its magnitude, into the systematics of the neighboring even–even isotopes. Also, interacting boson–fermion model predictions [35] by Arias, Frank and Van Isacker are in essential agreement with the experimental observations in ^{163}Dy.

We have recently started a search for the scissors mode in ^{167}Er in order to study the influence of the nucleon–core interaction on the transition strength and the fragmentation. Strong orbital M1 transitions are known in the neighboring nuclei ^{166}Er and ^{168}Er.

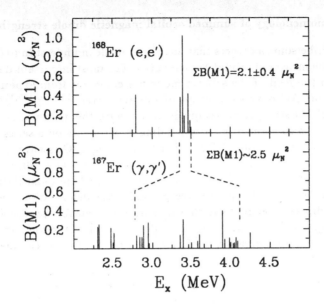

Fig. 10. Magnetic dipole strength distribution in ^{168}Er derived from (e,e$'$) measurements at the DALINAC (upper part) and measured dipole strength in ^{167}Er (converted into M1 strength) from (γ,γ') measurements at the S–DALINAC (lower part).

The combined results from nuclear resonance fluorescence spectra taken at the S–DALINAC at bremsstrahlung endpoint energies $E_0 = 3.5, 4.6$ and 5.8 MeV are presented in Fig. 10.

The mostly weak, but clearly identifiable transitions indicate a strong fragmentation of the measured strength. Under the (most likely very reasonable) assumption that the strength observed is indeed M1 strength the indicated three groups of transitions at $E_x \simeq 2.9, 3.5$ and 4.0 MeV yield a total strength of $\sum B(M1) \simeq 2.5\mu_N^2$. This value is comparable with the summed M1 strength observed in the neighboring nucleus ^{168}Er (upper part of Fig. 10 and also in ^{166}Er (not shown). However, the running sum of the strengths taken in the odd–A nucleus ^{167}Er saturates later than in the adjacent even–A nuclei indicating again a much larger fragmentation in the former nucleus.

In passing I note two possible improvements of the measurements of such weak transitions. Firstly, a Darmstadt/Cologne/Rossendorf collaboration will in the future at the low energy ($E_0 = 2.5 - 10$ MeV) photon scattering setup of the S–DALINAC use an EUROBALL CLUSTER–detector that has a high detection efficiency and the necessary background suppression needed to detect weak transitions. Secondly, a 180°–(e,e$'$) high–resolution scattering facility which has just been installed at the S–DALINAC by a Catholic University of Washington/Darmstadt collaboration is essentially a "spin–filter" whereby the multipolarity of transverse excitations can be determined rather quickly.

3.2.3 Phenomenology of summed orbital magnetic dipole strengths

I now describe some advances that have been made in understanding the physical origin of the measured orbital M1 strengths. The most important discovery made recently in the field of the scissors mode has come out from experiments [36] at the S–DALINAC done on a chain of the even–even ^{148,150,152,154}Sm isotopes. The orbital M1 strength varies quadratically with the deformation parameter δ. This result – also verified in corresponding experiments on a series of even–even Nd isotopes with varying deformation by a Cologne/Giessen/Stuttgart collaboration [37] – has been anticipated in a systematic study [38] of M1 strength in the rare earth region within the Nilsson model where quantitatively a direct correlation between the quadrupole ground state deformation and the orbital magnetic dipole strength was shown. Since the neutron–proton interaction is mainly responsible for the quadrupole deformation of the nuclear ground state the experimental observation is of great interest for the development of nuclear models of deformation.

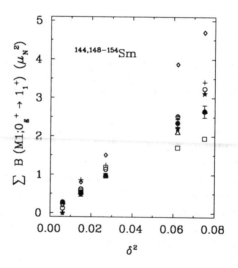

Fig. 11. The orbital M1 strength vs. the square of the ground state deformation parameter. The experimental points with error bars are compared to six theoretical predictions from refs. [39] ($\bigcirc$), [40] ($+$), [41] ($\diamond$), [42] ($\square$), [43] ($\triangle$) and [44] ($\star$), respectively.

A comparison of the original experimental data [36] with results from recent model calculations [39 – 44] is given in Fig. 11. As can be seen, practically all models – some of them after appropriate modifications of earlier versions – yield a strength more or less quadratic in the deformation parameter.

There are still discrepancies with the experimental results concerning the magnitude and the slope of the strengths, especially for the two most deformed isotopes. Two more predictions within the so called $n - p$ deformation model [45] and the IBM are not shown here but are treated in detail in [36]. It is clear that there seem to be still some fundamental problems with the various theoretical schemes, in particular towards a proper description of orbital M1 strength in truly rotational nuclei, i.e. ^{152}Sm and ^{154}Sm in our example. Given the elementary character of the excitation mode this is indeed surprising.

The obvious δ^2–dependence of the scissors mode strength can also be expressed in terms of a correlation plot [27] where the summed orbital M1 strengths in the nuclei indicated is plotted vs. the corresponding E2 strengths between the ground state and the first 2^+ state of the ground state band (Fig. 12). This striking manifestation of quadrupole collectivity in the magnetic dipole strength has also been looked at within the IBM–2 using a sum rule approach [27].

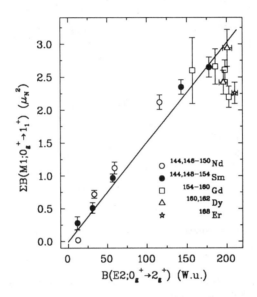

Fig. 12. Correlation plot of the summed orbital M1 strength below $E_x \leq 4$ MeV vs. the E2 strength from the g.s. to the first excited 2^+ state for the nuclei indicated in the figure. The solid line is a fit to the data points.

The strong M1/E2 correlation has first been discovered [46] when the respective M1 and E2 strengths were investigated in the frame of Casten's $N_p \cdot N_n$ scheme [47] in which a factor $P = N_p N_n / (N_p + N_n)$ with N_p and N_n being the number of valence protons and neutrons outside closed shells is considered. The factor P, a normalized form of $N_p \cdot N_n$, can be viewed as counting the average number of $p - n$ interactions compared to like–nucleon interactions. A correlation persists between the $B(\text{M1})$ and $B(\text{E2})$ values for the entire region of $0 \leq P \leq 8$.

An updated version [48] of the original plot [46] of the summed M1 strength vs. the P–factor is shown in Fig. 13. The rapid increase of transition strength for $4 < P < 5$ is correlated with the onset of deformation. It can be viewed in terms of increasingly dominant quadrupole interaction strength over the pairing strength. Furthermore, there is a saturation of the strength for $P \geq 6$ and the physical origin of this phenomenon is not immediately obvious. A number of suggestions have already been made in the literature (see e.g. [49, 50]) and for a rather extensive treatment of quadrupole collectivity in M1 transitions [27] and other references cited therein).

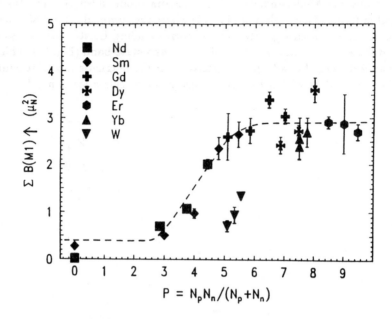

Fig. 13. Summed M1 strength vs. Casten's P-factor (this is an updated [40] version of a figure in [46]).

In the context of the observed M1/E2 correlation it is interesting to note that the IBM–2 fails to describe it [46] since for large boson numbers N the respective transition strengths behave like $B(M1) \simeq g^2 P$ and $B(E2) \simeq e_B^2 N^2$.

The physics content of Fig. 13 can be displayed slightly differently [51] if the summed orbital M1 strength is plotted vs. the mass number of the investigated nuclei (Fig. 14).

The figure shows the increase of strength towards midshell, then a saturation of the strength at about $3\,\mu_N^2$ and a drop for the heaviest rare earth nuclei, i.e. the three W isotopes. It is, of course, conceivable that this drop is a real nuclear structure effect but another possibility might just be that the strength in those nuclei is so fragmented that part of it might have just escaped detection.

The discussed experimental observations of the correlation of M1 and E2 strength and the saturation of those strengths at midshell are very interesting. Although theoretical interpretations of these facts are still uncertain one could still use a simple phenomenological approach in form of a sum rule that is essentially model independent and parameter free [43]. This sum rule is based on an expression of the M1 strength which, though first derived within the TRM is valid in a general context of the following sum rule of Lipparini and Stringari [52]:

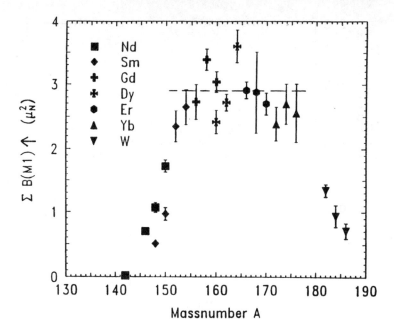

Fig. 14. Summed orbital M1 strengths in the nuclei indicated on the figure as a function of their mass number (data from a Cologne/Darmstadt/Stuttgart collaboration on nuclear resonance fluorescence experiments).

$$B(\text{M1})\uparrow \; \simeq \; \frac{3}{16\pi} \, \Theta_{sc} \, \omega_{sc} \, (g_p - g_n)^2 \, \mu_N^2 \tag{9}$$

Here ω_{sc} is the excitation energy of the scissors mode (called E_x before) and Θ_{sc} the mass parameter, which is very close to the moment of inertia. This latter quantity important in the isovector rotation can be estimated from the "classical" sum rule for E2 strength derived by Bohr and Mottelson long ago [53]. One arrives at an expression [43] for the summed strength

$$B(\text{M1})\uparrow \; \simeq \; \left\{ 0.0042 \, \frac{4NZ}{A^2} \, \omega_{sc} \, A^{5/3} \, \delta^2 \, (g_p - g_n)^2 \right\} \mu_N^2 \tag{10}$$

which is not only extremly simple and transparent but also contains the experimentally observed dependence of the M1 strength on the square of the deformation parameter.

As Fig. 15 from [51] shows convincingly that expression (10) works equally well for transitional and strongly deformed nuclei, and the ratio of the experimental over the calculated strength being unity for the majority of nuclei indicates that *all* the orbital magnetic dipole strength has been detected. This is certainly the second major discovery made recently in the field.

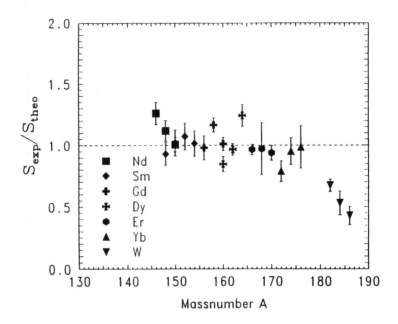

Fig. 15. Ratio of the summed experimental M1 strength and the one calculated from the expression of eq. (10). For the excitation energy of the scissors mode a value of 3 MeV was taken and for the gyromagnetic factors $g_p = 2Z/A$ and $g_n = 0$.

3.2.4 Some implications of those results: M1 strength and isotope shift

What can we learn with respect to nuclear monopole properties from the fact that the experimentally determined orbital M1 strength is related to the square of the quadrupole deformation of the ground state? Phenomenologically the nuclear isotope shift within a liquid drop model approach including a quadrupole shape deformation characterized by the usual deformation variable β is given by

$$\Delta\langle r^2 \rangle \;=\; \frac{4}{5}\, r_0^2\, A^{-1/3} \;+\; \frac{3}{4\pi}\, r_0^2\, A^{2/3}\, \Delta\langle \beta^2 \rangle, \tag{11}$$

and since (as shown *experimentally*) $\sum B(\mathrm{M1}) \sim \langle \beta^2 \rangle$, Iachello [54] and Otsuka [55] argue that nuclear monopole properties are also related to the summed orbital M1 strength. This is purely empirical though, and from such a relation variations of the nuclear radius can follow (or the other way around).

On the contrary, in a true IBM–2 approach [56] we can start from the non-energy weighted M1 sum rule of Ginocchio [50]

$$\sum B(\mathrm{M1}) \;=\; \frac{9}{4\pi}\, (g_\pi - g_\nu)^2\, \frac{P}{N-1}\, \langle 0_1^+ | \hat{n}_d | 0_1^+ \rangle \tag{12}$$

which relates the orbital M1 strength precisely to the d–boson expectation value in the nuclear g.s. using the concept of F–spin symmetry and in particular pure

F–spin for the ground state. Here P is Casten's P–factor introduced above. Since the E0 operator in the IBM is [54, 57]

$$\mathbf{T}(\text{E0}) = \gamma_0 \hat{n}_s + \beta_0 \hat{n}_d = \gamma_0 \hat{N} + \beta_0' \hat{n}_d \qquad (13)$$

and since

$$\Delta \langle r^2 \rangle = \gamma_0 + \beta_0' \, \Delta \langle \hat{n}_d \rangle \qquad (14)$$

we can rewrite the d–boson expectation value in eq. (12) and obtain

$$\sum B(\text{M1}) = \frac{9}{4\pi} (g_\pi - g_\nu)^2 \frac{1}{\beta_0'} \frac{P}{N-1} \left[\langle r^2 \rangle - \gamma_0 N \right] . \qquad (15)$$

So, while the phenomenological approach [54, 55] argues that the summed M1 strength is related to deformation and thus to the nuclear radius do we [56] start from a rather precise relation between the summed M1 strength and a particular measure of deformation, i.e. the expectation value of d–bosons in the nuclear ground state. Since in actual cases the $\gamma_0 N$ term is rather small we expect a linear relation between $\sum B(\text{M1})$ and $\langle r^2 \rangle$ with the factor $P/(N-1)$ determining the slope.

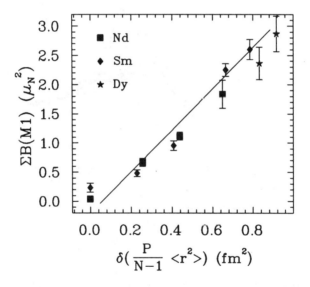

Fig. 16. Relation between the experimental orbital M1 strength and the quantity $\delta(\frac{P}{(N-1)} \langle r^2 \rangle)$ which is related to the isotope shift $\Delta \langle r^2 \rangle$.

In Fig. 16 the linear relation between the summed magnetic dipole strength and the isotope shift is indeed observed for the particular nuclei indicated in the figure. It appears that an almost constant β_0' "deformation" strength results for the particular region of rare earth nuclei considered ($86 < N < 96$). The decrease in slope when approaching the $N = 82$ closed shell spherical nuclei signals probably the breakdown of the simple expressions in eqs. (12) and (15).

The above arguments can be turned around. Large isotope shifts (as e.g. observed in the region of neutron deficient Au nuclei) should imply large M1 strengths. Furthermore, since $\sum B(M1) \sim \beta^2$ large M1 strengths might be observed in superdeformed nuclei.

One further implication where the transition strength and the excitation energy of the scissors mode play the role of a major scale is a test of F–spin symmetry discussed in [59] in detail.

3.2.5 Some remarks on summed spin magnetic dipole strengths

We have also in recent years started a comprehensive study of the spin magnetic dipole strengths in medium heavy and heavy deformed nuclei using both unpolarized and polarized protons. There is again no space here for extensive remarks on that topic and I will only list some basic features relevant to the scope of this lecture.

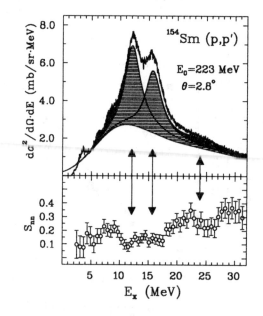

Fig. 17. Differential cross section and transverse spin–flip probability for inelastic polarized proton scattering on ^{154}Sm. The hatched areas show the double humped GDR. Visible on the low energy side of the GDR is the spin–flip M1 resonance between $E_x \simeq 5 - 12$ MeV and at $E_x = 23.4$ MeV the IVGQR. The arrows visualize the connection between the electric resonances and dips in the spin–flip probability.

One of the key nuclei we have looked at with unpolarized [33] and polarized [60] proton scattering at extreme forward angles and medium energies in a Darmstadt/Münster/TRIUMF collaboration is ^{154}Sm. As an example, the upper part of Fig. 17 shows the measured cross section for excitation energies $E_x \simeq 4 - 32$ MeV, the lower part the corresponding transverse spin–flip probability

S_{nn}. The relative maximum in S_{nn} near 8 MeV confirms the spin–flip character of the structures located at the low energy tail of the giant dipole resonance. From the angular distribution of the cross section the actual extracted spin–flip strength amounts to

$$B_\sigma(M1) \; = \; 10.5 \pm 2.0 \, \mu_N^2 \; . \tag{16}$$

A comparison of the experimental strength distribution with various theoretical predictions [44, 61, 62] is given in Fig. 18.

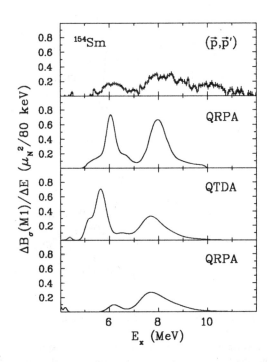

Fig. 18. Experimental and theoretical spin magnetic dipole strength distributions in ^{154}Sm. Underneath the experimental data [60] theoretical predictions from refs. [44, 61,- 62] are plotted.

The theoretical distributions have been folded with a Gaussian of variable width in order to facilitate the comparison. They were already calculated at a time when the experiment [33] revealed a double humped strength distribution. Now we find even more strength resting in a third bump which no calculation shows. From the comparison one can thus conclude that at present the agreement between the measured and the calculated distributions is still on a qualitative level. Furthermore, the theoretical predictions all use effective spin g–factors that are as usual quenched with respect to the bare ones.

Clearly, the major part of the spin magnetic dipole strength rests in the two peaks at lower excitation energies. Such a "two–peak structure" is long known from the famous M1 giant resonance problem in the spherical nucleus

^{208}Pb investigated experimentally in (e,e′) and (γ, γ') reactions. If the data of ^{208}Pb are combined with our inelastic proton scattering data (polarized and unpolarized, respectively) on the M1 giant resonance in ^{154}Sm and ^{238}U two remarkable observations occur (Fig. 19).

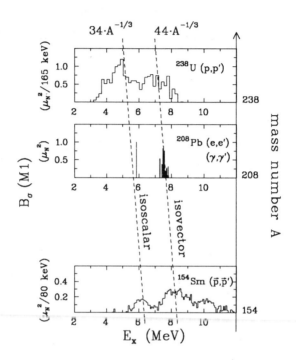

Fig. 19. Spin magnetic dipole strength distributions in ^{238}U, ^{208}Pb and ^{154}Sm. The center of gravity of the excitation energy of the two peaks representing the main strength follows a simple $A^{-1/3}$ law characteristic for spin–flip excitations between spin–orbit partner shells. The experimental strength distribution for ^{208}Pb has been combined from inelastic electron and photon scattering experiments [64, 65].

Firstly, the two open shell deformed nuclei show a much larger spreading (or fragmentation) of strength as compared to the spherical nucleus ^{208}Pb. This is certainly due to the high density of underlying quasiparticle states in ^{154}Sm and ^{238}U. Secondly, the center of gravity of the excitation energy of the two peaks obeys a simple $A^{-1/3}$ law characteristic for spin–flip excitations between shells of spin–orbit partners.

As is discussed in detail in [63], where also a full account of the present theoretical attempts to understand the spin M1 strength in the rare earth and actinide nuclei is given, this splitting can be attributed to the centroid energy difference of proton and neutron spin–flip excitations, respectivly. An analysis of the wave functions indicates that the first maximum in ^{154}Sm at 6 MeV corresponds to proton $1g_{9/2} \rightarrow 1g_{7/2}$ and $1h_{11/2} \rightarrow 1h_{9/2}$ spin–flip excitations while the second maximum is dominated by $1h_{11/2} \rightarrow 1h_{9/2}$ and $1i_{13/2} \rightarrow 1i_{11/2}$ neutron excitations, although in some states proton components are admixed. This situation is very reminiscent of the famous "isoscalar" and "isovector" neutron and proton M1–modes in the spherical nucleus ^{208}Pb that have fascinated the nuclear structure community for a long time [64, 65] and it is clear that we are dealing here in the deformed nuclei again with a collective state which has no good isospin, i.e. where "isoscalar" and "isovector" modes are mixed [66]. The interpretation in terms of a simple "isoscalar" and "isovector" picture [44] has therefore to be viewed with some caution, especially also in the light of a now experimentally found third peak in the strength distribution.

In summary, all the preceding evidence points clearly to the existence of a fairly collective spin magnetic dipole resonance in heavy deformed nuclei.

3.3 Orbital magnetic dipole mode and the isovector giant quadrupole resonance

As can be seen in Fig. 17 we also identified the isovector giant quadrupole resonance (IVGQR), about which our knowlegde is still rather scarce as compared to other giant resonances of low multipolarity, at $E_x = 23.4 \pm 0.6$ MeV in ^{154}Sm. It exhausts $76 \pm 11\%$ of the isovector E2 sum rule.

In the context of the still unsolved problems concerning the magnetic dipole modes in heavy deformed nuclei the deduced IVQGR properties play an important role. Firstly, it has been argued [67, 68] that the $K^\pi = 1^+$ component of this resonance is the real manifestation of the classical scissors mode, for which within a RPA approach an E2 ground state transition strength of 1380 e^2fm^4 has been estimated [67]. Considering that the contribution of the $K^\pi = 1^+$ component to the IVQGR is expected to be 40% of the total strength [69] we deduce 1050 ± 180 e^2fm^4 from our experiment and hence a value in fairly good agreement with the theoretical prediction. Secondly, the deduced IVGQR strength nearly completes the determined E2 strength in terms of sum rules. We are therefore able to test the completeness of the experimentally observed orbital M1 strength I spoke about in sect. 3.2.3 further by applying a recently formulated new energy weighted sum rule [70] by Moya de Guerra and Zamick to the data which connects the orbital M1 strength to the difference of the total isoscalar and isovector E2 strength.

$$\sum E_x B(M1)\uparrow = \frac{9}{16\pi} \chi \left\{ \sum_{isoscalar} B(E2)\uparrow - \sum_{isovector} B(E2)\uparrow \right\} \text{MeV} \mu_N^2 \quad (17)$$

where χ ist the strength of the quadrupole–quadrupole interaction and $B(E2)$ values are in units of e^2fm^4 (for further details, see [70]). From the experimental data for ^{154}Sm we get for the l.h.s. of eq. (17) 7.71 ± 0.44 μ_N^2 MeV and for the r.h.s. 9.32 ± 0.31 μ_N^2 MeV, i.e. we have detected below $E_x \leq 4$ MeV over 80% orbital M1 strength of the sum rule limit. This result is fully compatible with the one displayed in Fig. 15.

3.4 Summary

I have discussed an important example for an elementary excitation mode that has been and still is a major field of nuclear structure research at low energy. As I have already pointed out in [71], contrary to the well known E1 response of stable nuclei, the M1 response in a total of 12 open–shell nuclei – it is shown for six of them in Fig. 20 – could only in recent years be measured through a combination of different high–resolution probes, i.e. inelastic electron, photon and proton scattering. Remember that it took about thirty years after its discovery to understand the E1 giant resonance and it is no surprise that the data of Fig. 20 present a great challenge to nuclear structure theories.

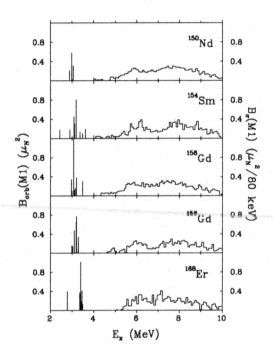

Fig. 20. Magnetic dipole response of several deformed rare earth nuclei determined by inelastic electron, photon and proton scattering.

The weakly collective *orbital* strength at around 3 MeV of excitation energy is called scissors mode strength made up from small angle vibrations of neutrons vs. protons. It is dependent on the square of the quadrupole deformation of the ground state and also strongly correlated to the E2 strength between the ground state and the first excited 2^+ state. Those facts are still not fully understood and the nature of the quadrupole–quadrupole force driving the nucleus to deformation (or alternatively the different neutron and proton deformation which might in the future be studied if pion beams of high resolution became available) needs to be clarified. Note that the $J^\pi = 1^+$ states excited from the ground state of open–shell nuclei through the scissors mode constitute still the most clear–cut examples for the mixed symmetry states of the IBM–2 and further searches for those states should be performed.

As is demonstrated beautifully in Fig. 20 the *spin* strength is removed from the orbital strength by the repulsive spin–isospin force up to a mean excitation energy of about 7.5 MeV. It is independent of deformation, strongly collective and it shows the phenomenon of quenching, which is still with us since about two decades and only partly explained. The spin strength furthermore is double humped and shows the "isoscalar–isovector" excitation picture reminiscent of the famous M1 strength problem in ^{208}Pb many people have been looking at for years. For a long time the question "Where has all the M1 strength gone?" has been with us. As Fig. 20 shows impressively: It is there, but quenched. This phenomenon will be treated further in my next lecture.

4 Looking for subnuclear effects

In the third lecture I will try to provide evidence for the fact that certain nuclear transitions at low energy as discussed in this course might well be influenced by subnuclear effects. As we will see in some detail such effects are uncovered in experiments with modern accelerators with high–quality beams and experimental detectors often through the simultaneous use of electromagnetic and hadronic probes. Such a combination has – as pointed out below – recently provided clear evidence for meson exchange current enhancement of isovector magnetic dipole strength. The conventional wisdom that the building blocks of nuclear structure are overwhelmingly provided by the nucleons is thereby not really shaken but naturally enlarged through the role which pions (and possibly even quarks) might play in addition.

4.1 The prime example for quenching of magnetic dipole spin–flip strength revisited: ^{48}Ca

Let us return to the simple case of the M1 response in nuclei and consider the classic example for a clear cut isovector excitation in the spherical nucleus ^{48}Ca which we discovered at the DALINAC now already over a decade ago [72] and which has since become the model for many more experiments and calculations. I will argue much along the same lines as before in refs. [1, 21] and show that the outstanding physical feature of the quenching of spin magnetism is still not understood at present. An (e,e') spectrum combined with (p,p') and (π, π') spectra [73, 74] taken at LAMPF, respectively, is shown in Fig. 21.

The backward angle (e,e') and the forward angle (p,p') reactions demonstrate the rather spectacular selectivity for the excitation of a real giant magnetic dipole state at 10.22 MeV, i.e. in a region of high level density in ^{48}Ca. From the signal to background ratio which is about 5 in the (e,e') and about 25 or better in the (p,p') spectrum we have another proof for the fact that the (p,p')

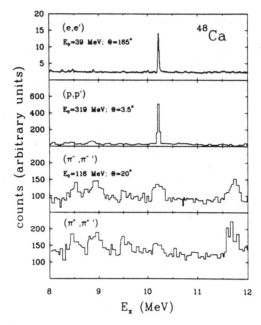

Fig. 21. Inelastic electron, proton and pion scattering spectra from the DALINAC and from LAMPF, respectively, all showing the excitation of the state at $E_x = 10.22$ MeV in ^{48}Ca (from [72, 73, 74]).

reaction is an ideal tool for investigating spin–flip strength, and indeed that it is what we have found already in some of the examples discussed in the second lecture on magnetic transitions. The pion spectra on the other hand do not show this selectivity since many more states besides the $J^\pi = 1^+$ state are strongly excited. Pions, however, have a real advantage over other probes. Since the π^-n interaction is stronger than the π^+n interaction pion scattering experiments have the power to test the validity of the assumption to which extent one is dealing here in ^{48}Ca with a pure $f_{7/2} - f_{5/2}$ neutron spin–flip transition [74].

The arguing is straightforward and I repeat here essentially what I have said in a previous lecture [1]. We know from the (e,e') experiment in which only about half of the strength predicted in the standard shell model has been observed – this very fact is called quenching – that for the reduced M1 *transition strength* the ratio

$$B(M1)_{EXP} / B(M1)_{TH} = \gamma^2 < 1. \qquad (18)$$

Similarly, if the measured $(\pi^-, \pi^{-\prime})$ *cross sections* (the data taken with negative pions are more extensive than the ones with positive pions) are compared to distorted wave impulse approximation (DWIA) calculations with sprectroscopic amplitudes taken from the same standard shell model calculation as in inelastic electron scattering, one has

$$\sigma_{exp}(\pi^-, \pi^{-\prime}) / \sigma_{DWIA}(\pi^-, \pi^{-\prime}) = \gamma^2 < 1. \qquad (19)$$

As we will see below, the quenching factor γ is roughly the same in (e,e')
and (π, π'). Since protons and neutrons contribute differently to the transition
strength in (π, π') and (e,e') the role of protons might be illuminated by combining both experiments. For that we make the following model. We decompose the
M1 matrix element in (e,e') – denoted here briefly as $\langle M1 \rangle$ – into an isoscalar
part $\langle IS \rangle$ and an isovector one $\langle IV \rangle$, i.e. we have

$$\langle M1 \rangle \;=\; \langle M1, IS \rangle + \langle M1, IV \rangle \tag{20}$$

and can relate the isoscalar and isovector single particle matrix elements to
the matrix elements for the excitation of a proton $\langle p \rangle$ or a neutron $\langle n \rangle$ in the
standard way:

$$\langle IS \rangle \;=\; (\langle p \rangle + \langle n \rangle)\,\sqrt{2} \tag{21}$$

$$\langle IV \rangle \;=\; (\langle p \rangle - \langle n \rangle)\,\sqrt{2} \tag{22}$$

Contrary we have:

$$\langle p \rangle \;=\; (\langle IS \rangle + \langle IV \rangle)\,\sqrt{2} \tag{23}$$

$$\langle n \rangle \;=\; (\langle IS \rangle - \langle IV \rangle)\,\sqrt{2} \tag{24}$$

The ratio $\langle IS \rangle / \langle IV \rangle$ is determined by the isoscalar and isovector gyromagnetic factors (as pointed out in the first lecture) and we have

$$\langle M1, IS \rangle / \langle M1, IV \rangle \;\approx\; 0.09 \tag{25}$$

i.e. isovector dominance in (e,e'). On the other hand - as one can look up in
text books – the matrix element for the excitation of the $\Delta(1232)$ resonance
in $(\pi^+, \pi^{-\,\prime})$ is three times as large for the π^- interaction with the neutron as
compared to the proton. We have therefore

$$\begin{aligned}
\langle \pi^-, IS \rangle / \langle \pi^-, IV \rangle \;&=\; (\langle p \rangle + \langle n \rangle) / (\langle p \rangle - \langle n \rangle) \\
&=\; (1 + 3) / (1 - 3) \\
&=\; -2 \,, \tag{26}
\end{aligned}$$

i.e. isoscalar dominance in (π, π').

As a *test* of this simple model let us take the open shell model $(fp)^8$ transition
amplitudes of McGrory and Wildenthal [75], abbreviated in the form

$$\langle 1^+, (fp)^8 \rangle \;=\; (\langle 1^+, IS \rangle - \langle 1^+, IV \rangle)\,\sqrt{2} \tag{27}$$

and rescale them by $(1+\delta_0)$ and $(1+\delta_1)$ such that the experimentally measured
transition strength is reproduced, i.e. in the form

$$\langle 1^+, EXP \rangle \;=\; [(1 + \delta_0)\langle 1^+, IS \rangle - (1 + \delta_1)\langle 1^+, IV \rangle]\,\sqrt{2} . \tag{28}$$

The rescaling factors are plotted in Fig. 22. The point ($\delta_0 = \delta_1 = 0$ or $1 + \delta_0 =
1 + \delta_1 = 1$, respectively) represents the original $(fp)^8$ model prediction.

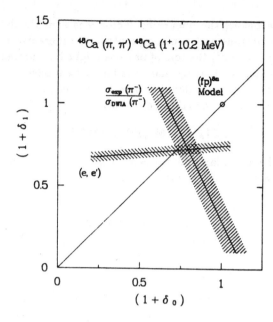

Fig. 22. Connection between the isovector $(1+\delta_1)$ and isoscalar $(1+\delta_0)$ rescaling factors in the excitation of the $J^\pi = 1^+$ state in ^{48}Ca at $E_x = 10.22$ MeV in inelastic electron and pion scattering. The given straight lines with error bands represent the values of the rescaling factors consistent with the description of the (e,e') and $(\pi^-, \pi^-{}')$ experiment. The diagonal stands for rescaling factors in case of a pure neutron transition.

Pure neutron transitions are characterized by the diagonal. In order to put the experimental values into Fig. 22, we rewrite eq. (28) into a linear equation of the form

$$(1+\delta_1) = \frac{\langle 1^+, IS\rangle}{\langle 1^+, IV\rangle}(1+\delta_0) - \frac{\sqrt{2}\langle 1^+, EXP\rangle}{\langle 1^+, IV\rangle} \qquad (29)$$

As has been noted in eq. (25), $\langle 1^+, IS\rangle/\langle 1^+, IV\rangle \approx 0.09$ in (e,e'), and we consequently have a slightly increasing slope with a positive intercept at $1+\delta_0 = 0$ because of $\langle 1^+, EXP\rangle \approx -\langle 1^+, IV\rangle/\sqrt{2}$. The scattering of negative pions yields (see eq. (26)) the steeply decreasing line in Fig. 22.

The crossing of the error bands of the (e,e') and $(\pi^-, \pi^-{}')$ experiments near the diagonal line representing a pure neutron transition is a confirmation that the $J^\pi = 1^+$ state in ^{48}Ca is excited in a fairly pure neutron transition. Furthermore, the *consistency* of the approach in which (e,e') and $(\pi^-, \pi^-{}')$ experiments have been combined lies in the fact that both experiments yield the same rescaling or quenching factor (see eqs. (18) and (19)), i.e. $\gamma^2 = (1+\delta_0) = (1+\delta_1) \approx 0.53$. It is clear from our arguments, that $(\pi^+, \pi^+{}')$ experiments would yield a line with increasing slope in Fig. 22. Such measurements would therefore check our conclusion of a fairly pure neutron transition further.

After this transparent exercise how pions, by their nature, in combination with electrons are able to test the role of neutrons and protons in a transition on a quantitative level let us return to the basic nature of the quenching by looking at the reduced M1 strength distribution from the ^{48}Ca(e,e') experiment (Fig. 23). There is little spreading of the strength. The strongest state at $E_x = 10.22$ MeV carries a strength of $B(\text{M1}) \uparrow = 3.9 \pm 0.3 \mu_N^2$ while the total observed strength between $E_x \approx 8 - 17$ MeV amounts to $5.3 \pm 0.6 \mu_N^2$ (see [76]).

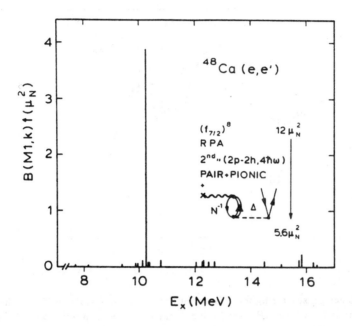

Fig. 23. Magnetic dipole strength distribution in ^{48}Ca (from [76]). The insert shows how the standard shell-model value of $12 \mu_N^2$ for the strength is reduced by various mechanisms down to $5.6 \mu_N^2$ close to the experimental value (from [77]).

As the insert of Fig. 23 shows, the measured strength is strongly quenched compared to the shell model value of $12 \mu_N^2$. There is a wealth of articles in the literature all trying to explain this quenching. One of the most recent very careful approaches to the problem has been the one by Arima and collaborators [77], using a new method of calculating the response function. Their ordinary RPA, second order RPA (including $2p - 2h$ states and excitations up to $4\hbar\omega$), pair plus pionic current and delta–hole current treatment results in a transition strength of $5.6 \mu_N^2$ close to the experimental value of $3.9 \mu_N^2$ for the strongest state. They conclude that the mixing of $2p - 2h$ states is more important than the Δ–hole state in the quenching phenomenon. The mixing of $2p - 2h$ states reduces the strength more than 20%, while the total exchange current (pair, pionic and $\Delta - h$ currents) decreases the strength of the not well-known $NN \rightarrow N\Delta$ transition potential.

The calculated quenching is still not enough to explain the experimental strength. This is also reflected by the form factor shown in Fig. 24. The RPA prediction (which is equivalent to a standard nuclear configuration mixing prediction with wave functions from [75]) greatly overshoots the first form factor maximum by a large amount. The situation is much improved by the configuration space calculation which has all exchange current contributions included, but there is still a difference to the experimental data.

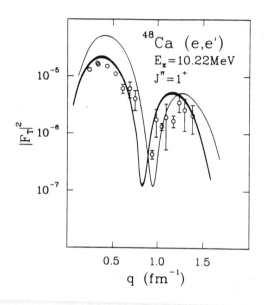

Fig. 24. Electron scattering form factor of the 10.22 MeV state in ^{48}Ca. The data [76] are compared to a $0\hbar\omega$ RPA prediction (thin line) and a prediction [77] in the $4\hbar\omega$ configuration space with all the exchange currents included (thick line).

Those findings are essentially in agreement with those from similar model approaches [78, 79], and it is very interesting to see how the upcoming large space fp-shell model calculations of Poves et al. [80] will be able to account for the experimental observations.

Clearly, the origin of the quenching of spin magnetism – as demonstrated here in some detail at the model case of ^{48}Ca – is presently still not understood.

4.2 Evidence for meson exchange current enhancement of isovector magnetic dipole strength in complex nuclei

It is well known that a purely nucleonic description of nuclei is insufficient to account for elctroweak observables in nuclei. Meson–exchange and Δ–isobar currents modify transition probabilities in Gamow–Teller (GT) β decay, and also lowest–order magnetic properties such as magnetic moments, M1 transition probabilities, and magnetic form factors [81, 82]. The identification of meson–exchange currents (MEC's) is straightforward in very light nuclei where there is very little uncertainty in the nucleonic wave function. In heavier nuclei it is difficult to isolate MEC or isobar effects because of uncertainties in the many-body nucleonic calculations. For example the quenching of GT strength observed in many nuclei [83] can be attributed to Δ–isobar currents, or to higher–order configuration mixing, and the relative importance of these two contributions is open to debate, as we have just discussed in the first part of this lecture.

Here, in the second part of this lecture which follows closely the presentation given earlier [84], data on isovector M1 and GT transitions to a triad of 1^+, $T = 1$ analog states in open–shell $A = 24$ nuclei are presented which allow a relatively clean identification of MEC contributions to the isovector M1 operator (for details see [84]). The experiments determine M1 and GT strength distributions from (e,e') and nucleon scattering experiments, respectively. For a self–conjugate $T_3 = 0$ target nucleus such as ^{24}Mg, and for M1 and GT transitions to 1^+, $T = 1$ final states, one may write approximately [81]

$$B(\text{M1}) \;=\; \frac{3(\mu_p - \mu_n)^2}{8\pi}\left[M(\sigma) + M(l) + M_\Delta + M_V^{MEC}\right]^2, \qquad (30)$$

$$B(\text{GT}) \;=\; \left[M(\sigma) + M_\Delta + M_A^{MEC}\right]^2, \qquad (31)$$

where the numerical factor in the $B(\text{M1})$ expression is 2.643 μ_N^2 and the ratio of coupling constants $(g_A/g_V)^2$ is not included in the definition of $B(\text{GT})$. The nucleonic spin matrix elements $M(\sigma)$ and the isobar contributions M_Δ are the same in both expressions. The MEC contributions are dominated by pion exchange and are predicted to be large for isovector and for axial–vector (GT) currents because of conservation of G parity. Unfortunately, the M1/GT comparison tends to be complicated by the (nucleonic) orbital contribution $M(l)$ to the M1 matrix element. The combined effects of orbital and the MEC contributions are measured by ratio

$$R(\text{M1/GT}) \;=\; \frac{\sum B(\text{M1})/2.643\mu_N^2}{\sum B(\text{GT})}. \qquad (32)$$

In their absence $R(\text{M1/GT})$ is unity, irrespective of the complexity of the nucleonic wave functions and of the exact magnitude of Δ–isobar contributions. Thus the sensitivity to uncertainties in the dominant nucleonic spin contributions is greatly reduced in $R(\text{M1/GT})$. The orbital contributions can be reliably predicted in the sd–shell using the unified sd–shell (USD) effective interaction of Wildenthal [85] which has been tested against a large body of experimental data. For the target nucleus ^{24}Mg, MEC contributions are expected to dominate the orbital contributions, especially when the M1 strength can be summed over a large region of excitation. The running sum of $B(\text{M1})$ strength in ^{24}Mg is shown in Fig. 25 (cross–hatched area) together with the USD predictions using free–nucleon values for spin and orbital g–factors in the M1 single–particle operator (dotted line).

We quote $\sum B(\text{M1})$ for two values of E_{max}, the upper limit in the running sum. The first value, $(4.85 \pm 0.36)\mu_N^2$ for $E_{max} = 11.4$ MeV, is dominated by the well established states at 9.96 and 10.71 MeV, whereas the second value $(5.84 \pm 0.40)\mu_N^2$ for $E_{max} = 15$ MeV, relies more heavily on the extraction of

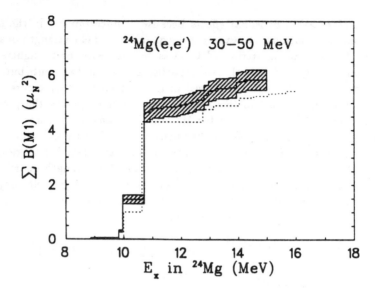

Fig. 25. Running sum of M1 strength in ^{24}Mg from the present experiment (cross-hatched area) and prediction of the USD shell model with free–nucleon g–factors (dotted curve), from [84].

relatively weak M1 strength. The corresponding M1 enhancement factors relative to the USD predictions, 1.13 ± 0.08 and 1.11 ± 0.08, show very satisfactory agreement.

The best estimates of total GT strengths in nuclei are at present obtained from (p,n), (p,p') and (n,p) reactions at energies between 120 and 500 MeV. The GT analogs of the isovector M1 transitions are driven by the strong $(\sigma\tau)$ part of the NN interaction. In the limit of vanishing energy transfer ($\omega = 0$) and momentum transfer ($q = 0$), the cross sections to members of a $1^+, T = 1$ isospin triad are related by $\sigma_{pn} = \sigma_{np} = 2\sigma_{pp'}$, where the factor of 2 arises from isospin coupling coefficients for the projectile. The cross sections can be converted to $B(GT)$ using the "unit cross section" $\hat{\sigma} = \sigma(q = 0, \omega = 0)/B(GT)$.

With these unit cross sections the result of the four experiments are shown in Fig. 26. The dotted lines correspond to the USD predictions of GT strength with free–nucleon values for the spin g factors. For E_{max} corresponding to 11.4 and 15 MeV in ^{24}Mg the $\sum B(GT)$ values are 0.74 ± 0.09 and 1.20 ± 0.17, respectively. This implies quenching of the GT strength relative to the USD predictions by factors of 0.72 ± 0.09 and 0.71 ± 0.10, respectively. The proposed M1/GT comparison of experimental data and calculations (quantified in Table I of [84]) is shown in Fig. 27.

We have concentrated on the total strengths which demonstrate the phenomenon of MEC most clearly. The USD interaction [85] gives the best available estimate of configuration mixing. With bare g–factors it lies below the experimentally determined ratio in the excitation energy ranges. The respective

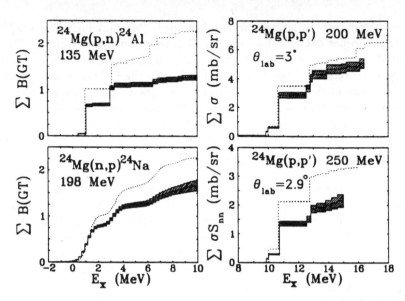

Fig. 26. Running sums of the GT strength from recent (p,n), (p,p′) and (n,p) experiments. The dotted lines represent predictions of the USD shell model with free–nucleon g-factors from [84].

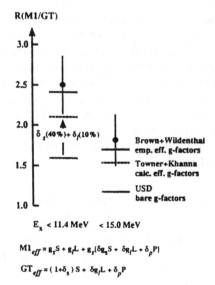

Fig. 27. Comparison of isovector M1 and GT strengths in ^{24}Mg. It is seen that in order to account for the experimental R (full dots with error bars) in the two indicated excitation energy ranges effective g-factors in the operators listed in the bottom of the figure have to be used in the calculations [86, 87]. Compared to the USD prediction with bare g-factors the $R(M1/GT)$ ratio is enhanced about 40% by the δ_s contribution and this is the effect of the enhanced vector over the axial vector MEC.

ratios predicted with empirical effective operators [86] that include the effects due to higher–order configuration mixing, Δ–isobar admixtures and MEC's are in essential agreement with the data. A quantitative analysis shows that the necessary enhancement of the $R(M1/GT)$ ratio over the USD calculations is due to the spin part of the effective operator, i.e. δ_s has to be enhanced by about 40%. As is seen in [87] this enhancement in the spin operator is essentially due entirely to the difference between MEC contributions to M_V^{MEC} and M_A^{MEC} discussed above and is the effect we are looking for. There is an additional 10% enhancement in $R(M1/GT)$ due to the δ_l contribution to the orbital part of the $B(M1)$. In the Towner–Khanna calculations this term originates from a strong cancellation between effects due to MEC's and those due to higher–order nuclear configuration mixing. However, the δ_l contribution is a factor 4 less important than the δ_s contribution in this case. The δ_p contribution is negligible.

Thus, the $R(M1/GT)$ ratio clearly is a senitive and direct measure of the MEC correction to the spin operator. The experimental value of $R(M1/GT)$ is in excellent agreement with expectations based on both [86, 87] effective operators, and hence directly confirms the importance of MEC's in nuclei, as has been shown with the example of ^{24}Mg. This has become possible through high precision electromagnetic and hadronic cross sections and their combined analysis, an improved knowledge of the effective nucleon–nucleon interaction in nuclei, and the existence of reliable many–body wave functions and effective operators for sd–shell nuclei. It will be interesting in the future to extend this type of comparison to other regions of the sd–shell, and we are working presently on a similarly precise electromagnetic and hadronic data set in the $A = 28$ nuclei.

In order to get already a feeling for the quality of the electromagnetic data I am rather proud to present in Fig. 28 a first inelastic electron scattering spectrum taken on ^{28}Si at a 180°–(e,e') high–resolution scattering facility which has just been installed at the S–DALINAC by a Catholic University of Washington / TH Darmstadt collaboration. This spectrometer is essentially a "spin-filter" whereby the multipolarity of transverse excitations and the transitions can be determined rather quickly and reliably.

4.3 Search for Δ–isobar components in the nuclear wave function

At the end of my lectures I will briefly revisit an old problem: What is the probability of Δ–isobar components in the nuclear wave function?

In the workshop at Scillac (France) from June 27 – July 1, 1988 "Hadronic Physics in the '90 with Multi–GeV Electrons" Torleif Ericson said in his introductory remarks [88]:

"It is obviously crucial to reach a proper knowledge of the building blocks and their interactions. Only then is it possible to master the effective interactions between these constituents to such a degree that we are perfectly certain we understand the physics and the quantitative predictions. Here is much work to be done, in spite of the spectacular individual successes we have achieved. In this picture the Δ–isobar enters in the physics of nuclei as a component with the same status as the nucleon. Thus, its interactions with other baryons, its aver-

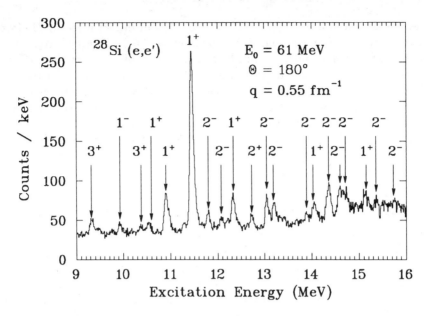

Fig. 28. First inelastic electron scattering spectrum taken at $\theta = 180°$ in ^{28}Si at the S-DALINAC. The spins and parities of all states shown can be determined straightforwardly in form factor measurements at two or three momentum transfers q.

age interaction in nuclear matter as well as its components in the nuclear wave function are of basic importance to our field."

We have thus decided to use the electromagnetic response of nuclei to possibly constrain the probability P_Δ of a Δ-component in the nuclear ground state. Since the 3-body system is completely treatable in a nonrelativistic approximation and since it is thus a good compromise between a "real" nucleus and tractability we wish to measure P_Δ in ^{3}He.

The ^{3}He ground state (g.s.) wave function is well known by precise measurements of the charge form factor of ^{3}He. These data are showing a significant deviation from a pure nucleonic ground state wave function, indicating contributions from baryonic resonances which are already present in the nucleus. This leads to a wave function of the form

$$|^3\text{He}\rangle \; = \; \alpha\,|ppn\rangle \; + \; \beta\left(\frac{1}{\sqrt{2}}|\Delta^{++}nn\rangle - \frac{1}{\sqrt{3}}|\Delta^{+}pn\rangle + \frac{1}{\sqrt{6}}|\Delta^{0}pp\rangle\right) \qquad (33)$$

where α and β are the amplitudes of the pure nucleonic and Δ-isobar pieces, respectively. Note the large Δ^{++} excess and the absence of a Δ^--component which plays a major role in the discussion that follows.

To my knowledge the only nucleus where experiments have yielded Δ-isobar components in the ground state wave function is deuterium and a probability of about $0.3 - 1.0\%$, depending on the model used, has been estimated [89, 90]. For

^{3}He the Δ component might be about 2% [91] and in fact a recent ^{3}He(γ,p$\pi^\pm$) experiment – performed along the same lines as the ^{3}He(e,e'$\pi^\pm$) experiment to be discussed in a moment – ended up at a Δ component less than 2% at the 90% confidence level [92]. In heavy nuclei, the Δ admixture into excited $J^\pi = 1^+$ states was also investigated and a value of about 3% has been deduced from the quenching of Gamow–Teller and M1 strength, that I have discussed in some detail in these lectures, with the help of the constituent quark model, however, in a very model dependent way [93, 94]. The problem has achieved additional significance recently when it was pointed out that a non–negligible value of P_Δ will even have an influence on the shell structure of nuclei [95].

4.3.1 The ^{3}He(e,e'$\pi^\pm$) reaction, Δ photoproduction and Δ knock–out

The ^{3}He(e,e'$\pi^\pm$) reaction which we have studied at the continuous wave electron accelerator MAMI within the A1 collaboration has been proposed by Lipkin and Lee as a sensitive test for the presence of Δ's in nuclei [96]. The idea is to use the diffent signatures of detecting the two charged pions for the two different reaction mechanisms, photoproduction of Δ and Δ knock–out. In the following I will use their arguments to describe how to obtain experimentally the probability P_Δ.

At first some assumptions are made about the properties of the Δ. Although it is really unclear whether a Δ present in the nuclear ground state can be described as an elementary fermion, as composed by three quarks or as a pion–nucleon resonance, Lipkin and Lee start from the following properties:

- The Δ is an object with isospin 3/2.
- Isospin is conserved in the strong interaction which mixes a Δ into a nuclear wave function.
- The amplitude for the interaction of a photon with a Δ is proportional to the charge of the Δ.

The first two assumptions are valid in almost any model and the third is correct if the Δ is either an elementary or a composite object in a totally symmetric wave function, as predicted by the standard quark model. The third assumption means furthermore, that a Δ^0 cannot be knocked out by a photon, but all other Δ–charge states can.

If a Δ is present in a nucleus it may be possible to knock it out with a photon and look for its effects in the final πN multiparticle state, but it is clearly necessary to distinguish between Δ's already present before the collisions and Δ's produced. Since, however, photoproduction of a Δ by a real or virtual photon from a single nucleon is a magnetic dipole transition involving transverse photons only, looking for the Δ knock–out in the longitudinal channel (where the Δ seems to be conspiciously absent [97]) is certainly favourable.

Let us briefly consider the two competing processes in some detail. In Δ photoproduction on a single proton always a Δ^+ is photoproduced, on a single neutron a Δ^0. The Δ^{++} cannot be photoproduced on a single nucleon, likewise Δ^-. Because of the nearly total symmetry of the ^{3}He ground state wave function in space, the photoproduction on a neutron or a proton has nearly equal probability. So we find for the relative probabilities for photoproducing the different charge states of Δ:

$$P_n \left(\gamma^3\text{He} \to \Delta^+ X\right) / P_n \left(\gamma^3\text{He} \to \Delta^0 X\right) = 2 \tag{34}$$

$$P_n \left(\gamma^3\text{He} \to \Delta^{++} X\right) = P_n \left(\gamma^3\text{He} \to \Delta^- X\right) = 0 \tag{35}$$

Taking into account the decay probabilities of Δ into the different πN–states, we obtain for the relative probabilities P_n to detect a specific charge state of a pion if the Δ was photoproduced:

$$P_n \left(\gamma^3\text{He} \to \pi^+ X\right) : P_n \left(\gamma^3\text{He} \to \pi^0 X\right) : P_n \left(\gamma^3\text{He} \to \pi^- X\right) = \frac{2}{9} : \frac{6}{9} : \frac{1}{9} \tag{36}$$

For Δ *knock–out* at first we must consider the relative probabilities of different charge states in the ^{3}He ground state. They are given by the Clebsch–Gordan coefficients for coupling the Δ to the two remaining nucleons and are found to be

$$P_k(\Delta^{++} nn) : P_k(\Delta^+ np) : P_k(\Delta^0 pp) = \frac{1}{2} : \frac{1}{3} : \frac{1}{6} \tag{37}$$

The third assumption of Lipkin and Lee above states that the Δ–photon coupling is proportional to the charge of the Δ. Therefore the Δ^{++} knock–out process is expected to be four times larger than the Δ^+ knock–out and a Δ^0 cannot be knocked–out because it does not couple to the photon. So we obtain the relative knock–out probabilities as

$$P_k \left(\gamma^3\text{He} \to \Delta^{++} X\right) / P_k \left(\gamma^3\text{He} \to \Delta^+ X\right) = 6 \tag{38}$$

$$P_k \left(\gamma^3\text{He} \to \Delta^0 X\right) = P_k \left(\gamma^3\text{He} \to \Delta^- X\right) = 0 \tag{39}$$

Again, with the decay probabilities into the different πN–states, we get the relative probabilities to detect a specific charge state of a pion if the Δ was pre–existing in the nucleus:

$$P_k \left(\gamma^3\text{He} \to \pi^+ X\right) : P_k \left(\gamma^3\text{He} \to \pi^0 X\right) : P_k \left(\gamma^3\text{He} \to \pi^- X\right) = \frac{19}{21} : \frac{2}{21} : 0 \tag{40}$$

Thus, comparing the results of the model for Δ photoproduction in eq. (36) and Δ knock–out in eq. (40) one observes very different signatures of the two processes, i.e. a skewness to positively charged pions in the knock–out process which results from the large probability of Δ^{++} over Δ^+ and Δ^0 (due to the couplings $4e^2, e^2$ and 0, respectively). In our ^{3}He$(e, e'\pi^{\pm})$ experiment we search just for this skewness in the π^+/π^- production ratio in the longitudinal channel at favourable kinematics where the incoming electron is scattered quasi–free on a "pre–existing" delta. The salient features behind the experimental are summarized again in Fig. 29.

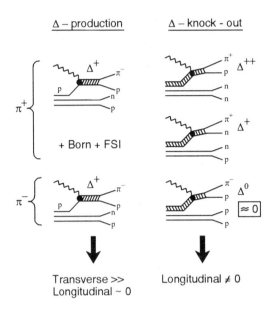

Fig. 29. The relevant Δ production and Δ knock-out graphs in experimental search for Δ-components in the ^{3}He ground state.

In the conceptually very simple Lipkin–Lee model only electroproduction of Δ's and Δ knock–out were taken into account, but nonresonant contributions, charge exchange, pion rescattering and other final state interactions (FSI) are ignored. They all can be, however, and have already been calculated for our experiment by Laget [98] and served for expected count rate estimates. It is clear that we are looking for an effect contained in only about 5% of the total number of produced pions.

We have used the 855 MeV/c electron beam from MAMI and the two big magnetic spectrometers A and B for pion and electron detection, respectively, in coincidence. The energy transfer ranged between 365 and 435 MeV which is slightly beyond that necessary for the electoproduction of the Δ-resonance. The pions from Δ-decay in flight were detected in the direction of the virtual photon. We worked in longitudinal kinematics; energy and momentum transfer, respectively, were kept constant for a Rosenbluth separation. Positively and negatively charged pions in coincidence with inelastically scattered electrons were detected under identical kinematical conditions with just the field of the pion spectrometer reversed.

The experiment has been performed just before easter 1994 and I am happy to show in Fig. 30 the first, still very preliminary results. In the upper part the $(e,e'\pi^+)$ channel is illustrated leading to a sharp peak that characterizes the coherent pion production into the ^{3}He g.s. and into the nd and nnp continuum. Naturally, the coherent pion production peak is absent in the $(e,e'\pi^-)$ channel

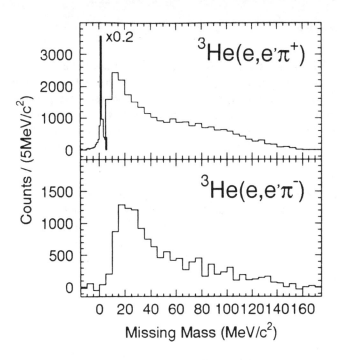

Fig. 30. First preliminary pion electroproction spectra on ^{3}He taken under the kinematical conditions indicated in the figure. The calculation of Laget (solid line) has been normalized to the continuum contribution in the $(e,e'\pi^+)$ spectrum. Note the sharp peak due to coherent pion production in the upper part which is absent in the lower part.

shown in the lower part. If one integrates the number of pions going into the continuum the ratio $N(\pi^+)/N(\pi^-)$ is about 2 indicating that the majority of the detected pions must come from delta production (see eq.(36)).

Laget's preliminary calculations, after being properly normalized on the peak in the π^+ coincidence spectrum, leave a surplus of pions in which the effect of Δ knock-out we are looking for might infact be hidden. We are presently performing the necessary Rosenbluth separation of the cross sections to isolate the wanted longitudinal channel.

In passing I note that we are in the process of preparing an extension of the present ^{3}He$(e,e'\pi^\pm)$ experiment in which we wish to detect electron, pions and protons in a triple coincidence (using in addition the third spectrometer at MAMI) in the ^{3}He$(e,e'p\pi^+)$nn and ^{3}He$(e,e'p\pi^-)$pp channels which are isospin symmetric in undetected nucleon pairs. It is hoped thereby that the weak signal which we are looking for in the present experiment will stand out stronger above the background of direct pion production. In a further increased step of sensitivity the ratio of charged pion asymmetries in electroproduction using a longitudinally polarized electron beam and a polarized ^{3}He target might also be looked at [99].

5 Conclusions

In these three lectures I have tried to highlight some of the experiments which were undertaken to study the nuclear response with electromagnetic probes at low energy. By way of specific examples I have discussed the power of inclusive single–arm photon, electron and proton and exclusive (e,e'x) and (p,p'x) coincidence reactions in evaluating particular features of low multipolaritiy electric and magnetic transitions and of small pieces in nuclear wave functions. An important reason for attempting to measure particular transition strengths with high precision and modern continuous electron accelerators like the S–DALINAC and MAMI and the simultaneous use of hadronic probes to look at the same subject has not only increased our understanding of the nucleus itself on a pure nucleonic level but has also thaught us about the existence of subtle subnucleonic effects. Undoubtely, many more special cases – some of them I have already indicated during the course of the lectures – will be uncovered as we continue to pursue the advantages of studying nuclear physics with electromagnetic probes.

Acknowlegements

I am very grateful to my many collaborators at the S–DALINAC and elsewhere for discussion and their many valuable contributions to the material of these lectures. Special thanks are due to P. von Brentano, B.A. Brown, C. Campos, H. Diesener, O. Engel, D. Frekers, H.-D. Gräf, O. Häusser, K. Heyde, U. Kneissl, N. Lo Iudice, M. Kuss, C. Lüttge, P. von Neumann-Cosel, C. Rangacharyulu, G. Schrieder, E. Spamer, S. Strauch, T. Suda, H. Wörtche, A. Zilges and the members of the A1–collaboration at MAMI for sharing their insight into the various topics treated in these lectures. Last but not least I thank Friedrich Neumeyer not only for his physics contribution but also for his great help and skill in preparing the manuscript of the lectures.

References

1.   A. Richter: Prog. Part. Nucl. Phys. **13** 1 (1985)
2.   H. Miska, H.D. Gräf, A. Richter, D. Schüll, E. Spamer and O. Titze: Phys. Lett. **B59** 441 (1975)
3.   H.D. Gräf, V. Heil, A. Richter, E. Spamer, W. Stock and O. Titze: Phys. Lett. **B72** 179 (1977)
4.   S.M. Kormanyos, R.J. Peterson, J. Applegate, J. Beck, T.D. Averett, B.G. Ritchie, C.L. Morris, J. McGill and D.S. Oakley: Phys. Rev. **C48** 250 (1993)
5.   R.J. Peterson: private communication
6.   Y. Torizuka, M. Oyamada, K. Nakahara, K. Sugiyama, Y. Kojima, T. Terasawa, K. Itoh, A. Yamaguchi and M. Kimura: Phys. Rev. Lett. **22** 554 (1969)
7.   B.A. Brown: private communication
8.   E.K. Warburton and B.A. Brown: Phys. Rev. **C46** 923 (1992)

9. H.L. Harney, A. Richter and H.A. Weidenmüller: Rev. Mod. Phys. **58** 607 (1986)
10. H. Diesener, U. Helm, G. Herbert, V. Huck, P. von Neumann-Cosel, C. Rangacharyulu, A. Richter, G. Schrieder, A. Stascheck, A. Stiller, J. Ryckebusch and J. Carter: Phys. Rev. Lett. **72** 1994 (1994)
11. P. von Neumann-Cosel, H. Diesener, U. Helm, G. Herbert, V. Huck, A. Richter, G. Schrieder, A. Stascheck, A. Stiller, J. Carter, A.A. Cowley, R.W. Fearick, J.J. Lawrie, S.J. Mills, R.T. Newman, J.V. Pilcher, F.D. Smit, Z.Z. Vilakazi and D.M. Whittal: Nucl. Phys. **A569** 373c (1994)
12. F. Zwarts, A.G. Drentje, M.N. Harakeh and A. van der Woude: Phys. Lett. **B125** 123 (1983); Nucl. Phys. **A439** 117 (1985).
13. D. Brajnik, D. Jamnik, G. Kernel, U. Miklavzic and A. Stanovnik: Phys. Rev. C **9** 1901 (1974)
14. Th. Kihm, K.T. Knöpfle, H. Riedesel, P. Voruganti, H.J. Emrich, G. Fricke, R. Neuhausen and R.K.M. Schneider: Phys. Rev. Lett. **56** 2789 (1986)
15. J. Ahrens, H. Borchert, K.H. Czock, H.B. Eppler, H. Gimm, H. Gundrum, M. Kröning, P. Riehm, G. Sita Ram, A. Zieger and B. Ziegler: Nucl. Phys. **A251** 479 (1975)
16. A. Veysièrre, H. Beil, R. Bergère, P. Carlos, A. Lepêtre and A. de Miniac: Nucl. Phys. **A227** 513 (1974)
17. R. Benz: Diploma thesis, Technische Hochschule Darmstadt (1984), unpublished
18. S. Kamerdzhiev, J. Speth and G. Tertychny: KFA Jülich Annual Report (1993) p. 157; and submitted to Phys. Rev. Lett.
19. W. Knüpfer: private communication
20. U. Helm: Ph.D. thesis, Technische Hochschule Darmstadt (1990), unpublished
21. A. Richter: Nucl. Phys. **A522** 139c (1991)
22. A. Richter, in: "The Building Blocks of Nuclear Structure", ed. A. Covello (World Scientific, Singapore, 1993) p. 335
23. D. Bohle, A. Richter, W. Steffen, A.E.L. Dieperink, N. Lo Iudice, F. Palumbo and O. Scholten: Phys. Lett **B137** 27 (1984)
24. N. Lo Iudice and F. Palumbo: Phys. Rev. Lett. **41** 1532 (1978)
25. F. Iachello: Phys. Rev. Lett. **53** 1427 (1984)
26. A. Arima, T. Otsuka, F. Iachello and I. Talmi: Phys. Lett. **B66** 205 (1977)
27. K. Heyde, C. De Coster, A. Richter and H.-J. Wörtche: Nucl. Phys. **A549** 103 (1992)
28. R. Nojarov, A. Faessler, P. Sarriguren, E. Moya de Guerra and M. Grigorescu: Nucl. Phys. **A563** 349 (1993)
29. R. Nojarov: Nucl. Phys. **A571** 93 (1994)
30. A.A. Raduta, I.I. Ursu and N. Lo Iudice: Nucl. Phys. **A551** 73 (1993)
31. D. Zawischa and J. Speth: Nucl. Phys. **A569** 343c (1994)
32. M. Sambataro, O. Scholten, A.E.L. Dieperink and G. Piccitto: Nucl. Phys. **A413** 333 (1984)
33. D. Frekers, H.J. Wörtche, A. Richter, R. Abegg, R.E. Azuma, A. Celler, C. Chan, T.E. Drake, R. Helmer, K.P. Jackson, J.D. King, C.A. Miller, R. Schubank, M.C. Vetterli and S. Yen: Phys. Lett. **B244** 178 (1990)
34. N. Huxel, W. Ahner, H. Diesener, P. von Neumann-Cosel, C. Rangacharyulu, A. Richter, C. Spieler, W. Ziegler, C. de Coster and K. Heyde: Nucl. Phys. **A539** 478 (1992)
35. I. Bauske, J.M. Arias, P. von Brentano, A. Frank, H. Friedrichs, R.D. Heil, R.-D. Herzberg, F. Hoyler, P. Van Isacker, U. Kneissl, J. Margraf, H.H. Pitz, C. Wesselborg and A. Zilges: Phys. Rev. Lett. **71** 975 (1993)

58   A. Richter

36.   W. Ziegler, C. Rangacharyulu, A. Richter and C. Spieler: Phys. Rev. Lett. **65** 2515 (1990)

37.   J. Margraf, R.D. Heil, U. Maier, U. Kneissl, H.H. Pitz, H. Friedrichs, S. Lindenstruth, B. Schlitt, C. Wesselborg, P. von Brentano, R.-D. Herzberg and A. Zilges: Phys. Rev. **C47** 1474 (1993)

38.   C. De Coster and K. Heyde: Phys. Rev. Lett. **63** 2797 (1989)

39.   I. Hamamoto and C. Magnusson: Phys. Lett. **B260** 6 (1991)

40.   E. Garrido, E. Moya de Guerra, P.Sarriguren and J.M. Udias: Phys. Rev. **C44** R1250 (1991)

41.   R.R. Hilton, W. Höhenberger and H.J.Mang: Phys. Rev. **C47** 602 (1993)

42.   K. Heyde and C. De Coster: Phys. Rev. **C44** R2262 (1991)

43.   N. Lo Iudice and A. Richter: Phys. Lett. **B304** 193 (1993)

44.   P.Sarriguren, E. Moya de Guerra, R. Nojarov and A. Faessler: J. Phys. **G19** 291 (1993)

45.   S.G. Rohozinsky and W. Greiner: Z. Physik **A332** 271 (1985)

46.   C. Rangacharyulu, A. Richter, H.J. Wörtche, W. Ziegler and R.F. Casten: Phys. Rev. **C44** R949 (1991)

47.   R.F. Casten, D.S. Brenner and P.E. Haustein: Phys. Rev. Lett. **58** 658 (1987)

48.   P. von Brentano, A. Zilges, U. Kneissl and H.H. Pitz: preprint, Universität zu Köln, (Nov. 1993)

49.   L. Zamick and D.C. Zheng: Phys. Rev. **C44** 2522 (1991)

50.   J.N. Ginocchio: Phys. Lett. **B265** 6 (1991)

51.   P. von Brentano, A. Zilges, R.-D. Herzberg, U. Kneissl, J. Margraf and H.H. Pitz: Nucl. Phys. **A**, in press

52.   E. Lipparini and S. Stringari: Phys. Lett. **B130** 139 (1983)

53.   A. Bohr and B.R. Mottelson: "Nucleare Structure. Vol.II" (Benjamin, New York, 1975) ch. 0

54.   F. Iachello: Nucl. Phys. **A358** 89c (1981)

55.   T. Otsuka: Hyperfine Ints. **74** 93 (1992)

56.   K. Heyde, C. De Coster, D. Ooms and A. Richter: Phys. Lett. **B312** 267 (1993)

57.   F. Iachello and A. Arima: "The Interacting Boson Model" (Cambridge University Press, New York, 1987)

58.   E. Otten, in: "Treatise on Heavy–Ion Science, Vol.8", ed. D.A. Bromley (Plenum, New York, 1989) p. 517

59.   O. Engel, A. Richter and H.J. Wörtche: Nucl. Phys. **A565** 596 (1993)

60.   H.J. Wörtche, C. Rangacharyulu, A. Richter, D. Frekers, O. Häusser, R.S. Henderson, C.A. Miller, A. Trudel, M.C. Vetterli and S. Yen: to be published

61.   D. Zawischa and J. Speth: Phys. Lett. **B252** 4 (1990)

62.   C. De Coster and K. Heyde: Phys. Rev. Lett. **66** 2456 (1991)

63.   C. De Coster, K. Heyde and A. Richter: Nucl. Phys. **A542** 375 (1992)

64.   S. Müller, G. Küchler, A. Richter, H.P. Blok, C.W. de Jager, H. de Vries and J. Wambach: Phys. Rev. Lett. **54** 293 (1985)

65.   R.M. Lazsewski, R. Alarcon, D.S. Dale and S.D. Hoblit: Phys. Rev. Lett. **61** 1710 (1988)

66.   E. Lipparini and A. Richter: Phys. Lett. **B144** 67 (1984)

67.   N. Lo Iudice and A. Richter: Phys. Lett. **B228** 291 (1989)

68.   D. Zawischa and J. Speth: Z. Phys. **A339** 97 (1991)

69.   D. Zawischa, J. Speth and D. Pal: Nucl. Phys. **A311** 445 (1978)

70.   E. Moya de Guerra and L. Zamick: Phys. Rev. **C47** 2604 (1993)

71.   A. Richter: Nucl. Phys. **A553** 417c (1993)

72. W. Steffen, H.-D. Gräf, W. Gross, D. Meuer, A. Richter, E. Spamer, O. Titze and W. Knüpfer: Phys. Lett. **B95** 23 (1980)

73. S.K. Nanda, C. Glashauser, K.W. Jones. J.A. McGill, T.A. Carey, J.B. Celland, J.M. Moss, S.J. Seestrom-Morris, J.R. Comfort, S. Levenson, R. Segel and H. Ohnuma: Phys. Rev. **C29** 660 (1984)

74. D. Dehnhard, D.H. Gay, C.L. Blilie, S.J. Seestrom-Morris, M.A. Franey, C.L. Morris, R.L. Boudrie, T.S. Bhatia, C.F. Moore, L.C. Bland and H. Ohnuma: Phys. Rev. **C30** 242 (1984)

75. J.B. McGrory and B.H. Wildenthal: Phys. Lett. **B103** 173 (1981)

76. W. Steffen, H.-D. Gräf, A. Richter, A. Härting, W. Weise, U. Deutschmann, G. Lahm and R. Neuhausen: Nucl.Phys. **A404** 413 (1983)

77. K. Takayanagi, K. Shimizu and A. Arima: Nucl. Phys. **A481** 313 (1988)

78. A. Härting, M. Kohno and W. Weise: Nucl. Phys. **A420** 399 (1984)

79. M.G.E. Brand, K. Allaart and W.H. Dickhoff: Nucl. Phys. **A509** 1 (1990)

80. A. Poves: Lectures presented at this school and private communication.

81. I.S. Towner: Phys. Rep. **155** 263 (1987)

82. D.O. Riska: Phys. Rep. **181** 209 (1989)

83. C.D. Goodman: Comments Nucl. Part. Phys. **10** 117 (1981)

84. A. Richter, A. Weiss, O. Häusser and B.A. Brown: Phys. Rev. Lett. **65** 2519 (1990)

85. H. Wildenthal: Prog. Part. Nucl. Phys. **11** 5 (1984)

86. A. Brown and B.H. Wildenthal: Nucl. Phys. **A474** 290 (1987)

87. S. Towner and F.C. Khanna: Nucl. Phys. **A339** 334 (1983)

88. T.E.O. Ericson: Nucl. Phys. **A497** 3c (1989)

89. H. Arenhoevel: Z. Phys. **A275** 189 (1975)

90. R.B. Wiringa, R.A. Smith and T.L. Ainsworth: Phys. Rev. **C29** 1207 (1984)

91. B. Baier, W. Bentz, C. Hajduk and P.U. Sauer: Nucl. Phys. **A386** 460 (1982)

92. T. Emura, S. Endo, G.M. Huber, H. Ito, S. Kato, M. Koike, O. Konno, B. Lasiuk, G.J. Lolos, K. Maeda, T. Maki, K. Maruyama, H. Miyamoto, K. Niki, C. Rangacharyulu, A. Sasaki, T. Suda, Y. Sumi, Y. Wada and H. Yamazaki: Phys. Lett. **B306** 6 (1993)

93. C. Gaarde, in: "Gamow–Teller resonances in Nuclear Physics", eds. C.H. Dasso, R.A. Broglia and A. Winther (North Holland, Amsterdam, 1982) p. 347.

94. W. Knüpfer, M. Dillig and A. Richter: Phys. Lett. **B122** 7 (1983)

95. R. Cenni, F. Conte and U. Lorenzini: Phys. Rev. **C39** 1588 (1989)

96. H.L. Lipkin and T.-S.H. Lee: Phys. Lett. **B183** 22 (1987)

97. C. Marchand, P. Barreau, M. Bernheim, P. Bradu, G. Fournier, Z.E. Meziani, J. Miller, J. Morgenstern, J. Picard, B. Saghai, S. Turck-Chieze, P. Vernin and M.K. Brussel: Phys. Lett. **B153** 29 (1985)

98. J.M. Laget: private communication.

99. R.G. Milner and T.W. Donnelly: Phys. Rev. **C37** 870 (1988)

Probing Nucleon and Nuclear Structure with High-Energy Electrons

Bernard Frois

C.E.A. Saclay, DAPNIA/SPhN, 91191 Gif sur Yvette, France.

Abstract: These lectures are an introduction to experimental nuclear physics using high energy electrons as a probe. The first lecture is an overview of available facilities dedicated to nuclear physics, Recent results on the neutron electric form factor and the N-Δ transition illustrate the development of modern techniques and in particular the importance of polarization. The second lecture discusses our understanding of the spin structure of the nucleon. The last lecture presents future perspectives and briefly reviews the physics program of the 15–30 GeV continuous beam electron accelerator, ELFE, proposed by the European Community. The specific example of color transparency is discussed.

1 Introduction

Electrons are pointlike and their interaction with other elementary particles is precisely known. Electron (or muon) scattering is a simple process. It is dominated by the exchange of a single virtual photon which probes the electromagnetic structure of the hadron. The incident electron or muon with energy E emits a photon that carries four-momentum $q = (\nu, \mathbf{q})$. This photon interacts in a known way with the charged constituents of the hadronic target. Because the energy and momentum transfer can be varied independently one can separately tune the excitation of the hadronic system and the spatial resolution with which one probes the system. Incident electron energies of the order of 1 GeV allow a spatial resolution down to the order of 1 fm, which is ideally suited to study nucleon distributions in nuclei [1,2]. With higher energy, $E \geq 10$ GeV, electrons and muons resolve the structure of nucleons and see the atomic nucleus as made of interacting quarks and gluons [3].

A few years ago, a major limitation for nuclear studies with electromagnetic probes was the duty factor of the accelerator. One used beam bursts only a fraction of the time (duty factor) with very high peak currents causing large backgrounds in coincidence experiments. The duty factor was at best of the order of 1%. Most data were obtained without the coincident detection of particles produced in the final state. The use of short beam bursts was due to heat dissipation problems in accelerating structures. This situation has recently changed due to technical developments. Several electron accelerators have now continuous electron beams. A wealth of new coincidence experiments is now in progress.

2 New Facilities

The experimental facilities dedicated to nuclear studies with electrons are shown in Fig. 1. A detailed review of these facilities can be found in ref. [4].

Accelerators at Bonn (ELSA), Cambridge, USA (MIT-Bates) and Amsterdam (NIKHEF-AMPS) have been recently upgraded. Coincidence experiments are explored at ELSA with a continuous beam of relatively low intensity up to $\simeq$ 3.5 GeV. AMPS is a ring added to an existing linac at NIKHEF. The ring is filled with short electron bursts until the beam becomes quasi continuous. The technique is called "pulse stretcher ring". AMPS supplies a 900 MeV high current beam for experiments either in an extracted mode or in a storage mode with internal gas targets. A similar project SHR (South Hall Ring) has been completed at MIT-Bates Laboratory in the United States at 1 GeV. Two new continuous beam electron accelerators have been recently completed, MAMI, a 800 MeV microtron at Mainz and CEBAF, a 4 GeV superconducting linear accelerator at Newport News (Virginia, USA). MAMI is a well equipped facility for a complete program of photon and electron studies below 1 GeV. The continuous acceleration of electrons at CEBAF is made possible due to a technological breakthrough in superconducting cavities. It will combine for the first time a continuous beam of high intensity and high energy resolution up to 4 GeV. One will have access to a new research domain on nucleon and nuclear structure. Three distinct experimental end stations are designed to be simultaneously used. ELFE is a European project [5] of 15–30 GeV continuous beam electron accelerator that is presented in the last part of these lectures.

Progress in polarized electron sources and polarized targets has been considerable in the last few years. This progress and the availability of continuous electron beams focus experimental techniques in new directions.

- Coincident detection of multiparticles.
- Polarization experiments.
- Non coplanar detection geometry.
- Large solid angle detectors.
- Internal targets.

Polarization degrees of freedom and non coplanar detection geometry give access to new information on nucleons and nuclei. Besides the longitudinal and transverse response functions, new interference terms can be measured [6]. The following sections illustrates of these new experimental techniques.

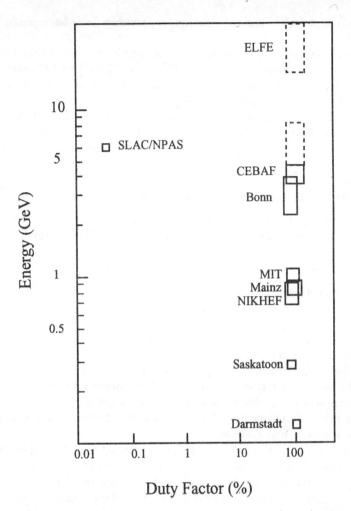

Fig. 1. Electron accelerators dedicated to nuclear studies.

3 Nucleon Form Factors

Elastic electron-nucleon scattering is specified by the electric and magnetic form factors $G_E(Q^2)$ and $G_M(Q^2)$, or the Dirac and Pauli form factors $F_1(Q^2)$ and $F_2(Q^2)$. These form factors are directly related to the distribution of charge and magnetization currents inside the nucleon. They are the basic observables related to the structure of nucleons that one would like to understand in terms of fundamental constituents.

For the proton, cross section measurements have been performed up to 32 GeV^2, where $G_M^p(Q^2)$ dominates. $G_E^p(Q^2)$ and $G_M^p(Q^2)$ have been separately determined up to 8 GeV^2. For the neutron, separations have been made up to $Q^2 = 2$ GeV^2.

To obtain a complete picture of the electromagnetic structure of the nucleon, it is fundamental to measure $G_E^p(Q^2)$ and $G_M^p(Q^2)$, $G_E^n(Q^2)$ and $G_M^n(Q^2)$. The usual method for the proton is to use the Rosenbluth separation technique but, with increasing Q^2, the extraction of $G_E^p(Q^2)$ becomes very difficult.

G_M is usually much larger and thus better known that G_E. Using polarization techniques, one can isolate and measure the interference term $G_E \cdot G_M$ in the cross section. This term is significantly larger than the G_E^2 terms. Its measurement will yield a much more precise determination of G_E. The elastic electron-proton scattering cross section will be measured with a longitudinally polarized electron beam and either a polarized target or a recoil polarimeter.

- Using a polarized target (solid NH_3 for example), one measures the cross section asymmetry A for two different relative orientations of the beam and target spins.
- Using an unpolarized target, one measures the recoil proton polarization.

The cross section asymmetry A is defined as:

$$A = \frac{\sigma_+ - \sigma_-}{\sigma_+ + \sigma_-} \tag{1}$$

In this expression, σ_+ and σ_- are the cross sections for scattering of longitudinally polarized electrons or muons on polarized nucleons, for positive and negative helicities of the incident electrons. The helicity of the incident beam can be easily reversed at a frequency of $\simeq 100$ Hz with modern polarized electron sources. Data are usually taken simultaneously for positive and negative helicities in order to minimize systematic uncertainties.

Experimental data are also taken for different orientations of the spins of the polarized nucleons. This allows one to isolate specific amplitudes in the nuclear response. Experiments using either a polarized target or a recoil polarimeter are complementary; experimental constraints and sources of systematic uncertainties are different. With a polarized target, the beam intensity has to be reduced because polarization in the target is degraded if the incident electron beam is too intense. Coincident detection of the recoil proton is essential for a clean separation of the elastic (hydrogen) events from the quasielastic (Nitrogen) ones. With a recoil polarimeter, full beam intensity can be used on a high power liquid H_2 target. However, this method uses a double scattering and the number of events decreases significantly at high Q^2.

Essentially similar ideas are used to measure the neutron electric form factor. Since there is no free neutron target, experiments are more complex. One uses either a deuteron or a ^{3}He polarized targets. Several new results have been recently obtained at Bonn, MIT-Bates, Mainz, NIKHEF and SLAC. A recent review by Milner [7] discusses these results and compares the different experimental techniques. A first result from Mainz [8] for the electric form factor of the neutron is illustrated in Fig. 2. The experiment uses polarized quasielastic electron scattering $(e, e'n)$ from $^3\boldsymbol{H}e$ to determine G_E^n. In this experiment, the electron and the recoil neutron are detected in coincidence. The goal of this experiment is to determine G_E^n in the momentum transfer range $Q^2 = 5-15$ fm^{-2}. The first result demonstrates the feasibility of the experiment and the agreement with previous data [9]. The experimental detection system will be upgraded this year and the precision of the measurements should significantly increase next year. Several new results on the neutron magnetic form factor have also been published recently [10-13].

These experiments have just started and it will take a few years to complete these efforts. It is already clear that the use of polarization techniques will lead to significantly more precise data on the nucleon form factors.

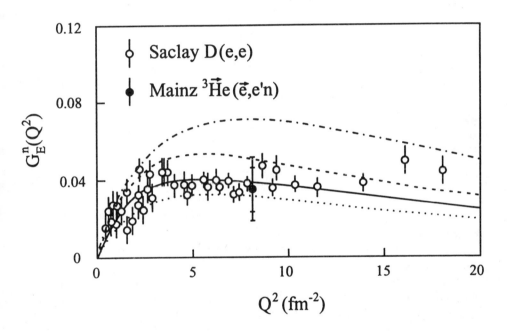

Fig. 2. The electric form factor of the neutron. The solid circle is the new result from Mainz obtained by polarized electron scattering from polarized ^{3}He [8]. The open circles are previous data from Saclay [9].

4 The $N \to \Delta$ Transition and the Nucleon Deformation

In the naive quark model, the three quarks are moving in the lowest s state of the confining potential both in the ground state N(940) and in the first excited state, the Δ $P_{33}(1232)$ resonance. In this model the $\Delta \to N + \gamma$ transition is a pure magnetic M1 transition corresponding to the spin-flip of a quark in the stretched configuration $J = 3/2$ of the Δ to the configuration $J = 1/2$ of the nucleon where two quarks couple their spin to 0. Theory shows that the spherical picture of the nucleon is only an approximation. There exists a deformation caused by the color magnetic interaction due to the one gluon exchange at short distances. This deformation creates an E2 component in the $N \to \Delta$ transition. The measurement of the E2/M1 ratio allows one to determine the deformation of the system of three quarks in the nucleon.

Different methods have been proposed to measure this transition. Experiments are in progress at Bonn, Brookhaven, Mainz, MIT-Bates and in the future at CEBAF. The difficulty of these measurements is to isolate resonant from non-resonant terms. Recent results have been reported by the Bonn group obtained by a measurement of the $H(e, e'\pi^0)$ reaction at $Q^2 = -0.127$ GeV^2 [14]. The π^0 decay was measured in the range $15° < \theta_{\pi^0} < 20°$ by observing the highest decay energy photon in coincidence with the scattered electron in a lead glass counter. The multipoles were extracted from the data by assuming that the resonant magnetic dipole amplitude, $M1_+$ is dominant. At the maximum of the resonance:

$$\frac{Re(S_{1+}^* M_{1+})}{|M_{1+}|^2} = -13 \pm 3 \qquad (2)$$

This is a relatively large value that has to be confirmed by a precise separation of the non resonant multipoles. Recent data on the $N(\gamma, \pi)$ reaction were obtained at the LEGS facility at Brookhaven using polarized photons to measure the E2 transition amplitude in the $\gamma N \to \Delta$ [15,16]. The results demonstrate that unpolarized data have little sensitivity to the E2 amplitude. Polarization data are imperative. Since most existing data have been obtained without polarization, this probably explains why existing multipole analysis are not sufficient to describe the data. Experiments with polarized photons are also in progress at Mainz.

The separation of the resonant and non-resonant amplitudes is the goal of an important research program at MIT-Bates laboratory. Four out-of-plane spectrometers (OOPS) have been recently assembled and will be used in the near future to measure the $H(e, e'p)\pi^0$ and $H(e, e'p)\gamma$ for a non-coplanar detection geometry. The separation by asymmetry measurements [17] appears to be very promising [18,19]. Another proposal at MIT-Bates laboratory will provide complementary information. Different response functions will be measured using a focal plane polarimeter.

In a few years, at CEBAF, the different helicity amplitudes will be separated by using polarized electron scattering from a polarized nucleon target with a 4π

detector, CLAS. This new detector represent a major breakthrough in this field, all the reaction channels will be measured simultaneously. For example, CLAS will allow for the first time to determine in a single measurement a complete angular distribution of the reaction $e + p \to n + \pi^+$ with a high precision. The N^* program proposes to measure 10^9 events in the final state for the electroproduction of pions, vector mesons and kaons.

For a pure M1 transition the cross section asymmetry is $A = -1/2$. Existing data are shown in Fig. 3 together with several theoretical predictions. The sensitivity of these experiments is not sufficient

to determine a small effect of the order of a few percents.

At CEBAF the experimental precision will be sufficient to discriminate between the various models of the nucleon. A major progress will be the possibility of studying the Q^2 dependence of the $E2/M1$ ratio. For $Q^2 \to \infty$, helicity conservation implies that the $E2/M1$ ratio $\to 1$. The study of the Q^2 dependence is thus a way of exploring the transition from non-perturbative to perturbative QCD.

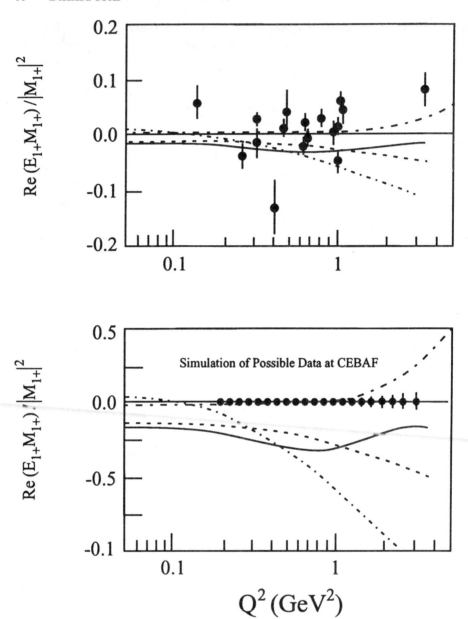

Fig. 3. Electroproduction of the Δ resonance. The top figure is a comparison between various theoretical predictions and existing data. The figure below is an estimate of the precision of possible measurements at CEBAF with the CLAS detector. Note that the scale is enlarged in the lower figure.

PART I: STRUCTURE OF THE NUCLEON

Nucleons are made of quarks and gluons. Specific models of QCD give a reasonable description of many nucleon properties, but their formal derivation from QCD is not yet resolved. The best-known of such models is the "non-relativistic quark model", in which a nucleon is composed of three valence "constituent quarks" of two kinds, the up (u) and the down (d). A proton, with an electric charge of +1, is a bound state of uud and a neutron, which has no electric charge, is a bound state of udd. Apart from the charge, the proton and neutron are essentially identical. In both nucleons the individual and total orbital angular momentum of quarks is zero. The basic concepts of this model are similar to those used to describe atomic structure in terms of electron orbits, or nuclei in terms of nucleon orbits within the shell model. It is a static approximation in which quantum fluctuations are frozen. Constituent quarks are the "effective fields" describing the low energy properties of QCD. Their mass is about 1/3 of the nucleon mass and their binding energy is small. The interaction between constituent quarks in principle takes into account the existence of a "sea" of virtual particles produced by the very short-lived quantum fluctuations, during which a quark emits and absorbs a gluon, or a gluon transforms into a quark-antiquark pair $(q\bar{q})$.

In addition to the u and d quarks there are heavier quarks which can be produced in high energy collisions: s, c, b and t. Heavy quarks are always unstable and decay to u or d via weak interactions.

Although it is a somewhat naïve picture, the constituent quark model has provided up to now a good description of "global" nucleon properties, such as masses and magnetic moments. The central question, is to what extent it describes the internal structure of nucleon.

I.1 Seeing Quarks in the Nucleon

One of the most important probes of nucleon structure is the deep inelastic scattering (DIS) of high energy leptons (electrons, muons or neutrinos) from nucleons. In these experiments one "takes a snapshot" of the nucleon, as if freezing the state of the nucleon's constituents at a given instant in time. Due to the weakness of the color interaction at high energies, the effective reaction mechanism involves a hard scattering on a single quark. Pointlike constituents have been discovered in the nucleon that have spin 1/2 and can be presumably identified with the elementary fields of QCD, the quarks. Indirect arguments indicate that the mass of the quarks is small. From the angular distribution of scattered leptons one can infer the distribution of momentum carried by quarks just prior to the collision. These distributions are usually given as function of the variable x, the fraction of nucleon momentum carried by a given quark, $0 \leq x \leq 1$. It is known today that quarks carry about 50% of nucleon's momentum and the rest is carried by gluons.

I.2 The Nucleon Spin Structure

The spin of a composite particle is the vector sum of the spins and angular momenta of its elementary constituents. The constituents of a nucleon are spin-$\frac{1}{2}$ quarks and spin-1 gluons. The constituent quark model suggests that the spin of the nucleon is built up from the spins of the constituent quarks. Therefore, the measurement of the nucleon spin fraction carried by quarks is a crucial step towards understanding the nucleon structure. In what follows, we identify spin with its longitudinal component, also called (helicity). Experiments with polarized electron or muon beams measure polarized structure functions $g_1^p(x)$ and $g_1^n(x)$ [20] as a function of the Bjorken scaling variable x, where x is the fraction of nucleon momentum carried by a given quark.

I.2.1 Bjorken and Ellis-Jaffe Sum-Rules

Theorists can not yet calculate directly from QCD the proton and neutron spin structure functions. There are, however, theoretical prediction for the integrals Γ_1^p and Γ_1^n.

$$\Gamma_1^p = \int_0^1 g_1^p(x)dx \tag{3}$$

$$\Gamma_1^n = \int_0^1 g_1^n(x)dx \tag{4}$$

The "Bjorken sum rule" relates the difference between proton and neutron spin structure to a well-known coupling constant measured in β-decay. This result was first derived by the Bjorken in 1966 [21] from symmetry properties of strong interactions, prior to be birth of QCD, and has since been shown to be a rigorous consequence of QCD. A significant experimental deviation from the Bjorken sum rule would therefore imply a breakdown of the foundations of QCD. Richard Feynman wrote that its "verification or failure would have a most decisive effect on the direction of future high-energy theoretical physics."

For $Q^2 \to \infty$,

$$\Gamma_1^p - \Gamma_1^n = \int_0^1 g_1^p(x)dx - \int_0^1 g_1^n(x)dx \tag{5}$$

$$= \frac{1}{6}\left|\frac{g_A}{g_V}\right|[1 - \mathcal{O}(\alpha_s)] \tag{6}$$

Bjorken predicted only the difference between proton and neutron spin structure functions. Separate predictions for the proton and neutron spin structures can be made, at the price of introducing additional dynamical assumptions. Shortly before the first polarized scattering experiments were performed, Ellis and Jaffe [22] wrote down a sum rule for the polarized proton spin structure functions $g_1^p(x)$ and $g_1^n(x)$. They used the constituent quark model as a guide, and assumed that there was no contribution to the spin structure from virtual strange–antistrange quarks in the "sea".

$$\Gamma_1^p = \frac{1}{2}(F - D) \tag{7}$$

$$\Gamma_1^n = \frac{1}{2}(\frac{2}{3}F - \frac{4}{9}D) \tag{8}$$

where F and D are universal weak decay constants.

I.2.2 The EMC Spin Surprise

This simple picture seemed to be confirmed by the results of the first two DIS experiments [23] in which a high-energy polarized electron beam was scattered on a polarized proton target, i.e. the spins of the electrons, and the spins of the nucleons in the target, were aligned either parallel or antiparallel to the beam direction. These experiments were proposed at SLAC in 1971 by Vernon H. Hughes at Yale as the first exploration of the spin structure of the nucleon. They were carried out between 1976–1983, by a Bielefeld-SLAC-Tsukuba-Yale collaboration, and used polarized electrons with energies between 6 and 21 GeV, covering the kinematic range $0.1 \leq x \leq 0.7$. Everything seemed to be as prescribed by the naive constituent quark model.

Particle physicists were therefore taken by surprise when a completely different picture of the nucleon emerged in 1987–1988 from the European Muon Collaboration (EMC) measurements at CERN [24]. Exploration of a wider kinematic region, using muons of higher energy, indicated that quarks carry only around 20% of the proton spin. The muon data also suggested that virtual strange quarks in the sea contributed to the spin of the nucleon.

I.2.3 Theoretical Interpretation

One suggested way to explain these surprising results is to take into account a quantum effect known as the "axial anomaly" of QCD (by which quantum effects destroy a classical symmetry) [25]. The anomaly makes it possible for spin carried by gluons to mix with spin carried by quarks, thus modifying the structure of the quark "sea". Effectively, if the gluon polarization is big, i.e. if the amount of spin carried by the gluons is large and positive, the fraction of the nucleon spin carried by quarks will appear to be smaller than the actual fraction carried by them. Thus, the existence of a large anomaly effect would explain the smallness of the apparent quark contribution to the nucleon spin. Quantum anomalies are very subtle, but they play a major role in modern physics, from electromagnetic decay of a pion to the quantum Hall effect in condensed matter physics. The EMC data raised the possibility that the anomaly may also be significant for the interpretation of DIS data, although this still needs to be confirmed experimentally.

An alternative way of explaining the small quark spin fraction is within an approximate model of QCD in which the neutron or proton corresponds to a sort of "knot" in the field of pions (a π^+ meson comprises $u\bar{d}$; a π^- a $d\bar{u}$, and a π^0 a mix of $u\bar{u}$ and $d\bar{d}$) [26]. This idea can be illustrated by considering an analogy with waves propagating on a vibrating string [27]. When a string is fixed at both ends, waves can be excited by applying an external force at some point along the string. The waves propagate and eventually dissipate to zero. In QCD, pions are the analog of such waves. In addition to exciting waves, a knot can be tied on the string, before fixing the ends; the knot can then be moved along the string, but it cannot be untied. If the string has some rigidity, there is a finite amount of energy associated with the knot. Such a knot can propagate and carry energy; it is thus a non-dissipating wave since there is no continuous deformation of the string through which the knot can be untied. Such a knot is called a "topologically stable soliton". Solitons appear in many fields of physics and often behave like particles for all practical purposes. An interesting aspect of the soliton model is that the smallness of the spin carried by quarks is independent of the details of the model and only depends on the symmetry properties of QCD, which are built into models of this type.

The soliton and the anomaly explanations involve fundamentally different theoretical concepts and differ in some of their predictions. Both involve subtle theoretical ideas about the nature of the vacuum. Quantum field theory tells us that the vacuum is far from being "empty". It is populated by virtual particles, thus making it into an interacting many-body system. A future generation of polarized DIS experiments is likely to resolve the question which is the right explanation, thus providing new insight into profound issues in non perturbative QCD. This fact and the surprise provided by the EMC measurement of the proton spin structure gave urgency to the need of measuring the analogous structure of the neutron and re-checking the proton data. An extensive experimental program was therefore immediately proposed to test the validity of the Bjorken sum rule and to measure the fraction of nucleon spin carried by quarks. Several experiments are presently being carried out or in preparation at CERN, SLAC and DESY in Hamburg, to measure the spin structure function of the neutron and to repeat the proton measurement with improved accuracy.

After five years of intense experimental activity, exciting new data have been recently measured for the proton, the neutron and the deuteron. Since free neutrons are unstable, the neutron measurement actually uses a target of deuteron or ^{3}He. These data have been obtained with electron and muon beams, in Europe by the Spin Muon Collaboration at CERN [28] and in the United States by the E142 and E143 collaborations at SLAC [29,30]. Both groups are international collaborations with many institutions. Spain in particular participates in the SMC experiment with a strong group from the University of Santiago de Compostela.

I.2.4 Polarized Deep Inelastic Lepton Scattering

If unpolarized lepton beams and targets are used in scattering experiments, then an average over all the possible orientations of quark spin in the nucleon or nucleus is obtained. To measure the fraction of the nucleon spin carried by quark spins, one must therefore freeze both the orientation of the spin of the incoming lepton, and the spin of the target nucleon, so that the lepton and nucleon spin are either parallel or antiparallel to each other. 100% polarization means that the spins of all the particles in the beam or in the target are oriented in the same direction. One then measures the cross-section asymmetry, i.e. the difference in the probabilities of scattering leptons from nucleons with spin parallel or antiparallel to the spin of the incident lepton – these differences are very small (10^{-3}). The experiments are challenging and, as well as polarized high-energy beams, need complex polarized targets and detectors. Electron and muon beams have sufficient intensities to perform such measurements, but they are impossible with neutrino beams, because of the very small probability of neutrino interaction with nucleons. Experiments of this type started about 25 years ago, but the surprise provided by the EMC results focused international efforts on the study of nucleon spin structure.

From the experimental asymmetries, a quantity can be deduced, known as the spin-dependent structure function $g_1(x)$. This quantity contains direct information on the spin structure of the nucleon. The net spin carried by quarks and antiquarks of a given flavor is denoted by Δq. In the absence of effects due to axial anomaly, polarized lepton experiments measure directly a specific combination of the spin carried by the u, d and s flavors: for the proton this is $4\Delta u + \Delta d + \Delta s$. The relative weight of each flavor is proportional to the square of its electric charge. To determine the fraction of nucleon spin carried by quarks, $\Delta \Sigma \equiv \Delta u + \Delta d + \Delta s$, one needs two additional independent combinations of Δu, Δd and Δs, which can be obtained from measurements of the lifetimes of specific particles, such as neutrons and Σ-hyperons, that decay by beta emission.

I.2.5 Electrons or Muons?

Both electrons and muons are pointlike charged particles and both are used to measure the quark structure of nucleons. High intensity polarized electron beams of up to 30 GeV are now available at SLAC and at DESY in Germany. This allows the kinematic range $0.03 \leq x \leq 0.8$ to be explored, where the statistical accuracy of the data is higher than that obtained using muon beams. Muon beams are produced as secondary beams in proton accelerators, so their intensity is much smaller. However, their higher energy (100–200 GeV), gives access to significantly lower values of x, typically $\sim 5 \cdot 10^{-3}$. The high muon energy also guarantees that only a single quark is struck in the reaction, and their higher mass ensures that corrections due to electromagnetic radiation are also smaller. Electron and muon data are thus complementary: electron data have high accuracy in a limited x-range, while muon data have lower statistical accuracy but extend to significantly lower values of x and higher momentum transfer. Systematic uncertainties are similar.

The muon scattering experiments carried out by the Spin Muon Collaboration use muon beams of 100–200 GeV, a spectrometer to measure the scattered muons, and a beam polarimeter, in an experimental area the size of a football stadium. The spectrometer produces a magnetic field to bend scattered particles according to their momentum and includes about 150 planes of detectors of variable sizes, up to 3m×3m. To overcome the relatively small number of muons in the beam, a thick target is used. To cancel systematic uncertainties to a large extent, the target consists of two cells with opposite longitudinal polarizations, inserted in a superconducting magnet of 2.5 T. Both cells are 60 cm long and 5 cm in diameter and are separated by 30 cm. Nucleons are polarized by dynamic nuclear polarization and their spins are frozen in the direction of the beam at 60 mK using a ^{3}He/^{4}He dilution refrigerator. SMC measured the spin structure of the deuteron in 1992 and the proton in 1993. The measurements on the proton cover the range $0.003 \leq x \leq 0.7$ and improve the accuracy of the experimental uncertainties on the proton data significantly.

Several experiments are in progress at SLAC. In 1992 the E142 experiment used a 22 GeV polarized electron beam and a ^{3}He nucleus (two protons and one neutron) target with a typical polarization of 40%. To a good approximation, the two protons in the ^{3}He nucleus couple their spins to zero, so a polarized ^{3}He nucleus can be considered as a polarized neutron target. In the range $0.03 \leq x \leq 0.6$, the data have an impressive accuracy, due to the high electron flux available at SLAC.

Systematic uncertainties are reduced by measuring quasi- simultaneously the probabilities of electron scattering from nucleons with spin parallel or antiparallel to the incident electron spin. This is achieved by fast reversals (120 Hz) of the spin orientations of the electrons in the incident beam. The E143 experiment, which used NH3 and ND3 as proton and deuteron targets respectively, took data at the end of 1993, using targets built on the same principle as the SMC target. Both E142 and E143 experiments use magnetic spectrometers. Beam energy for E142 was 22 GeV and was 29 GeV for E143. The development of strained gallium arsenide crystals increased the polarization of the electron beam in 1993 from 40% to 90%, making it possible to reach a high statistical accuracy. So far, about 10^8 proton and deuteron events have been measured in the range $0.027 \leq x \leq 0.7$.

I.2.6 Outlook

The surprising results observed by the EMC collaboration in 1988 are now confirmed by the new more precise proton data of the SMC and E143 Collaborations [28,30]. Fig. 4 is a comparison of these data. An excellent agreement is observed between the various data sets.

Combined with the neutron results of SMC and E142, they corroborate the Bjorken sum rule when higher order QCD corrections are taken into account. This provides a beautiful confirmation of the theory, which accurately describes

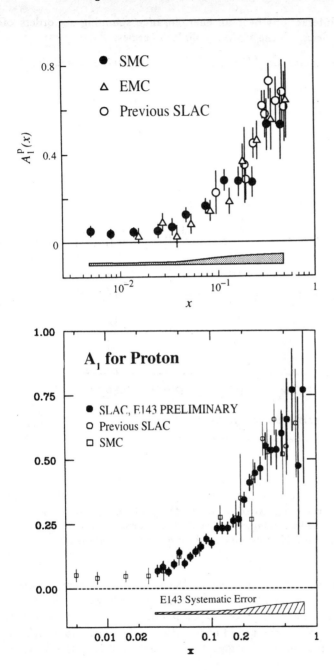

Fig. 4. Comparison of the CERN (SMC) and SLAC (E143) data

all aspects of strong interactions over an energy range spanning six orders of magnitude, from beta decay to the highest collider energies.

Figure 5 shows the results for $\Delta\Sigma$ extracted from all the published data.

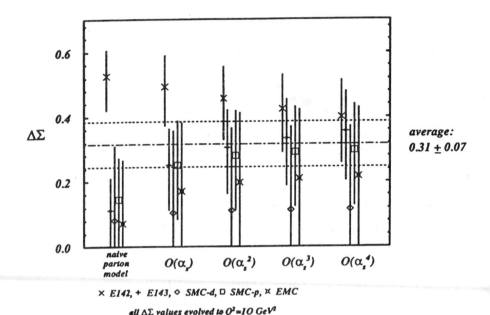

× E142, + E143, ◇ SMC-d, □ SMC-p, ⋈ EMC

all $\Delta\Sigma$ values evolved to Q^2=10 GeV²

Fig. 5. $\Delta\Sigma$, the fraction on nucleon spin carried by quarks, as extracted from all polarized lepton-nucleon scattering experiments (E143 results are preliminary). The horizontal axis represents the increasing orders of QCD perturbation theory employed in obtaining $\Delta\Sigma$ from the data. (J. Ellis and M. Karliner, to be published).

In the lowest order of perturbation theory there is no agreement between the neutron and proton experiments, but when higher and higher-order QCD corrections are taken into account, all the experimental results converge to the value $\Delta\Sigma = 0.30 \pm 0.11$. In the absence of effects due to the axial anomaly, this implies that the quark spins are responsible for only about 30% of the nucleon spin.

The origin of the missing fraction of the nucleon spin is not yet understood. We know it must come either from gluon spins or from the orbital angular momentum of quarks and gluons, but we do not know the mechanism which is responsible for the relative strength of the various contributions.

These recent results disagree with the Ellis–Jaffe sum rule and imply, if one neglects the effect of the axial anomaly, that the fraction of the nucleon spin carried by strange quarks is appreciable and negative, $\Delta s = -0.1 \pm 0.04$. This means that on average, the spins of strange quarks in the sea are pointing in the opposite direction to the direction of the nucleon spin. This result is of particular interest, since a non-zero value of Δs arises purely due to virtual $s\bar{s}$ pairs, i.e. a non-valence component of the nucleon, contradicting the intuition based on the quark model.

The apparent smallness of the quark contribution to the nucleon spin remains a mystery. The fact that quarks carry only a small fraction of nucleon spin seems contrary to the intuition based on the constituent quark model. The same applies to the non-zero spin carried by virtual $s\bar{s}$ pairs in the nucleon. Yet, the quark model has been very successful in describing the phenomenology of strongly interacting particles. It is crucial to reconcile this apparent paradox. Theoretical effort in this direction is now under way and some progress has already been made.

We know that constituent quarks are composite objects made of a quark interacting with a sea of gluons and of virtual quark–antiquark pairs. Polarized electron and muon scattering experiments show that the role played by the sea is a significant effect. The possibility of a $s\bar{s}$ polarization in the nucleon now needs to be determined with the best accuracy possible. Experiments planned in the near future at CERN, DESY and SLAC aim to increase their precision and take data at the highest Q^2 possible, where theoretical predictions are on a firm basis.

The complete E143 results on the accurate comparison of the proton and deuteron are expected to be published later this year. SMC plans to take new data on the deuteron this year and on the proton next year. At SLAC, two new experiments, E154 and E155, will measure the proton, the deuteron and ^{3}He spin structure in late 1995 with a 50 GeV electron beam, the highest available electron energy. The experimental approach proposed by the HERMES collaboration at DESY involves a novel technique based on an atomic gas jet target of polarized nucleons, injected into a beam of polarized electrons circulating in the HERA electron–proton collider. The main advantage of this experiment is the high purity of the atomic jet target, very good statistics and particle identification in the final states. HERMES plans to take its first data in 1995. An important issue is the size of the contribution of the "axial anomaly" and the closely related question of gluon polarization The various proposed theoretical ideas differ considerably in their predictions of the amount of spin carried by gluons. Several experiments have been proposed to measure the amount of gluon polarization, such as the study of polarized proton–proton scattering at the new Relativistic Heavy Ion Collider (RHIC) being constructed at Brookhaven in the US. This future polarized scattering experiments should resolve the enigma of the proton spin. Measurements of the Drell-Hearn-Gerasimov sum-rule [31,32] and of the strange quark form factors (discussed by J.D. Walecka in his lectures) will bring important information on the strange quark content of the nucleon.

PART II: NUCLEON DISTRIBUTIONS

The shell model considers that each nucleon moves independently in a mean field. The success of this picture was unexpected and came historically as a spectacular surprise. In a nucleus, there is a priori no central field, only strong two-body interactions. The importance of nucleon-nucleon correlations, the distribution of the mass between nucleons and the saturation of nuclear densities seem to discourage the idea of treating nucleons as independent particles moving in an average potential. This is now well understood: many-body theory has shown that in the nuclear medium the two-body effective interaction is significantly modified at short distances since nucleons cannot scatter into already occupied states. The Pauli principle demands that each nucleon occupy a distinct quantum eigenstate or "single-particle orbit" in the average potential (mean field) generated by the interactions with the other nucleons, thus generating a shell structure.

Mean-field theory is only the leading contribution in a series expansion in terms of multiparticle correlations. The remarkable success of the shell model is attributed to the fact that nucleons can scatter out of their mean-field orbits for only short times, within the limits allowed by the uncertainty principle, because of energy conservation and the Pauli exclusion principle. As a result of these short-lived fluctuations, nucleons do not occupy their single-particle orbits all of the time. The occupation probability of single-particle orbits provides a quantitative measure of the inadequacy of mean field theory.

Electron scattering plays an important role in the study of the nuclear many-body problem. Several reviews on theory and experimental results can be found in ref. [2].

II.1 Charge Distributions

In the shell model the charge density distribution is just the sum of the squares of the proton single particle wave functions folded by the finite size of the proton. Thus the charge distribution of a nucleus is the observable that gives the most detailed information on its ground state wave function. Nuclear charge distributions are determined by elastic electron scattering from nuclei. Electrons probe the whole nuclear volume with a spatial resolution which is related to their momentum transfer. Incident electron energies of the order of 500 MeV typically yield a spatial resolution of the order of 0.8 fm, which corresponds to the nucleon size and thus is ideally suited to the study of nucleon distributions.

Electron scattering data, combined with measurements of muonic X-ray transitions, have determined the charge densities $\rho(r)$ in the nuclear interior of many nuclei to an accuracy of the order of 1 %. The charge densities of medium and heavy nuclei are now known with sufficient accuracy to provide stringent tests of the most complete many-body calculations. Typical is the case of ^{208}Pb, where the cross section has been measured over 13 orders of magnitude.

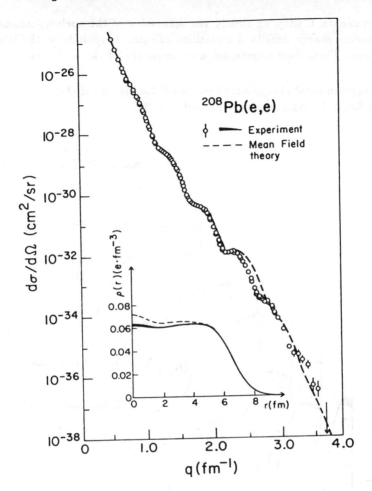

Fig. 6. Elastic electron scattering cross sections from ^{208}Pb. [33]. The inset depicts the experimental ^{208}Pb charge density (solid curve) together with a mean-field prediction (dashed curve) [34].

Cross section has been measured over 13 orders of magnitude. Figure 6 shows the experimental cross sections for ^{208}Pb at an incident energy of 500 MeV. The charge density of ^{208}Pb is determined by fitting the parameters of a model density to the elastic electron scattering cross sections and to the very precise measurements of muonic X-ray transitions.

The inset in Fig 6 shows the charge density deduced from these data [33] together with the prediction of a mean field calculation [34]. The experimental precision, depicted by the thickness of the solid curve, is of the order of 1 % even in the center of the nucleus. In the central region of the nucleus the density is almost constant, showing clearly the saturation of the nucleon-nucleon force. One observes charge density fluctuations of small amplitude in the interior of the nucleus. These fluctuations are a consequence of the shell structure of the nucleus.

The experimental charge densities for all doubly closed shell nuclei are compared in figure 7 with mean-field predictions [34].

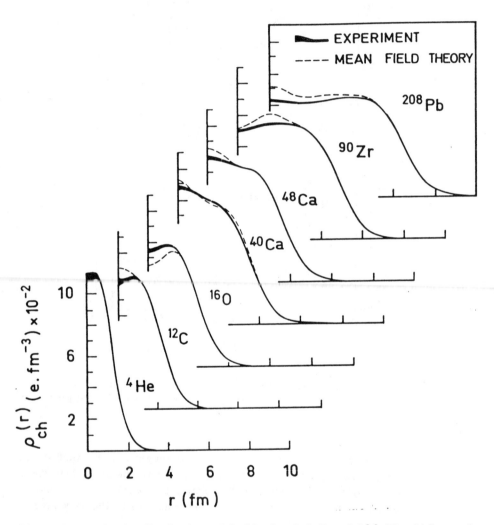

Fig. 7. Charge density distributions of doubly closed shell-nuclei [1]. The thickness of the solid curve depicts the experimental uncertainty. The dashed curves are mean-field predictions [1,34].

These calculations use the same finite range density dependent force for all nuclei. Although the experimental charge densities are reasonably well described by theory, the precision of experimental results clearly show that there is a significant discrepancy between experiment and theory in the nuclear interior.

II.1.1 Mean Field Predictions

Because of the complexity of the nucleon-nucleon interaction, the nuclear many-body problem can be solved only for systems with only few nucleons. For medium and heavy nuclei, drastic simplifications are required in order to find approximate solutions. Mean field theory is one of the most successful of these approximations. In this microscopic approach, the interaction of a nucleon with the other nucleons in the nucleus is described by an effective interaction which is different from the interaction between two free nucleons. The physics of correlations is included in the effective interaction that one uses with simple uncorrelated single-particle wave functions. The wave functions and the mean-field potential are determined by self-consistent calculations using a variational approach. The effective interaction is strongly density dependent. In principle the G-matrix theory of Brueckner should allow to build the effective force from a realistic nucleon-nucleon force. Unfortunately, despite many years of effort, it is still technically too difficult to calculate a heavy nucleus as ^{208}Pb starting from a realistic interaction between free nucleons. Charge densities predicted in the center of medium and heavy nuclei by Brueckner theory are too large, showing that the saturation of the nucleon-nucleon force is not correctly described. One has not succeeded to start from a realistic nucleon-nucleon interaction and build an effective force which has correct saturation properties without introducing a phenomenological density dependence. This correction accounts for three- and higher many-body correlations and the effect of many-body forces. The most reliable calculations are performed with finite range interactions derived from G-matrix theory.

Mean field theory gives a good general description of nuclear charge density distributions, in particular of the nuclear surface. The radial moments of charge densities are very well described. The major disagreement between experiment and theory concerns charge density fluctuations.

II.1.2 The 3s Charge Distribution

The narrow structure predicted by theory in the center of ^{208}Pb is due to the two 3s protons occupying the valence orbit. The absence of this structure in the measured charge density indicates that the 3s proton distribution is significantly modified by correlations beyond the mean-field approximation.

The detailed study of the interior of the charge distributions of two nuclei differing by one 3s proton, ^{206}Pb $(\nu 3p_{1/2}^{-2} \otimes |0^+\rangle_{208})$ and ^{205}Tl $(3s_{1/2}^{-1} \otimes |0^+\rangle_{206})$ offers the possibility of learning about the shape and strength of the proton 3s orbital.

Figure 8 shows the charge difference between ^{206}Pb and ^{205}Tl [35]. The characteristic shape of the 3s proton orbit is observed. Experiment and theory also have the same remarkable

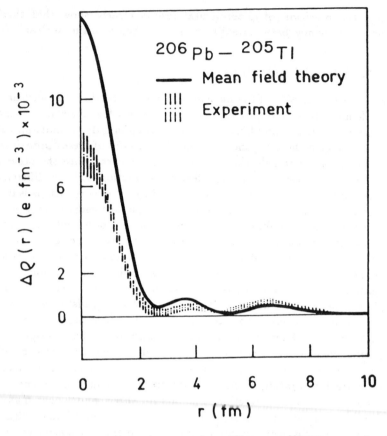

Fig. 8. The charge density difference of ^{206}Pb and ^{205}Tl [34,35,36].

similarity in configuration space and in momentum space. This result demonstrates that the concept of an independent orbit is valid in the central region of the nucleus. This result provides for the first time the spatial distribution of a particle in a quantum orbit in a dense medium.

These results suggest that the explanation of the damping of nuclear charge density fluctuations is due to many-body correlations in which a nucleon scatters to states higher than the Fermi energy and produces a depletion of the normally fully occupied states of the Fermi sea. This was first conjectured by Brown, Gunn and Gould [37] as a consequence of the experimental nuclear level density. At the Fermi surface, the level spacing is reproduced by a static potential, so the effective nucleon mass m^* is given by $m^* = m$, while for levels far from the Fermi surface the increase in level spacing requires $m^* = 0.7\, m$. This enhancement of the effective mass at the Fermi surface can be related to a depletion of valence orbits. For valence orbits, the quasiparticle strength Z has the same interpretation as the spectroscopic factor of a single-particle transition. Z is related to $\tilde{m}/m$, where $\tilde{m}$ is the energy dependent part of the effective mass, by the equation

$$Z = \left(\frac{\tilde{m}}{m}\right)^{-1} \tag{9}$$

It is a dynamical effect associated with the time (i.e. energy) dependence of the mean-field which is not taken into account in the Hartree-Fock approximation. This effect exists both in nuclear matter and in finite nuclei. It is also observed in the behavior of the specific heat of liquid ^{3}He at $0°$ K and $2°$ K. This is now understood as a characteristic property of many-fermions systems.

II.2 Correlated Wave Functions

Recent progress has been achieved many-body calculations. One hopes that in the near future it will be possible to calculate correlated wave functions for heavy nuclei. An example of recent development is the Correlated Basis Function (CBF) theory and Fermi HyperNetted Chain (FHNC) technique of correlated nuclear matter (NM). This method has been generalized to perform ab–initio calculations on complex nuclei [38,39].

For $A \leq 6$ nuclei, $< H >$ and other quantities of interest may be calculated using Monte Carlo techniques to sample the necessary many-body integrals. This is not feasible in larger nuclei and in nuclear matter. In this last system, FHNC theory has been successfully applied [40,42].

The FHNC equations have been explicitly solved in doubly closed shell nuclei, for semirealistic interactions (without tensor components) and Jastrow correlations. In such nuclei the energy per particle and the momentum distribution have been computed [38,39] and the accuracy of the method has been satisfactorily verified against variational MonteCarlo calculations, when available.

At present, the FHNC technique i)is being extended to deal with the full correlation operator and ii) to treat nuclei with N≠Z. Results, in particular for the last case, are expected in the near future.

The FHNC scheme has been recast in a form which makes use of analogous NM quantities, at different densities, in the spirit of the Local Density Approximation (LDA). This scheme, called FHNC/LD, has the property of converging to the exact solution in a very few iterations, particularly for large nuclei [41]. FHNC/LD has been checked with the exact FHNC solutions for the above mentioned model nuclei; it has been also used, at its lowest order, with realistic correlations to compute the momentum distribution in ^{16}O and ^{40}Ca.

The Jastrow correlations enhance the momentum distribution at large momenta, with respect to IPM. A further enhancement is provided by the operatorial part of the correlation operator. The agreement of the FHNC/LD estimate with the Monte-Carlo one in ^{16}O is very promising.

II.3 Momentum Distributions

Many-body theory predicts that the probability of finding a particle with momentum k and energy E is split in two parts, the single-particle component and a multi-particle component. The single particle component corresponds to the one-hole strength which is well localized in energy; it is this component which is measured by the $(e, e'p)$ reaction. The multi particle component turns out to be distributed over a very large energy range and has not yet been observed by the $(e, e'p)$ reaction. This explains the apparently low value of the single particle strength observed by the $(e, e'p)$ reaction if one compares this value to n(k), while there is a good agreement with theoretical predictions for Z.

By extending the existing ^{208}Pb$(e, e'p)$ data [43,44] to higher momenta, quasi-hole orbitals can be extracted with sufficient precision to distinguish between the different models and determine the effect of correlations on proton momentum distributions. This allows one to use the $(e, e'p)$ probe and determine orbital functions deep into nuclear interior [45].

With the continuous electron beam from the Amsterdam Pulse Stretcher (AmPS) at NIKHEF the first measurements were performed of high momentum components in the reaction ^{208}Pb$(e, e'p)^{207}$Tl [46]. Momentum distributions were deduced up to $p_m = 500$ MeV/c for the transitions to the first four excited states in ^{207}Tl. Fig. 9 shows the new data for the $3s_{1/2}$ proton knock-out.

The solid curves result from independent particle wave functions calculated in a Woods-Saxon potential and adjusted to describe the low p_m data. The dashed (dotted) curves result from modifications of these wave-functions to take into account correlation effects. It should be noted that such modifications leave the low p_m components of the momentum distributions essentially unaffected.

The new NIKHEF data indicate the sensitivity of the momentum distribution at large p_m to different theoretical hypotheses on the effect of correlations. Similar data will also be available from Mainz in the near future. These data demonstrate that with the luminosity and kinematic flexibility available at CEBAF it will be possible to probe the momentum distributions to even larger p_m with high statistics and to separate their longitudinal and transverse components. This longitudinal/transverse separation is the fundamental breakthrough possible only at CEBAF. At such high values of the nucleon momentum, various effects, final-state interactions and meson exchanges might also contribute. Such processes will contribute in a different way in the longitudinal and transverse response functions and thus can be separated. In particular the effect of meson-exchange currents is expected to be large in the transverse response while in the longitudinal one they are expected to be small.

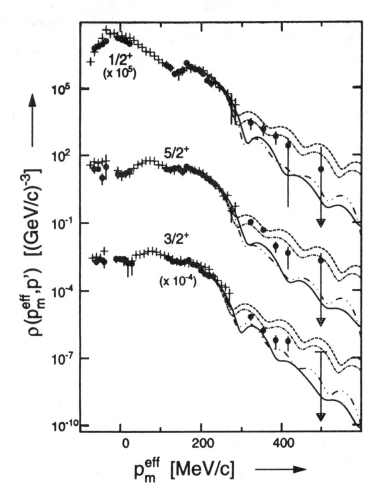

Fig. 9. Proton momentum distribution for the 3*s* and 2*d* orbitals measured at NIKHEF. The curves correspond to various theoretical predictions [46].

PART III: FUTURE PERSPECTIVES

In the past two decades, we have seen the emergence of a theory that identifies the basic constituents of matter and describes the strong interaction. The elementary building blocks of atomic nuclei are colored quarks and gluons. The theory describing their interactions is Quantum Chromodynamics (QCD) which has two special features, asymptotic freedom and color confinement. Asymptotic freedom means that color interactions are weak at short distances. Color confinement results in the existence of hadrons and in the impossibility to observe quarks and gluons as single particles.

Asymptotic freedom and color confinement imply the existence of two regimes. At short distances, well below the nucleon size ($\simeq 1$ fermi $= 10^{-15}$ m), quarks and gluons are in the so called "perturbative regime" and behave in essence as free particles. At large distances color interactions are strong and confine quarks and gluons into hadrons (mesons and baryons). The most familiar baryons are the nucleons, i.e. protons and neutrons.

A coherent description of nuclei at the fermi scale has been achieved in terms of nucleons interacting through the exchange of mesons. However, at shorter distances, more and more mesons are exchanged and the mesonic approach becomes untractable. One must explicitly take the internal structure of baryons and mesons into account.

The formation of hadronic matter in the universe is clearly one of the important questions in contemporary physics, a challenge for nuclear, particle and astrophysics. The exploration of the structure of hadrons requires to single out specific quark and gluon configurations. The study of their spatial structure and time evolution involves both nucleons and nuclei as hadronic targets. The nucleus plays both the role of a target and a detector. It is necessary to detect in scattering experiments all the produced particles (exclusive process) or some selected ones (semi-exclusive process). This program needs a sufficient energy and momentum to access quarks and gluon dynamics but it does not need an extremely high energy. As it will be shown below it can most appropriately be done with a $15 \div 30$ GeV electron accelerator with high duty cycle, high intensity, high resolution and polarized beams.

Complementary experiments to study the quark structure of matter have been proposed with ultra high energy heavy ions. Thus, the experimental strategy follows two distinct paths.

- Study of the evolution from quasi free quarks or correlated quark systems to massive particles (hadrons) where they are confined. This will require a new facility, a 15-30 GeV continuous beam electron accelerator to study specific reaction channels to well identified final states.
- Study of quark deconfinement in heavy ion collisions at ultra high energies to discover the quark gluon plasma. The United States are now building the relativistic heavy ion collider (RHIC) at Brookhaven. In Europe, an exploratory program has started at CERN. An ambitious program proposes to use the future large hadron collider (LHC) at CERN.

III.1 The Quark Structure of Nuclei

Experiments with high energy accelerators have established the validity of QCD in the perturbative regime. The next step is the exploration of the non-perturbative domain which is of interest to both nuclear and particle physics communities.

Nearly all existing data on quark distributions in hadrons have been obtained by inclusive scattering of high energy particles. In such reactions, one strikes quarks with considerable energy and reconstructs quark distributions from scattering data. These experiments only give access to an average over all the possible quark configurations in the nucleon or nucleus. In order to get precise information on specific configurations, to study their spatial structure and to follow their evolution controlled by the confinement mechanism, one needs different kind of data.

This is the domain of exclusive reactions where all produced particles are observed in coincidence with the scattered electron.

This necessitates a high duty cycle. Moreover exclusive reactions at high momentum transfer have a small probability and therefore require high luminosity ($10^{35} \div 10^{38}$ nucleons/cm^2/s). Finally, experiments with polarized beams are needed to disentangle specific reaction channels through interference measurements.

The study of exclusive reactions is not possible with existing accelerators because of their technical limitations. Muons and neutrino beams have too low an intensity to obtain the precision and high statistics required. Existing high energy electron accelerators have too small a duty factor (< 0.1 %). The possibility of high polarization levels (~ 80 %) with high currents is a technical breakthrough of which the feasibility only recently has been shown.

III.2 The ELFE Project

The ELFE (an Electron Laboratory For Europe) project is presented in detail in the book [5]. It is a new facility dedicated to the study of the quark and gluon structure of matter. The proposals discussed in ref. [5], form an extensive research program of exclusive and semi-exclusive reactions on hadronic targets in order to probe the correlated quark and gluon structure. In this research domain there is only a limited amount of experimental data with poor statistics. A number of puzzling effects have been observed in high energy collisions on nuclei: EMC effect, color transparency in proton-nucleus collision, J/Ψ production in hadronic collisions, etc ... They are related to the interface between short- and long-range physics. They have triggered a large number of theoretical speculations which remain inconclusive due to the lack of significant data. This is due to the limitations of existing accelerators. The ELFE project is a dedicated facility for a detailed investigation of the dynamics of quarks and gluons in hadronic matter. The research program will focus on the following research topics:

- *Exclusive processes.* Exclusive electroproduction processes, including polarization experiments, are needed to study the spatial structure of hadrons. Because they require coherent scattering of the quarks, exclusive observables are sensitive to the quark gluon wave function of the hadrons. Typical examples are virtual Compton scattering and form factors of mesons or baryons.
- *Nucleus as a detector.* A central idea is to use the nucleus as a microscopic detector to determine the time evolution of the elementary quark configurations in the building up of hadrons. A typical example of this research

program is color transparency in quasi-elastic reactions and in charmonium production. Another one is hadronization in the nuclear medium. For these processes, the nucleus is used as a medium of *varying length*.

- *Heavy Flavors* The study of the production and the propagation of strangeness and charm provides us with an original way to understand the structure of hadronic matter. The corresponding reactions do not involve the valence quarks of the target and probe its sea quark (intrinsic strange or charm content) and gluon distributions.
- *Short Range Structure of Nuclei*. At short distances, nuclear structure cannot be reduced to nucleons or isobar configurations. To unravel such exotic configurations dedicated experiments (large x structure functions and ϕ production) are proposed.

III.3 Accelerator and Detector Requirements

The choice of the energy range of 15 to 30 GeV for a new electron accelerator, dedicated to nuclear physics, is fixed by three constraints:

- Hard electron-quark scattering: one must have sufficiently high energy and momentum transfer to describe the reaction in terms of electron-quark scattering. The high energy corresponds to a very fast process where the struck quark is quasi-free. High momentum transfers are necessary to probe short distances.
- Nuclear sizes: The energy of the incident electron beam is determined to match the characteristic interaction time τ to the diameter of the nucleus. Starting from the rest frame time $\tau_o \sim 1$ fm/c and taking into account a typical Lorentz dilation factor $\gamma = E/M$ this means a time τ of several fm/c's in the laboratory. If the energy transfer is too large, the building-up of hadrons occurs outside the nucleus which can then no longer be used as a microscopic detector.
- Charm production requires a minimum electron beam energy of 15 GeV to have reasonable counting rates.

Exclusive and semi-inclusive experiments are at the heart of the experimental program. To avoid a prohibitively large number of accidental coincident events a high duty cycle is imperative. This simultaneously requires a high luminosity because of the relatively low probability of exclusive processes. Finally a good energy resolution is necessary to identify specific reaction channels. A typical experiment at 15 GeV (quasielastic scattering for instance) needs a beam energy resolution of about 5 MeV. At 30 GeV the proposed experiments require only to separate pion emission. The characteristics of the accelerator are summarized in table 1.

Table 1. ELFE Accelerator Parameters

Beam Energy	$15 \div 30$ GeV
Energy Resolution FWHM	3×10^{-4} @ 15 GeV
	10^{-3} @ 30 GeV
Duty Factor	$\simeq 100$ %
Beam Current	$10 \div 50\ \mu$A
Polarized Beams	P > 80 %

With existing facilities available for particle physics, such as SLAC and HER-A (HERMES program), one can only perform a limited set of exclusive experiments. In Fig. 1, the ELFE project is compared to the existing electron facilities dedicated to nuclear physics.

ELFE will be the first high energy electron beam beyond 10 GeV
with both high intensity and high duty factor.

The various components of this experimental program put different requirements on the detection systems that can be satisfied only by a set of complementary experimental equipment. The most relevant detector features are the acceptable luminosity, the particle multiplicity, the angular acceptance and the momentum resolution. High momentum resolution (5×10^{-4}) and high luminosity (10^{38} nucleons/cm^2/s) can be achieved by magnetic focusing spectrometers. For semi-exclusive or exclusive experiments with more than two particles in the final state, the largest possible angular acceptance ($\sim 4\pi$) is highly desirable. The quality and reliability of large acceptance detectors have improved substantially in the last two decades. The design of large acceptance detectors proposed for the ELFE project uses state of the art developments to achieve good resolution and the highest possible luminosity.

III.4 Color Transparency: A Typical Example

Examples of exclusive reactions and quantities which will be studied with ELFE are: electromagnetic form factors of nucleons and mesons, form factors of electromagnetic transitions from the nucleon to excited nucleons such as the Δ or the S_{11}, real and virtual Compton scattering off nucleons and electroproduction of mesons. A detailed picture of hadrons will emerge from these studies.

A major challenge of the research program on exclusive reactions will be to understand both the precocious scaling and the helicity violation. Quark-quark correlations have been proposed as an explanation but the question is still open.

Color transparency illustrates the power of exclusive reactions to isolate simple elementary quark configurations. This concept was originally developed by Mueller [47] and Brodsky [48]. The experimental technique to probe these configurations is the following:

- For a hard exclusive reaction, say electron scattering from a proton, the scattering amplitude at large momentum transfer Q^2 is suppressed by powers of Q^2 if the proton contains more than the minimal number of constituents. This is derived from the QCD based quark counting rules, which result from the factorization of wave-function-like distribution amplitudes. Thus protons containing only valence quarks participate in the scattering. Moreover, each quark, connected to another one by a hard gluon exchange carrying momentum of order Q, should be found within a distance of order $1/Q$. Thus , at large Q^2 one selects a very special quark configuration: all connected quarks are close together, forming a small size color neutral configuration sometimes referred to as a *mini hadron*. This mini hadron is not a stationary state and evolves to build up a normal hadron.

- Such a color singlet system cannot emit or absorb soft gluons which carry energy or momentum smaller than Q. This is because gluon radiation — like photon radiation in QED — is a coherent process and there is thus destructive interference between gluon emission amplitudes by quarks with "opposite" color. Even without knowing exactly how exchanges of soft gluons and other constituents create strong interactions, we know that these interactions must be turned off for small color singlet objects.

An exclusive hard reaction will thus probe the structure of a *mini hadron*, i.e. the short distance part of a minimal Fock state component in the hadron wave function. This is of primordial interest for the understanding of the difficult physics of confinement. First, selecting the simplest Fock state amounts to the study of the confining forces in a colorless object in the "quenched approximation" where quark-antiquark pair creation from the vacuum is forbidden. Secondly, letting the mini-state evolve during its travel through different nuclei of various sizes allows an indirect but unique way to test how the squeezed mini-state goes back to its full size and complexity, i.e. how quarks inside the proton rearrange themselves spatially to "reconstruct" a normal size hadron. In this respect the observation of baryonic resonance production as well as detailed spin studies are mandatory.

To the extent that the electromagnetic form factors are understood as a function of Q^2, $eA \rightarrow e'(A-1)p$ experiments will measure the color screening properties of QCD. The quantity to be measured is the transparency ratio T_r which is defined as:

$$T_r = \frac{\sigma_{Nucleus}}{Z \sigma_{Nucleon}} \tag{10}$$

At asymptotically large values of Q^2, dimensional estimates suggest that T_r scales as a function of $A^{\frac{1}{3}}/Q^2$. The approach to the scaling behavior as well as the value of T_r as a function of the scaling variable determine the evolution from the pointlike configuration to the complete hadron. This highly interesting effect can be measured in an $e, e'p$ reaction that provides the best chance for a quantitative interpretation. The first experiment to investigate Color Transparency was performed at Brookhaven by comparing (p,2p) reactions on hydrogen and on a nucleus [49]. Ralston and Pire [50] have shown that there are ambiguities in the interpretation of these results. These ambiguities should be smaller in (e,e'p) reactions because the electron-nucleon interaction is weaker and better understood that the nucleon-nucleon interaction. A first experiment at SLAC (NE-18) has performed a preliminary exploration of Color Transparency with the (e,e'p) reaction for H, C, Fe and Au, in the Q^2 range of 1–7 GeV^2 [51,52]. In these kinematics, no effect was observed. A new experiment at CEBAF is planned in the near future. This experiment is expected to increase the precision of the NE-18 data.

The interplay between the perturbative and non-perturbative aspects of QCD cannot be easily explored by existing high energy machines. The SLAC electron machine is of a suitable energy, but its 10^{-4} duty factor is too low for high statistics coincidence measurements. CEBAF is capable of delivering the required beam characteristics, but its energy is probably too low to observe a significant effect in transparency experiments.

III.5 Conclusions

There is a bright future for probing the structure of nucleons and nuclei with high energy electrons. These lectures have just sketched a few facets of this field. An intensive research program is carried out in Europe with new continuous beam facilities at Bonn, Mainz and NIKHEF. New results from the electric neutron form factor and the proton momentum distribution in ^{208}Pb illustrate the power of these new facilities. Many experiments are in progress and new results will be available in the near future.

Soon, a completely new experimental domain will be explored with CEBAF in the United States. This new facility presented by J.D. Walecka in his lectures will provide three simultaneous continuous electron beams with energies between 0.4 GeV and 4 GeV and currents up to 200 μA. The three experimental halls will be equipped with complementary detectors with state-of-the art instrumentation. A large community of users is actively preparing an exciting experimental program.

NuPECC is now preparing the future of nuclear physics in Europe. A new facility, the ELFE project is now being discussed. ELFE will be the first high energy electron beam beyond 10 GeV with both high intensity and high duty factor. The ELFE research program lies at the border of nuclear and particle physics. Most of the predictions of QCD are only valid at very high energies where perturbation theory can be applied. In order to understand how hadrons are built, however, one is in the domain of confinement where the coupling is strong. It is fundamental to guide theory by the accurate, quantitative and interpretable measurements obtained by electron scattering experiments, in particular in exclusive reactions. This research domain is essentially a virgin territory. There is only a limited amount of experimental data with poor statistics. It is not possible to make significant progress in the understanding of the evolution from quarks to hadrons with the available information.

This lack of data explains to a large extent the slow pace of theoretical progress. The situation will considerably improve due to technical breakthroughs in electron accelerating techniques. ELFE will be the first high energy machine offering the high luminosity and high duty cycle demanded by the exclusive reaction program.

A few exploratory experiments can be achieved by existing or planned facilities at the price of considerable efforts. This is for example the case of the proton electric form factor at SLAC. Also the proton transverse spin structure function can be studied at RHIC through dilepton pair production in polarized proton-proton collisions. The exploratory program on color transparency at Brookhaven with protons and at SLAC with electrons did strengthen the need for dedicated experiments with high energy resolution and high duty cycle electron beam. The HERMES program at HERA proposes a first detailed study of semi-inclusive reactions. A dedicated facility with a continuous beam electron accelerator of 15 to 30 GeV would increase counting statistics by orders of magnitude. Such new information would allow for the first time a detailed study of color neutralization.

The formation of hadronic matter in the universe is clearly one of the important questions in contemporary physics, a challenge for nuclear, particle and astrophysics. Nuclear physics is now going to use the tools that have been forged by twenty years of research in QCD, to elucidate the central problem of color interaction: color confinement and the quark and gluon structure of matter. The goal is to understand how quarks and gluons build up hadrons and nuclei.

Europe is well equipped for nuclear physics studies by intermediate energy electron scattering at Bonn, Mainz, NIKHEF. Beyond 1995 the performances of European accelerators will be superseded by the quality of the facilities available at CEBAF. Nuclear physics in United States will have access to a completely new domain of high energy resolution and coincidence experiments. Up to now the experimental situation has been well balanced between Europe and the United States. This friendly competition was one of the sources of success of this field. The situation has changed because of the high cost of experimental facilities. No European nation alone has enough financial and human resources available to build alone a facility for the next century. We will need to focus our efforts with NuPECC at a European level to make a coherent effort and build an exciting future.

Acknowledgements

It is for me a great pleasure to give these lectures at La Rabida. I like the wonderful atmosphere of Spain and I am very grateful to the organizers of this school for their kind invitation. These lectures are based on work done in collaboration in particular with Marek Karliner, Costas Papanicolas and Bernard Pire. I would like to give them my warmest thanks.

References

1   B. Frois and C.N. Papanicolas, Ann.Rev Nucl. Part.Sci. 37 (1987) 133.
2   *Modern Topics in Electron Scattering*, B. Frois and I. Sick, World Scientific 1991, Singapore.
3   T. Sloan, G. Smadja and R. Voss, Phys. Rep. 162 (1988) 46.
4   S. Kowalski, Proceedings of the Conference on the Intersections of Particle and Nuclear Physics, 1994 St. Petersburg, Florida, to be published.
5   *The ELFE project, an electron laboratory for Europe*, Proceedings of the conference on Physics with a 15-30 GeV High Intensity Continuous Beam Electron Accelerator, Mainz, 7-9 October 1992, edited by J. Arvieux and E. De Sanctis, published by the Italian Physical Society, Bologna, Italy.
6   T.W. Donnelly and A.S. Raskin, Annals of physics, 169 (1986) 247.
7   R. Milner, MIT preprint 1994, LNS 94/71.
8   M. Meyerhoff et al., Phys.Lett. B327 (1994) 201.
9   S. Platchkov et al., Nucl.Phys. A510 (1990) 740.
10 P. Markowitz et al., Phys.Rev. C48 (1993) R5.
11 J. Jourdan, Few-Body Conference, 1994 Williamsburg, Virginia.
12 T. Reike, PhD. Thesis Bonn University.
13 H. Gao et al., Caltech preprint 1994, OAP-729.
14 F. Kalleicher, Thesis, University of Mainz, 1993.
15 G. Blanpied et al, Phys. Rev. Lett. 69 (1992) 1880.
16 M.A. Khandaker et al., to be published.
17 C.N. Papanicolas et al., Nucl.Phys. A497 (1989) 509.
18 S. Dolfini, PhD Thesis, University of Illinois (1993).
19 J.B. Mandeville, PhD Thesis, University of Illinois (1993).
20 V.W. Hughes and J. Kuti, Ann. Rev. Nucl. Part. Sci. 33 (1983) 611.
21 J.D. Bjorken, Phys. Rev. 148 (1966) 1467; Phys. Rev. D1 (1970) 465; ibid. D1 (1970) 1376.
22 J. Ellis and R.L. Jaffe Phys. Rev. D9 (1974) 1444; Phys. Rev. D 10 (1974) 1669.
23 E-80 (SLAC), M.J. Alguard et al., Phys. Rev. Lett. 37, (1976) 1261; ibid. 41 (1978) 70; E-130 Collaboration (SLAC), G. Baum et al., Phys. Rev. Lett. 51 (1983) 1135; G. Baum et al., Phys. Rev. Lett. 45 (1980) 2000.
24 EMC Collaboration (CERN), J. Ashman et al. Phys. Lett., B206 (1988) 364; Nucl. Phys. B328 (1989) 1.
25 For a pedagogical discussion of the anomaly see G Altarelli 1990 in Proc. 27th Int. School of Subnuclear Physics 1989 A Zichichi (ed) (Plenum Press, New York).
26 J. Ellis, S.J. Brodsky and M. Karliner Phys. Lett. B 206 (1988) 309.
27 For an introduction to soliton interpretation see M. Karliner 1990 in Proc. 1989 Trieste Summer School J.C. Pati et al. (editors) World Scientific, Singapore.
28 SMC Collaboration (CERN), B. Adeva et al. 1993 Phys. Lett. B302 (1993) 533; D. Adams et al. Phys. Lett. B329 (1994) 399 and references therein.
29 E142 Collaboration (SLAC) P.L. Anthony et al., Phys. Rev. Lett. 71 (1993) 959.
30 E143 Collaboration (SLAC), R. Arnold et al. 1994 preliminary results presented at Conference on the Intersections of Particle and Nuclear Physics St. Petersburg, Florida.
31 S.D. Drell and A.C. Hearn, Phys.Rev.Lett. 16 (1966) 908;
32 S.B. Gerasimov, Yad. Phys. 2 (1965) 598; Sov.J. Nucl.Phys. 2 (1966) 430.

33   B. Frois et al., Phys. Rev. Lett. 38 (1977) 152.

34   J. Dechargé and D. Gogny, Phys. Rev. C21 (1980) 1568

35   J.M. Cavedon et al., Phys. Rev. Lett. 49 (1982) 978.

36   B. Frois et al., Nucl. Phys. A396 (1983) 409.

37   G.E. Brown, J.H. Gunn and P. Gould, Nucl. Phys. 46 (1963) 598.

38   G.Co, A.Fabrocini, S.Fantoni and I.E.Lagaris, Nucl. Phys. A549 (1992) 439.

39   G.Co, A.Fabrocini and S.Fantoni, Nucl. Phys. A586 (1994) 73.

40   V.R. Pandharipande and R.B.Wiringa, Rev. Mod. Phys. 51 (1979) 821.

41   S.C. Pieper, R.B.Wiringa and V.R. Pandharipande, Phys. Rev. C46 (1992) 1741.

42   S.Rosati, in "From nuclei to particles", Proc.Int.School E.Fermi, course LXXIX, ed. A. Molinari (North Holland, Amsterdam, 1982).

43   E.N.M. Quint et al., Phys. Rev. Lett. 57 (1986) 186.

44   E.N.M. Quint et al., Phys. Rev. Lett. 58 (1987) 50

45   L. Lapikas, Proceedings of the International Conference on Nuclear Physics, Wiesbaden 1992.

46   I. Bobeldijk et al., NIKHEF preprint 1994, submitted to Phys. Rev. Lett.

47   A.H. Mueller, Proceedings of the XVII Rencontres de Moriond, J. Tran Thanh Van Editor, Editions Frontières, Gif-sur-Yvette, France, 1982, p.13.

48   S.J. Brodsky, Proceedings of the Thirteen International Symposium on Multi-particle Dynamics, edited by W. Kittel, W. Metzger, and A. Stergion, World Scientific, Singapore, 1982, p.963.

49   A.S. Carroll et al., Phys. Rev. Lett. 61 (1988) 1698.

50   J.P. Ralston and B. Pire, Phys.Rev.Lett. 61 (1988) 1823.

51   N.C.R. Makins et al., Phys. Rev. Lett. 72 (1994) 1986.

52   O'Neill et al., Caltech Preprint 1994 OAP-731, submitted to Phys. Rev. Lett.

Relativistic Theory of the Structure of Finite Nuclei*

Peter Ring

Physikdepartment der Technischen Universität München,
D-85747 Garching, Federal Republic of Germany

Abstract. Relativistic Mean Field (RMF) theory and several of its recent applications to problems of nuclear structure at low energies are discussed, such as isotope shifts in the Pb region, rotational bands in superdeformed nuclei, the problem of identical bands, a relativistic theory of pairing correlations in nuclei and dynamical investigations of the time-dependent relativistic mean field equations for the description of giant resonances.

1 Introduction

Since the early days of nuclear physics one has known, that the spin plays an important role in our understanding of nuclear spectra, and that, in principle, one should use a relativistic theory for a complete description of the nuclear system even in the low energy domain. Early attempts in this direction of Teller et al [1], however, have been forgotten and it was much later – in the seventies – that Walecka[2, 3] pointed out the power, the simplicity and the elegance of a phenomenological relativistic description of the nuclear system. In recent years these methods have become very popular. They have been extended so far, that they now can be applied not only to nuclear matter and to the ground state of spherical doubly closed shell nuclei, but also to the entire region of the periodic table, to exotic nuclei with large neutron excess as well as to deformed and superdeformed nuclei and even to rotating and vibrating nuclei.

Relativistic Mean Field (RMF) theory is a phenomenological theory. It is conceptually similar to density dependent Hartree-Fock theory with Skyrme forces, which is nowadays standard for the microscopic description of nuclear properties over the entire periodic table. It shares with this theory, that it contains only a few parameters which are adjusted to data of nuclear matter and of and a few closed shell nuclei. All the other nuclei are described with one parameter set.

*supported by the BMFT under the project 06 TM 733.

So far it also shares with Skyrme theory the disadvantage that pairing is taken into account only in a very primitive way, the so-called *constant gap approximation*, which uses experimental gap parameters to determine BCS occupation probabilities for the calculation of the densities in each step of the iteration. This method works rather well for the ground states of even-even nuclei, where experimental gaps are available. Its predictive power, however, is limited in the region of exotic nuclei far from the stability line, where no experimental gap parameters are known and in the high spin region, where pairing is quenched and where the finite range of the pairing interaction should be taken into account.

In this series of lectures a overview is given of a number of recent investigations of problems of low energy nuclear structure within the framework of RMF theory. In section 2 we give a short summary of the basic facts of relativistic mean field theory. In section 3 we discuss the problem of isotope shifts in the Pb region, which could not be explained so far in non-relativistic density dependent Hartree-Fock calculations and which shows that a proper treatment of the spin-orbit term plays a crucial role for the understanding of many details in nuclear structure. In section 4 we deal with RMF theory in the rotating frame and the very successful description of identical bands in superdeformed nuclei. In section 5 we present a relativistic theory of pairing in nuclei and first investigations in this direction. In section 6 we discuss recent applications of time-dependent RMF theory for the description of giant resonances in nuclei and for Coulomb dissociation and in section 7 we give a short summary and an outlook.

2 Relativistic Mean Field Theory

Relativistic Mean Field (RMF) theory is a classical relativistic field theory similar to classical electrodynamics. Only the nucleons are treated as quantum mechanical Dirac-particles with the four component wave functions ψ_i moving with relativistic dynamics in several classical meson fields. As in any mean field theory, these fields describe in an average way the interaction produced by the exchange of the corresponding mesons. The theory is phenomenological similar as the density dependent Hartree-Fock theories with Skyrme or Gogny forces. Therefore the number of mesons entering this theory is limited and the corresponding parameters, the meson masses and the coupling constants between mesons and nucleons are neither derived from a more fundamental theory, nor taken from the properties of these mesons in the vacuum, rather they are adjusted to data of the correlated nuclear many-body system. The mean field of single pions is neglected. It breaks parity, a fact not observed in nature. Pions are, however, taken into account indirectly by the mean field of the so-called σ-meson. This particle has not been observed experimentally so far as a resonance, but it has been interpreted as a complicated bound state of two pions in the medium. On the phenomenological level we cannot decide whether this interpretation is the proper one. Many experimental observations, however indicate the importance of a large scalar field in the nucleus. This scalar field is taken into account by the phenomenological σ-meson in RMF-theory. Since relativistic covariance leads to an attractive fields for scalar particles and to repulsive fields for vector particles, we need in addition a vector meson, the ω-meson. This particle has been observed in nature as a resonance at $m_\omega = 783$ MeV. σ- and ω-meson carry both isospin 0. For nuclei with large neutron excess we need in addition a meson carrying isospin 1, the ρ-meson, which is also observed in nature as a resonance at $m_\rho = 763$ MeV. Finally we need the Coulomb-force, i.e. the photon.

Apart from the Dirac wave functions $\psi_i(x)$ $(i = 1 \ldots A)$ for the nucleons we therefore have one field for the scalar σ-meson $\sigma(x)$ and 4 fields for each Lorentz-vector particle, i.e. for the ω-meson: $\omega^\mu(x)$, for the three iso-vector ρ-mesons $\vec{\rho}^\mu(x)$ and for the photon $A^\mu(x)$.

In order to keep the number of adjustable parameters small, we use the experimental values for the masses of the nucleons m_N and the masses of the ω- and the ρ-meson. In this most primitive version of the theory we have only four parameters, the mass of the σ-meson m_σ and the three coupling constants g_σ, g_ω and g_ρ. The rest is more or less given by Lorentz invariance and the principle of simplicity (minimal coupling). It leads to the following Lagrangian density:

$$\mathcal{L} = \bar{\psi} \left(\not{p} - g_\omega \not{\omega} - g_\sigma \sigma - m_N \right) \psi$$
$$+ \frac{1}{2} \partial_\mu \sigma \partial^\mu \sigma - \frac{1}{2} m_\sigma^2 \sigma^2 - \frac{1}{4} \Omega_{\mu\nu} \Omega^{\mu\nu} + \frac{1}{2} m_\omega^2 \omega_\mu \omega^\mu, \qquad (1)$$

where $\Omega_{\mu\nu} = \partial_\mu \omega_\nu - \partial_\nu \omega_\mu$ is the field tensor for the vector mesons. For simplicity we neglect in this and in the following equations the other vector mesons, the ρ-mesons $\vec{\rho}^\mu$ and the photons A^μ, because they are treated in a similar way as the ω. In all the calculations, however, these fields are fully taken into account.

Although this simple version of the Walecka-model is able to give a qualitative description of saturation and many nuclear properties[3], it fails to describe properly the experimental surface properties. Boguta and Bodmer[4] therefore extended the model by including a nonlinear self-coupling of the scalar mesons, i.e. the mass term $\frac{1}{2} m_\sigma^2 \sigma^2$ in the Lagrangian (1) is replaced by a polynomial of forth order $U(\sigma) = \frac{1}{2} m_\sigma^2 \sigma^2 + \frac{1}{3} g_2 \sigma^3 + \frac{1}{4} g_3 \sigma^4$ containing two additional parameters g_2 and g_3. Conceptually this is equivalent to the modification of the original Skyrme force with its linear density dependence by a modified density dependence of the type ρ^α. All the great quantitative achievements of RMF-theory in recent years were only possible in this nonlinear version of the theory. It is nowadays standard and rather uniform parameter sets have been adjusted in the literature.

Using the fields ψ_i, σ and ω as dynamical variables Hamiltons variational principle allows to derive the Euler-equations as equations of motion: the Dirac equation for the nucleon spinors and Klein-Gordon equations for the meson fields:

$$\{\gamma_\mu(i\partial^\mu + g_\omega \omega^\mu) + (m_N + g_\sigma \sigma)\} \psi_i = 0, \qquad (2)$$
$$\Box \sigma + U'(\sigma) = -g_\sigma \rho_s \qquad (3)$$
$$\{\Box + m_\omega^2\} \omega^\mu = g_\omega j^\mu \qquad (4)$$

The scalar density $\rho_s = \langle \bar{\psi}\psi \rangle$, the baryonic current $j^\mu = \langle \bar{\psi}\gamma^\mu\psi \rangle$ are calculated from the solution of the Dirac equations by summing over all particle states with the appropriate BCS occupation probabilities v_i^2:

$$\rho_s = \sum_i v_i^2 \, \bar{\psi}_i \psi_i, \qquad j^\mu = \sum_i v_i^2 \, \bar{\psi}_i \gamma^\mu \psi_i. \tag{5}$$

In principle we should – in a relativistic theory – include into these sums also negative energy solutions of the Dirac equations, the anti-particle states in the Dirac sea, i.e. to allow for a polarization of the vacuum. In principle this can be done[5] using a suitable renormalization procedure. In such a case, however, the parameters of the Lagrangian have to be readjusted and so far no essential differences have been found for such a readjusted renormalized theory compared to a theory neglecting the vacuum polarization. Since one is dealing with an phenomenological theory, one therefore usually uses the *no sea approximation* and neglects these anti-particle contributions.

The quantities ρ_s and j^μ serve as sources in the Klein-Gordon equations (3) and (4), which determine the mesons fields. These meson fields, finally, enter the Dirac equation (2) and determine the motion of the nucleons in a self-consistent way. The whole system of equations is solved by iteration.

For ground state properties of even-even nuclei we have the static limit with meson fields constant in time and in addition time reversal invariance, i.e. the currents vanish. In this case the theory becomes extremely simple:

$$\{\alpha(p - g_\omega \omega) + g_\omega \omega_0 + \beta(M + g_\sigma \sigma)\}\, \psi_i \;=\; \epsilon_i \psi_i, \tag{6}$$

$$-\Delta\sigma + U'(\sigma) \;=\; -g_\sigma \rho_s, \tag{7}$$

$$\{-\Delta + m_\omega^2\}\, \omega^0 \;=\; g_\omega \rho_v. \tag{8}$$

In a relativistic theory we have to distinguish two types of densities, the scalar density $\rho_s = \langle \bar{\psi}\psi \rangle$, and the vector density $\rho_v = \langle \psi^+\psi \rangle$, the time-like component of the current. The vector density is the usual baryon density. It contains the sum of the squares of large and small components, which are normalized to unity. The scalar density contains the difference of the square of large and small components and is not normalized. This purely relativistic effect prevents the theory from an attractive collapse: in the case of a very strong attractive σ-field the gap in the Dirac equations becomes relatively small. However, the theory remains stable, because in such a case the small components become relatively large too and the scalar density is quenched. Since this quantity is the source of the scalar field, the latter is quenched too in a self-consistent way and the collapse, which one would find in an equivalent non-relativistic system, is prevented.

The six parameters m_σ, g_σ, g_ω m_ρ, g_2, and g_3 have been obtained by careful fitting of experimental data of nuclear matter and a few finite spherical nuclei. Pairing properties are taken into account in the constant gap approximation. In Table 1 we show standard parameter sets (the masses are given in MeV, the parameter g_2 in fm^{-1} and the other coupling constants are dimension-less) and some results for symmetric nuclear matter: the baryon density ρ_0 (in units of fm^{-3}), the binding energy per particle E/A, the incompressibility K, the

Parameter	HS	NL1	NL2	NL-SH
m_N	939.0	938.0	938.0	939.0
m_σ	520.0	492.25	504.89	526.059
m_ω	783.0	795.359	780.0	783.0
m_ρ	770.0	763.0	763.0	763.0
g_σ	10.47	10.138	9.111	10.444
g_ω	13.8	13.285	11.493	12.945
g_ρ	4.035	4.976	5.507	4.383
g_2	0.0	-12.172	-2.304	-6.9099
g_3	0.0	-36.265	13.783	-15.8337
ρ_0	0.148	0.151	0.146	0.146
E/A	-15.731	-16.426	-17.018	-16.346
K	546.3	211.1	399.2	355.4
m^*/m	0.541	0.573	0.670	0.597
J	34.9	43.5	45.1	36.1

Table 1: Various sets of meson parameters and results for asymmetric nuclear matter. Details are given in the text.

asymmetry parameter J (all given in units of MeV), and the effective mass m^*/m.

The standard set was for a long time NL1[6]. Since it produced a too large value for the asymmetry parameter J, recently, the new set NL-SH[7], has been more carefully adjusted to the isospin properties of nuclei. It is thus particularly suitable for calculations in the region of exotic nuclei far from the line of β-stability. Both sets share the problem of having a negative value for the non-linear parameter g_3 proportional to σ^4. The classical solutions of the field equations move only in a local minimum. For very large σ-values the theory becomes unstable even in the classical limit. It is always unstable, if quantum mechanical tunneling processes are taken into account. This fact seems at first a serious shortcoming of the theory. We have to keep in mind, however, that we are dealing with a phenomenological theory adjusted to the classical limit for not too high densities, where everything is stable. Already by this reason the theory is by no means a fully fledged quantum theory. We also give in Table (1) the parameter set NL2, which has a positive value g_3. It is successful for the description of light nuclei. It fails, however, completely for heavy nuclei. In addition we give the parameter set of Horowitz and Serot[8], which does not contain nonlinear terms. It produces a much to high value for the incompressibility K.

Using various parameter sets (as for instance NL1 and NL-SH given in Table 1) one has calculated the ground state properties of many nuclei in the peri-

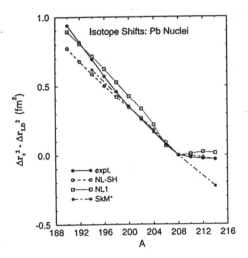

Figure 1: Isotope shifts of Pb nuclei obtained with the parameter set NL1 and NL-SH. The empirical values along with the values form NL-SH exhibit a conspicuous kink in the isotope shifts around N=126. The SkM* values show large deviations from the empirical data for heavier nuclei (from Ref. [12]).

odic table in recent years[9], in particular binding energies, nuclear radii, charge distributions and nuclear deformation parameters. One- and multidimensional energy surfaces have been produced by constraint Hartree calculations[10, 11]. Apart from the region of transitional nuclei, where the whole concept of a mean field theory breaks down, the agreement with experimental data is excellent. It is of the same quality as standard Skyrme and Gogny results.

3 Isotope Shifts in the Pb-Region

There are cases where the conventional non-relativistic theories fail. One of those are the famous isotope shifts in the Pb-region, which have been measured with extreme accuracy by atomic beam laser spectroscopy[13], and which could be reproduced in the framework of the Skyrme model[14]. If plotted as a function of the mass number as in Fig. 1, the experimental radii show a pronounced kink at the magic neutron number $N = 126$. This kink is neither reproduced by Skyrme nor by Gogny calculations. On the other hand relativistic calculations with the parameter sets NL1 and NL-SH show a remarkable agreement with the experimental data. A more careful analysis[15, 16] shows, that the origin for this deviations is the isospin dependence of the spin orbit term. The Skyrme model uses in principle an isospin independent two-body spin-orbit force. The exchange term to this force, however, induces a strong isospin dependence. In the relativistic theory the isospin dependence has its origin in the coupling constant g_ρ of the ρ-meson. Since it is the very large sum of the absolute values of the scalar and vector fields, which determine the strength of the relativistic spin-orbit term, the ρ-field plays only a minor role (less then 10 %) and the isospin dependence of relativistic calculations is rather week. In fact, in RMF theory the spin-orbit term is a pure single particle effect. It does not have an exchange contribution, which causes the strong isospin dependence of the spin-orbit term in Skyrme and Gogny calculations. A recent fit[16] of a new Skyrme parameter set with a two-body spin-orbit term without exchange, but allowing for a variable isospin dependence $V_{ls} \propto (1 + x_w P^\tau)$ through the parameter x_w leads to very small values of this parameter $x_w \approx 0.005$.

4 RMF Theory in Rotating Frame

In non-relativistic nuclear physics the cranking model[17] plays an important role in the description of rotating nuclei. It is the symmetry breaking mean field version of a variational theory with fixed angular momentum and can be derived as an approximate variation after angular momentum projection[18]. In the self-consistent version[19] it allows to include alignment effects[20] as well as polarization effects induced by the rotation, such as Coriolis-anti-Pairing or changes of the deformation. But already in the simplified version of the *Rotating Shell Model* with fixed mean fields[21] it is able to describe successfully an extremely large amount of data in the high spin region of deformed nuclei.

The cranking idea can be used for a relativistic description too[22, 23]: one simply transforms the coordinate system to a frame rotating with constant angular velocity Ω around a fixed axis is space assuming – as in non-relativistic nuclear physics – that this axis is perpendicular to the symmetry axis of the nucleus in its ground state. Such a transformation in Minkowski space is given in text books[24]:

$$
\begin{pmatrix} t \\ x \\ y \\ z \end{pmatrix} \implies \begin{pmatrix} \tilde{t} \\ \tilde{x} \\ \tilde{y} \\ \tilde{z} \end{pmatrix} = \begin{pmatrix} 1 & 0 & 0 & 0 \\ 0 & & & \\ 0 & & e^{it\Omega J} & \\ 0 & & & \end{pmatrix} \begin{pmatrix} t \\ x \\ y \\ z \end{pmatrix} \tag{9}
$$

According to the cranking prescription the absolute value of the angular velocity $|\Omega|$ will be determined after the self-consistent solution of the equations of motion in the rotating frame by the Inglis condition[17]:

$$
\langle \Omega J \rangle_\Omega = |\Omega|\sqrt{I(I+1)} \tag{10}
$$

Using the transformation properties of scalars, vectors, spinors, etc., we obtain in the rotating frame the following quantities

$$
\tilde{\sigma}(x) = e^{it\Omega L}\sigma(x) \tag{11}
$$

$$
\begin{pmatrix} \tilde{\omega}^0(x) \\ \tilde{\omega}(x) \end{pmatrix} = \begin{pmatrix} 1 & 0 \\ -\Omega \times \tilde{r} & 1 \end{pmatrix} \begin{pmatrix} e^{it\Omega L}\omega^0(x) \\ e^{it\Omega J}\omega(x) \end{pmatrix} \tag{12}
$$

$$
\tilde{\psi}(x) = e^{it\Omega J}\psi(x), \tag{13}
$$

where $L = -i(r \times p)$ is the orbital and $J = L + S$ is the total angular momentum containing the 4×4-matrices S for the spinor fields with spin $\frac{1}{2}$ and a 3×3-matrices S for vector fields with spin 1. For details see Ref. [22].

Using these quantities we obtain the following Lagrangian in the rotating frame

$$
\begin{aligned}
\tilde{\mathcal{L}} = & \ \bar{\tilde{\psi}} \left(\tilde{\gamma}^\mu (\tilde{D}_\mu + g_\omega \tilde{\omega}_\mu) + g_\sigma \tilde{\sigma} + m_N \right) \tilde{\psi} \\
& + \frac{1}{2} \tilde{\partial}_\mu \tilde{\sigma} \tilde{\partial}^\mu \tilde{\sigma} - U(\tilde{\sigma}) \\
& - \frac{1}{4} (\tilde{\partial}^\mu \tilde{\omega}^\nu - \tilde{\partial}^\nu \tilde{\omega}^\mu)(\tilde{\partial}_\mu \tilde{\omega}_\nu - \tilde{\partial}_\nu \tilde{\omega}_\mu) + \frac{1}{2} m_\omega^2 \tilde{\omega}^\mu \tilde{\omega}_\mu
\end{aligned}
\tag{14}
$$

where $\tilde{D}^\mu$ is the covariant derivative with respect to the rotating metric. Neglecting in the following the tilde sign, we derive the classical equations of motion, which are in the quasi-static limit of the form:

$$
\{\alpha(p - g_\omega \omega) + g_\omega \omega_0 + \beta(m_N + g_\sigma \sigma) - \Omega J\} \psi_i = \epsilon_i \psi_i \tag{15}
$$
$$
\{-\Delta + (\Omega L)^2\} \sigma + U'(\sigma) = -g_\sigma \rho_s \tag{16}
$$
$$
\{-\Delta + (\Omega L)^2 + m_\omega^2\} \omega^0 = g_\omega \rho_v \tag{17}
$$
$$
\{-\Delta + (\Omega J)^2 + m_\omega^2\} \omega = g_\omega j \tag{18}
$$

These equations are very similar to the RMF equations in the non-rotating frame. There are only three essential differences:

1. The Dirac equation (15) contains a Coriolis term ΩJ in full analogy to non-relativistic cranking.

2. The Klein-Gordon equations for the mesons contain terms proportional to the square of the corresponding Coriolis terms. It turns out, however, that they can be neglected completely for all realistic cranking frequencies, because (i) they are quadratic in Ω and (ii) mesons being bosons are to a large extend in the lowest s-states with only small d-admixtures

3. The Coriolis operator in the Dirac equation breaks time-reversal invariance. Currents j are induced, which form the source of magnetic potentials in the Dirac equation (*nuclear magnetism*). In this way the charge current j_c is the source of the normal magnetic potential A, the iso-scalar baryon current j_B is the source of the spatial components ω of the ω-mesons and the iso-vector baryon current j_3 is the source of the spatial component ρ_3 of the ρ-mesons. In contrast to the Maxwellian magnetic field A having a small electro-magnetic coupling, the large coupling constants of the strong interaction causes the fields ω and ρ to be important in all cases, where they are not forbidden by symmetries, such as time reversal. They have a strong influence on the magnetic moments[25] in odd mass nuclei, where time reversal is broken by the odd particle, as well as on the moment of inertia in rotating nuclei, where time reversal is broken by the Coriolis field[26].

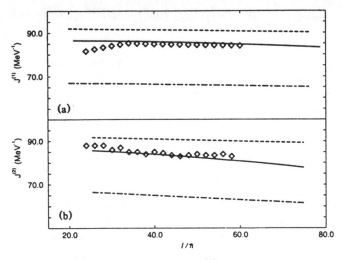

Figure 2: (a) Static ($\mathcal{J}^{(1)}$) and (b) dynamic ($\mathcal{J}^{(2)}$) moment of inertia for the lowest superdeformed band in the nucleus ^{152}Dy. The dashed line corresponds to the calculation without the spatial contributions of the vector mesons. In the dotted line such contributions are taken into account in a semi-classical way and the full line represents the self-consistent solution including these contributions fully (From Ref. [26]).

In order to investigate the applicability of Relativistic Mean Field theory for the description of rotating superdeformed nuclei, we chose as an example the well know superdeformed band in the nucleus ^{152}Dy. It was the first superdeformed band in rotating Rare Earth nuclei discovered experimentally by the Daresbury group[27].

The RMF equations in the rotating frame in Eqs. (15-18) were solved self-consistently by expanding the Dirac spinors as well as the meson fields in terms of eigenfunctions of a deformed oscillator, as discussed in details in Refs. [9] and [28]. In Fig. 2 we show the static and the dynamic moment of inertia for the lowest superdeformed band in the nucleus ^{152}Dy as a function of the angular momentum. It is clearly seen that a calculation without nuclear magnetism, i.e. without the spatial contributions of the vector meson-fields, which is in good agreement with the experimental quadrupole moments (see Ref. [29]) produces much too small moments of inertia. A semi-classical correction where these contributions, derived in Thomas Fermi approximation using a rigid rotor current, are taken into account in first order perturbation theory, overemphasized the moments of inertia by roughly 10 %. Only if one takes these contributions into account in a fully self-consistent way, is perfect agreement with experimental data achieved. In the region of small angular momenta one still observes very small deviations, which could possibly be understood as the influence of remaining pairing correlations in this region of intermediate spins.

We also find that nuclear magnetism practically has no influence on the shape of the nucleus. The *rms*-radii are practically identical for the calculation with and without nuclear magnetism and their dependence on the angular momentum

Band	E (MeV)	Q_0 (fm^2)	$\mathcal{J}^{(2)}$	$\mathcal{J}^{(1)}$	$\mathcal{J}_{rig}$
^{152}Dy	-1228.358	4287.7	82.544	86.41	93.35
^{151}Tb*$(+,+)$	0.53	-2.91	-1.554	-1.25	-1.69
^{151}Tb $(+,-)$	0.51	-3.17	0.654	0.35	-1.80
^{151}Tb*$(-,+)$	0.56	1.30	-0.001	0.10	-0.45
^{151}Tb*$(-,-)$	0.54	1.19	0.145	1.48	0.39

Table 2: Binding energy E, mass quadrupole moment Q_0, dynamic ($\mathcal{J}^2$), static ($\mathcal{J}^1$) and rigid body ($\mathcal{J}_{rig}$) moment of inertia at the angular momentum $I = 50\hbar$ for the superdeformed band in ^{152}Dy and relative changes of these values (given in %) several bands in the neighboring nucleus ^{151}Tb at the same angular momentum. They have the quantum numbers (PS) of parity (P) and signature (S).

is very week. The same is true for the deformations expressed by quadrupole- and hexadecupole moments. The mass quadrupole moments decrease in the spin range from 20 to 60 $\hbar$ only very little, running from 4350 to 4260 fm^2 and the corresponding hexadecupole moments change for the same region from 20600 to 19600 fm^4. The changes induced by nuclear magnetism are of the order of a few per mill. The average charge quadrupole moment if found to be 18.6 eb, which is in good agreement with a value of 18 eb obtained in a non-relativistic calculation[30], and the experimental value of 19 eb [31] From the quadrupole moments we can derive the Hill-Wheeler parameters $\beta = 0.72$ and $\gamma = 0.7°$ for the quadrupole deformations, which corresponds closely to a nearly prolate deformed nucleus with an axis ratio of 1:1.9, close to the standard value 1:2 of the harmonic oscillator model.

Let us now investigate the problem of identical bands. For this purpose we calculate, in a self-consistent way, bands in the neighboring nucleus ^{151}Tb by removing one proton from different levels in the ^{152}Dy core. The proton hole induces a polarization of the ^{152}Dy-core, which has two effects: it leads to changes of deformation and in addition to changes in the current distribution. In Table 2 we show the values obtained after solving in a fully self-consistent fashion the relativistic mean field equations for the odd system in the four lowest configurations. We show the values for several observables for the lowest superdeformed band in ^{152}Dy. The relative changes with respect to this reference band in the four bands of ^{151}Tb are given in per mill. According to the simple $A^{5/3}$-rule we expect changes in the moment of inertia by ≈ 11 per mill. The calculated values for the moments of inertia $\mathcal{J}^1$ and $\mathcal{J}^2$ for the band with the quantum numbers $(-+)$, which we we shall in the following call the *identical band*, are, however, at least an order of magnitude smaller. This is by no means trivial, because we find considerably larger changes in the quadrupole moments and in the rigid

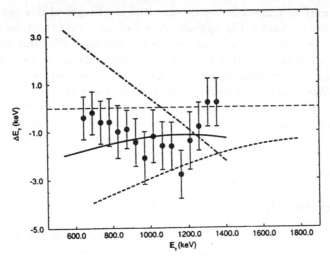

Figure 3: Differences ΔE_γ for the identical band with the quantum numbers $(-+)$. The fully self-consistent solution (full line) and solutions neglecting nuclear magnetism (dashed line) or polarization induced by the proton hole (dashed dotted line) are compared with the experiment (full dots).

body moments of inertia. In fact in most of the other bands the changes are also much larger.

In order to have a direct comparison with the experiment we show in Fig. 3 the difference $\Delta E_\gamma = E_\gamma(\text{Tb}) - E_\gamma(\text{Dy})$ between the transitions in the band with the quantum numbers $(-+)$ in the nucleus ^{151}Tb and in the lowest superdeformed band in ^{152}Dy. The agreement with the experimental values is excellent, the energy differences are of order of 1 keV. This band correspond to a hole in the orbit with the approximate Nilsson quantum numbers $[301]\frac{1}{2}-$. This orbit has a very small number of oscillator quanta along the z-axis (the symmetry axis), which yields nearly vanishing contributions to the moment of inertia. We are therefore in agreement with the qualitative argument put forward in Ref. [32]

This is, however, not the full story. In order to investigate the very good quantitative agreement, we have carried out two additional calculations in Fig. 3 for the *identical band* band with the quantum numbers $(-+)$: First we neglected the polarization induced by the proton hole, i.e we calculated the energy differences for wave functions for the nucleus ^{151}Tb obtained from the ^{152}Dy core by just removing one proton, without requiring self-consistency for the odd mass configuration. In this case we find the dashed dotted line in Fig. 3 which is in sharp disagreement with the experimental data. Next we took into account the polarization, but we neglected nuclear magnetism, i.e. the contributions of the spatial components of the vector meson fields and find the dashed line in Fig. 3, which is also in disagreement with experiment.

We therefore conclude, that a very delicate cancelation process occurs in identical bands in superdeformed nuclei. Polarization of the quadrupole moments and of the density alone would induce changes of the order of 5 – 10 per mill.

Neglecting nuclear magnetism would also lead to changes of this order of magnitude. Obviously both act in opposite direction, such that the finial differences are only in the order of 1 per mill. So far the precise mechanism for this cancelation is not fully understood. It requires definitely much more systematic investigations. Nonetheless it seems to us a very satisfying and surprising result, that without any free parameter, and simply using the set NL1 adjusted to nuclear matter and a few spherical nuclei, long before identical bands had been identified, we can obtain this degree of accuracy in the relatively simple minded relativistic mean field approach. We have to emphasize, however, that full self-consistency as well as the inclusion of the nuclear currents are very important in this context.

5 Relativistic Theory of Pairing

In relativistic mean field theory the mesons are treated as classical fields. In this case one obtains a relativistic single-particle operator for the nucleons containing only terms of the structure $\psi^+\psi$. Obviously, it is impossible to describe a superfluid behavior of the nuclear many-body system in such a classical framework. For this phenomenon we need either a two-body interaction of the form $\psi^+\psi^+\psi\psi$ or a generalized single-particle field $\psi^+\psi^+ + \psi\psi$. In order to derive those terms we have to quantize not only the nucleon but also the meson fields.

A very simple way to describe pairing correlations in a many-body system has been proposed in non-relativistic quantum mechanics by Gorkov[33]. It can be easily extended to the relativistic framework (for details see Ref. [34]). In addition to the normal Green's functions[†]

$$G_{ab} = -i\langle A|T\psi_a\bar\psi_b|A\rangle, \tag{10}$$

one introduces anomalous Green's functions

$$F_{ab} = -i\langle A+2|T\bar\psi_a\bar\psi_b|A\rangle \quad \text{and} \quad \tilde F_{ab} = -i\langle A|T\psi_a\psi_b|A+2\rangle \tag{20}$$

and derives an equation of motion for these quantities. It contains higher order Green's functions of the form $\langle A|T\psi\phi\bar\psi|A\rangle$ and $\langle A+2|T\bar\psi\phi\bar\psi|A\rangle$, where ϕ stands for the quantized meson fields σ or ω. Using the Klein Gordon equations, which hold also for the corresponding field operators they can be replaced by the two-particle Green's functions $\langle A|T\psi\bar\psi\psi\bar\psi|A\rangle$ and $\langle A+2|T\psi\bar\psi\bar\psi\bar\psi|A\rangle$. In order to close this system of equations several approximations are made:

- The two-particle Green's functions are factorized and expressed by the one-body functions (19) and (20) (*Gorkov factorization*):

$$\langle A|T\psi_e\bar\psi_d\psi_c\bar\psi_b|A\rangle = -G_{ed}G_{cb} + G_{cd}G_{eb} + \tilde F_{ec}F_{db},$$
$$\langle A+2|T\psi_e\bar\psi_d\bar\psi_c\bar\psi_b|A\rangle = -G_{ed}F_{cb} + G_{ec}F_{db} - G_{eb}F_{dc}. \tag{21}$$

[†]the index $a = (t, r, s)$ contains coordinates x^μ as well as spin indices (s)

- The difference between the wave functions $|A\rangle$ and $|A+2\rangle$ in the anomalous Greens'functions is neglected. This leads to violation of particle number in the final relativistic HFB theory.

- The retardation in the propagators for the mesons is neglected (*instantaneous approximation*). In this case one ends up with a two-body interaction local in time.

After some algebra one is able to derive a set of relativistic Hartree-Fock-Bogoliubov equations

$$
\begin{pmatrix} h-\lambda & \Delta \\ -\Delta^* & -h^*+\lambda \end{pmatrix} \begin{pmatrix} U \\ V \end{pmatrix}_\nu = \begin{pmatrix} U \\ V \end{pmatrix}_\nu E_\nu
\tag{22}
$$

with the Dirac operator

$$
h = \boldsymbol{\alpha p} + \beta \Sigma,
\tag{23}
$$

the mass operator and the pairing field

$$
\Sigma_{ac} = \delta_{ac} m_N + V_{adce}\rho_{ed} - V_{adec}\rho_{cd}, \qquad \Delta_{ac} = V_{acde}\kappa_{de}.
\tag{24}
$$

V_{abcd} are two-body matrix elements obtained from the meson exchange forces. The Latin indices $a = (\boldsymbol{r}, s)$ contain here no longer the time. ρ_{ab} and κ_{ab} are the normal and anomal density matrices:

$$
\begin{aligned}
\rho_{ab} &\equiv \rho_{\alpha\beta}(\boldsymbol{r}_a, \boldsymbol{r}_b) = \langle A|\psi_\beta^+(\boldsymbol{r}_b)\psi_\alpha(\boldsymbol{r}_a)|A\rangle \\
\kappa_{ab} &\equiv \kappa_{\alpha\beta}(\boldsymbol{r}_a, \boldsymbol{r}_b) = \langle A|\psi_\beta(\boldsymbol{r}_b)\psi_\alpha(\boldsymbol{r}_a)|A\rangle
\end{aligned}
\tag{25}
$$

The structure of the relativistic HFB equations (22) is very similar to the conventional HFB equations (see for instance Ref. [35]) with one-meson exchange potentials. However, it contains all the relativistic effects described in section 2.

We apply this theory first to nuclear matter at various densities. Since there are no data on the size of the pairing gap in this case, we compare with non-relativistic results based on the Gogny force[36], which is considered to yield a reliable and quantitative description of nuclear pairing properties in the entire periodic table.

It turns out, that pairing in nuclear matter has practically no influence on the binding energy and on the saturation behavior. Pairing effects are just several orders of magnitude smaller the other energies involved in this problem. This means that we can easily separate the Hartree-Fock from the BCS problem and to not have do solve the full HFB problem. For the Hartree-Fock problem we use the well established RMF theory with the parameter set NL1. Considering only $(S = 0, T = 1)$-pairing we are left in this case with the following gap equation in nuclear matter:

$$
\Delta(p) = -\frac{1}{4\pi^2} \int_0^\infty v_{pp}(p, k) \frac{\Delta(k)}{\sqrt{(\epsilon(k) - \lambda)^2 + \Delta^2(k)}} k^2 dk,
\tag{26}
$$

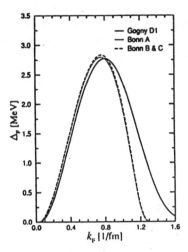

Figure 4: The pairing gap of symmetric nuclear matter at the Fermi $\Delta(k_F)$ as a function of the density expressed in terms of the Fermi momentum k_F for the Gogny force (full line) and various versions of the Bonn Potential.

where $\epsilon(k) = \sqrt{k^2 + m^{*2}}$ is the eigenvalue of the Dirac equation for the corresponding density. The ω-field contributes only a constant value, which is absorbed into the chemical potential λ, which is adjusted to the chosen density. $v_{pp}(p, k)$ is the relativistic version of the particle-particle interaction obtained from the one-meson exchange potentials.

In Eq. (26) on has for each momentum p to integrate over all k-values. In this integral there are large contributions from the high momentum region leading for the parameter sets of RMF theory to very unrealistic pairing properties, roughly a factor 3 larger than the values calculated with the Gogny force[34]. Pairing turns out to depend crucially on the high-momentum, i.e. on the short-distance behavior of the force, which is not suitably adjusted in RMF theory, where only the low momenta below the Fermi level are important.

In order to investigate this question further, we therefore used interactions adjusted to the nucleon-nucleon scattering data, i.e. several relativistic versions of the Bonn potential[37] and solved with those potentials the gap equation (26) in nuclear matter using for the effective mass m^* the values of the non-linear parameter set NL1.

In Fig. 4 we show the resulting gap parameter at the Fermi surface $\Delta_F = \Delta(k = k_F)$ as a function of the density represented by the Fermi momentum k_F for three versions of the Bonn potential (A, B and C) and compare it with the corresponding quantity obtained in a non-relativistic calculation[36] based on Gogny's force D1. Apart from the tail at larger densities, we find excellent agreement in both cases. In particular we find maximal pairing correlations of roughly 2.8 MeV at Fermi momentum $k_F \approx 0.8$ fm^{-1}, i.e. at roughly one fifth of nuclear matter density in both cases.

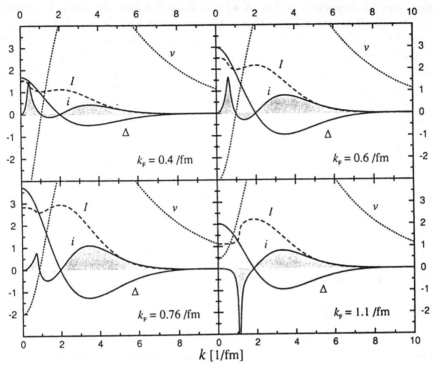

Figure 5: The pairing gap parameter $\Delta(k)$ as a function of k together with the potential $\nu = v_{pp}(0.8, k)$, the integrand $i(k)$ of the gap equation (26), and the total integral $I(k)$ in the gap equation integrated from the value k to infinity for the Bonn potential discussed in Fig. 4.

This is in agreement with the usual observation that pairing in nuclei is a surface phenomenon. In Ref.[36] it has been shown that using this density dependence of the gap parameter in semi-classical calculations the average pairing properties of finite nuclei can be reproduced rather well. Whether nuclear matter at saturation with a Fermi momentum of roughly 1.35 fm^{-1} is super-fluid is hard to decide. In any case it depends critically on small details of the interaction. For Gogny forces there seems to be a small pairing gap of roughly 0.5 MeV left, whereas for the Bonn potentials there is no pairing at this density. Obviously the three Bonn potentials A, B, and C produce nearly identical results.

We realize that also in this realistic calculation the high momentum region in the integral (26), where the repulsive part of the ω-mesons dominates, gives large contributions to the pairing gap. Within this model nuclear pairing is therefore produced to a large extend by repulsive forces.

In order to understand this astonishing result we have to consider the gap equation in more detail. In Fig. 5 we show the pp-interaction $v_{pp}(p, k)$ for $p = 0.8$ fm^{-1} – corresponding to the maximum of the gap at the Fermi surface in Fig.4 – as a function of k. For small momenta the force v_{pp} is attractive (mostly caused by σ-exchange), and for large values of k the repulsive contributions of the ω-mesons dominate. In Fig. 5 we also show the function $\Delta(k)$. It is positive only for small k-values. Together with the minus sign in Eq. (26) we obtain therefore positive contributions to the integral in this k-region. For large k-values the interaction v_{pp} as well as the gap-parameter Δ change sign leading again to positive contributions in the integral. It is not the strong attraction of the σ-exchange which produces the larger part of pairing correlations, but the strong repulsion of the ω-exchange. This is a case where repulsive forces can produce pairing (see also Ref. [38]).

The present considerations are based on the assumption, that renormalization effects of the effective pairing force caused by higher correlations as for instance Brueckner correlations are negligible. Definitely this assumption has to be investigated more carefully in the future. The simple fact, however, that the applications of bare nucleon-nucleon interactions in the BCS equations leads to very reasonable pairing correlations at the Fermi surface for nearly all densities, gives us confidence to believe that such renormalizations can be neglected in the $(S=0, T=1)$-channel and that the conclusion that repulsive forces have a major influence on nuclear pairing remains valid also for investigations going beyond the BCS approximation.

6 Time Dependent RMF Theory

So far we have treated only static problems in relativistic mean field theory. Only the relativistic cranking theory was not completely static, but the time dependence could be removed by a suitable transformation to a rotating frame. In this section we deal with really time dependent problems, i.e. we solve the time dependent relativistic mean field equations (2-4) for proper initial conditions. For that purpose we consider a heavy ion reaction with relativistic energies, as they are now available at modern accelerators as for instance at the GSI in Darmstadt. For a study of nuclear structure we use these highly energetic projectiles only for Coulomb excitation of the target. Because of the relativistic energy and the Lorentz-contraction the reaction time is extremely short and the target feels in practice only a δ-like Coulomb push, i.e. all the protons are suddenly pushed in one direction, and, since we are working in the center of mass frame, all the neutrons are pushed at the same time in the other direction.

Without studying the reaction process (for more details in this direction see the review article in Ref. [39]) we therefore assume the following initial condition: We start with a self-consistent solution of the static problem for the target nucleus as obtained by the solution of Eq. (8) and transform all the protons (neutrons) by a covariant Lorentz-boost to a frame moving with velocity $\pm v$ in z-direction against each other. Starting with this initial wave function we solve the Dirac equations (2) in time. At each time step the densities and currents are calculated and the corresponding Klein-Gordon equations (3) and (4) are solved neglecting retardation effects, i.e. neglecting the second time derivatives ∂_t^2 in these equations. This approximation is justified, because the large meson masses in these equations produce forces of very short range, where at the energies under consideration retardation effects are small. This can be seen even more

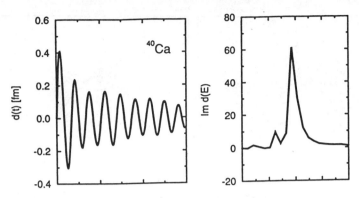

Figure 6: Time dependent dipole moments and the corresponding Fourier transforms of the nuclei ^{40}Ca for the initial velocity $v = \pm 0.035c$ using the parameter set NL-SH.

clearly in Fourier space, where can we neglect the energy contribution k_0 in the denominator $k_0^2 - k^2 - m_{meson}^2$ of the meson propagators. At each time step we therefore assume, that the nucleons feel instantaneously the static fields produced by the densities at this point in time.

Such calculations require a large numerical effort. For a nucleus with A particles we have to solve $8 \times A$ coupled partial differential equations of Dirac type for the four complex components of the spinors. In additions at each time step the meson fields have to be determined by the solution of the Klein-Gordon equations. Since currents are essential in this case, we have 13 fields (one σ-field and in each case four ω^μ- ρ^μ- and A^μ fields). Only the 3-component of the isospin is taken into account for the ρ-meson, because we do not consider charge transfer processes and the pionic fields are neglected by the argument of reflection symmetry at a plain containing the direction of the velocity v (the z-direction). The parity- or time-reversal symmetry arguments are no longer valid in this case. Further details to these calculations are given in Ref. [40].

We are now able to study the self-consistent motion of the nuclear system under this sudden Lorentz push, which drives protons and neutrons in opposite directions. In Fig. 6 we display for ^{40}Ca the time-dependent dipole moments $d(t)$ of the system for small initial velocity. We observe a rather harmonic – weakly damped – oscillation with small amplitude containing superpositions of different modes. It corresponds to the giant dipole resonance (GDR) of this nucleus. The corresponding Fourier transform give us the spectrum. We see two distinct peaks with excitation energies 12.3 and 18.6 MeV. The main peak is at 18.6 MeV and is in reasonable agreement with the experimental value of 19.8±0.5 MeV. We find a width of roughly 3.2 MeV as compared with the experimental value of 4 MeV. Since this model does not take into account two-body collisions we should not expect to find the full experimental width. These calculations, however, contain the coupling to the continuum (escape width), which plays for light nuclei the major role in our understanding of the width. This is also found in these calculations. This time-dependent investigations allow us therefore the study of collective excitations. Depending on the initial condition we find several

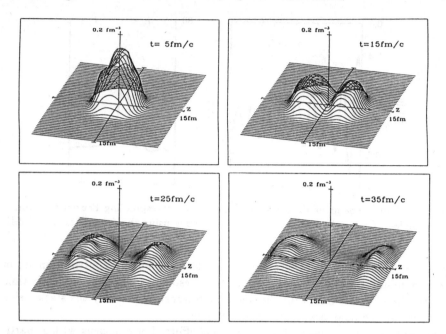

Figure 7: Time evolution of the total baryon densities of ^{48}Ca for the the large initial velocities $v_\pi = 0.4c$ and $v_\nu = \pm 0.286c$.

modes of different character. In Fig. 6 we studied the case of protons moving against neutrons and found the giant dipole resonance. Starting initially with wave functions deformed by a constrained Hartree calculation, we obtained other modes such as iso-scalar giant quadrupole resonance (ISGQR) or the iso-vector quadrupole resonance (IVGQR) at energies in agreement with experiment (for details see Ref. [40]).

So far we have only considered small amplitudes oscillations, because we started with initial conditions very close to the static solutions. In Fig. 7 we show a case with a very large initial velocity (which is probably impossibly to achieve in one Coulomb push). In this case the motion is no longer harmonic. A straight dissociation of the entire nucleus is observed. Two clusters are formed that consist of pure proton and neutron matter. The restoring force is not strong enough to pull back the two clusters and they separate. These are unbound systems and dissociate quickly into individual nucleon.

7 Conclusion and Outlook

Within recent years a large number of problems in nuclear structure physics have been tackled by relativistic mean field theory. It has turned out, that this theory is very successful in describing nuclear properties over the entire periodic table with a minimal set or parameters in a fully microscopic way. It is a phenomenological theory as all the other competing non-relativistic density dependent Hartree-Fock theories, but is goes beyond these theories in several respects: (a) it incorporates relativistic phenomena such as the large scalar attraction and the large vector repulsion in a consistent way, (b) it deals properly with the spin-orbit term in the nuclear medium, which has a great influence on many details in nuclear structure, (c) it uses the field-theoretical techniques and the language of nucleons and mesons. I is therefore much better suited to provide a connection to more fundamental theories derived from QCD, and (d) it is in many ways technically simpler, requiring less manpower, than Skyrme or Gogny theories, which are either non-local or have to take into account non-local effects by complicated momentum dependencies. Let us therefore conclude that modern theories of nuclear structure should be relativistic.

There is certainly a lot to be done. Relativistic Brueckner calculations seem to be very successful in dealing with the old Coester line. There should be a better connection of such theories to the purely phenomenological models discussed in this paper. In this case one probably will end up with density dependent meson masses and coupling constants. They might give us a better understanding of the nonlinear parameterizations used so far.

Acknowledgments

I would like to express my deep gratitude to all my collaborators, in particular to H. Berghammer, G.A. Lalazissis, W. Koepf, J. König, H. Kucharek, A. Rummel, M.M. Sharma, and D. Vretenar. Without their great enthusiasm and without their efforts these investigations would not have been possible.

References

[1] M.H. Johnson and E. Teller; Phys. Rev. **98** (1955) 783

[2] J.D. Walecka; Ann. Phys. (N.Y.) **83** (1974) 491

[3] B.D. Serot and J.D. Walecka; Adv. Nucl. Phys. **16** (1986) 1

[4] J. Boguta and A.R. Bodmer; Nucl. Phys. **A292** (1977) 413

[5] Z.Y. Zhu, H.J. Mang, and P. Ring; Phys.Lett. **254B** (1991) 325

[6] P.-G. Reinhard et al.; Z. Phys. **A323** (1986) 13

[7] M.M. Sharma, M.A. Nagarajan, P. Ring; Phys. Lett. **B312** (1993) 377

[8] C.J. Horowitz and B.D. Serot; Phys. Lett. **140B** (1984) 181

[9] Y.K. Gambhir, P. Ring, A. Thimet; Ann. Phys. (N.Y.) **198** (1990) 132

[10] W. Koepf and P. Ring; Phys.Lett 212B (1988) 397

[11] D. Hirata, H. Toki, I. Tanihata, and P. Ring; Phys. Lett. **B314** (1993) 168

[12] M.M. Sharma, G.A. Lalazissis, and P. Ring; Phys.Lett.B317 (1993) 9

[13] E.W. Otten; in *Heavy-Ion Science*, (Plenum, New York, 1989) **7**, p. 517

[14] N. Tajima et al.; Nucl. Phys. **A551** (1993) 434

[15] G.A. Lalazzisis, J. König, P. Ring, and M.M. Sharma; Contr.Int.Conf.on Shapes and Nuclear Structure at Low Energies, Antibes, France, June 1994

[16] M.M. Sharma, G.A. Lalazissis, and J. König; to be published

[17] D.R. Inglis; Phys. Rev. **96** (1954) 1059

[18] R. Beck, H.J. Mang, and P. Ring; Z. Phys. **231** (1970) 26

[19] P. Ring, R. Beck, and H.J. Mang; Z. Phys. **231** (1970) 10

[20] B. Banerjee, H.J. Mang, and P. Ring; Nucl. Phys. **A215** (1973) 366

[21] P. Ring and H.J. Mang; Phys. Rev. Lett. **33** (1974) 1174

[22] W. Koepf and P. Ring; Nucl. Phys. **A493** (1989) 61

[23] K. Kaneko, M. Nakano, and M. Matsuzaki; Phys. Lett. **B317** (1993) 261

[24] L.D. Landau and E.M. Lifshitz; *Course of Theoretical Physics*, Pergamon Press, Oxford (1959)

[25] Ulrich Hofmann and P. Ring, Phys. Lett. **214B**, 307 (1988)

[26] J. König and P. Ring; Phys. Rev. Lett. **71** (1993) 3079

[27] P.J. Twin et al.; Phys. Rev. Lett. **57**, 811 (1986)

[28] Y.K. Gambhir and P. Ring; Mod. Phys. Lett. **8** (1993) 787

[29] W. Koepf and P. Ring, Nucl. Phys. **A511**, 279 (1990)

[30] I. Ragnarsson and S. Aberg, Phys. Lett. **180B**, 191 (1990)

[31] M.A. Bentley et al.; Phys. Rev. Lett **59** (1987) 2141

[32] T. Byrski et al.; Phys. Rev. Lett. **64** (1990) 1650

[33] L.P. Gorkov; Sov. Phys. JETP **7** (1958) 505

[34] H. Kucharek and P. Ring; Z. Phys. **A339** (1991) 23

[35] P. Ring and P. Schuck; *The Nuclear Many-body Problem*, Springer Verlag, Heidelberg (1980)

[36] H. Kucharek, P. Ring, P. Schuck, R. Bengtsson, and M. Girod; Phys. Lett. **216B** (1989) 249

[37] R. Machleidt; Adv. Nucl. Phys. **19** (1989) 189

[38] F.J.W. Hahne and P. Ring; Phys. Lett. **259B** (1991) 7

[39] C.A. Bertulani and G. Baur; Phys. Rep. **163C** (1988) 300

[40] D. Vretenar, H. Berghammer, and P. Ring; Nucl. Phys. in print

Semiclassical Description of the Relativistic Nuclear Mean Field Theory

X. Viñas

Departament d'Estructura i Constituents de la Matèria
Facultat de Física, Universitat de Barcelona
Diagonal 647, E-08028 Barcelona, Spain

Abstract: Semiclassical relativistic particle and energy densities for a set of fermions submitted to a scalar field and to the time-like component of a four-vector field are presented in the Wigner–Kirkwood and extended Thomas-Fermi mean field theories. The semiclassical approach is then applied to the non-linear $\sigma - \omega$ model and the resulting variational equations are solved for finite nuclei and semi-infinite symmetric nuclear matter.

1 Introduction

The conventional approximation to nuclear physics is based on a non-relativistic formulation although the mean value of the nucleon velocity in the nuclear medium can be as large as forty per cent of the velocity of the light. Twenty years ago, and combining relativity and Yukawa's theory of the nucleon-nucleon interaction described by exchange of mesons, Walecka and coworkers developed the so-called quantum hadrodynamics (QHD) which is a quantum field theory for the nuclear many-body problem [1,2]. In this formalism the nucleons interact through exchange of virtual mesons. In the simplest version only a vector field accounting for the short-range repulsion and a scalar field responsible for attraction are needed to describe the nuclear saturation. The scalar field couples to the mass of the nucleon resulting that the nucleon has, in the nuclear medium, an effective mass m^* smaller than its bare mass. The vector field is associated with the exchange of the ω meson and the scalar field is due to σ meson that mocks up the two pion exchange.

The full QHD theory is complicated and in practice one deals with phenomenological approach. The coupling constants of the meson-nucleon Lagrangian and the mass of the σ meson are taken as free parameters that are fitted to reproduce the properties of the nuclear matter as well as of the finite nuclei. Most calculations have been done in the mean field approach which neglects the exchange terms and the negative energy states. This relativistic mean field theory (RMFT) gives reliable results along the whole Periodic Table with the advantage of incorporating automatically the spin-orbit force and the finite range effects as well as density dependence and non-local effects. The quality of this theory is comparable with the one obtained using Skyrme forces [3] in the conventional non-relativistic mean field approximation.

In many applications where one is only interested in the average value of some properties like binding energies or nuclear densities, the semiclassical methods have become a useful tool. In particular, the Density Fuctional Theory (DFT) is widely used to treat inhomogeneous Fermi Systems in many fields of physics and chemistry [4-7]. The theoretical justification of the DFT is provided by the Hohenberg-Kohn theorem [8] which states that the ground-state energy of a set of interacting fermions can be expressed through a unique functional of the local density ρ from where it is determined making use of the variational principle. Due to the fact that the exact functional is not known, one has to work with approximate methods. In the non-relativistic nuclear case the most popular and succesful of them is the so-called Extended Thomas-Fermi method [9] mainly used together with Skyrme forces. It is based in a semiclassical expansion in powers of $\hbar$ of the density matrix. This method gives reasonable binding energies and densities in comparison with the full quantal ones calculated in the Hartree-Fock approach especially if the $\hbar^4$ corrections are included [10,11].

Although the relativistic Thomas-Fermi approximation was introduced in the early thirties [12], the inclusion of the $\hbar$ corrections (Relativistic Extended Thomas Fermi RETF) has only been developed very recently. This has been done by Centelles et al. by using the Wigner transform of the propagator associated to the Dirac hamiltonian [13], by Weigel et al. employing Green's functions together with the Wigner representation [14] and Dreizler et al. who have proved the Hohenberg-Kohn theorem in the relativistic domain and derived the RETF functionals through Green's functions and Fourier transforms [15].

2 Theory

In order to derive the RETF approach to the RMFT we start from a single -particle time independent Dirac Hamiltonian containing a scalar field $S(\mathbf{r})$ and a time-like component of a four-vector field $V(\mathbf{r})$.

$$\hat{H} = -i\boldsymbol{\alpha} \cdot \nabla + \beta\, m^*(\mathbf{r}) + I V(\mathbf{r}) \tag{1}$$

where $m^*(\mathbf{r}) = m + S(\mathbf{r})$.

The standard techniques used to get the $\hbar$ expansion of the density matrix [9,16-18] are not useful to treat the relativistic problem. It is due to the structure of the Dirac Hamiltonian constituted by non-commuting matrices.

Our method [13] starts from the propagator associated to the Dirac Hamiltonian (1) :

$$\hat{G}(\eta) = \exp\left(-\eta \hat{H}\right). \tag{2}$$

that fulfills the symmetrized Bloch equation :

$$\frac{\partial \hat{G}}{\partial \eta} + \frac{1}{2}\left(\hat{H}\hat{G} + \hat{G}\hat{H}\right) = 0. \tag{3}$$

The Wigner transform (WT) [16] of equation (3) reads :

$$\frac{\partial G_w}{\partial \eta} + \frac{1}{2}\left\{H_w, \exp\left(\frac{i\hbar}{2}\overleftrightarrow{\Lambda}\right), G_w\right\} = 0, \tag{4}$$

where A_w denotes the WT of the operator $\hat{A}$ and the shorthand notation $\{A, C, B\}$ stands for $ACB + BCA$. Use has been made of the following property of the WT of a product of operators [16]:

$$\left(\hat{A}\hat{B}\right)_w = A_w \exp\left(\frac{i\hbar}{2}\overleftrightarrow{\Lambda}\right) B_w, \tag{5}$$

with $\overleftrightarrow{\Lambda} \equiv \overleftarrow{\nabla}_r \cdot \overrightarrow{\nabla}_p - \overleftarrow{\nabla}_p \cdot \overrightarrow{\nabla}_r$ (the arrows indicate the direction in which the gradients act). In particular , it should be pointed out that the WT of the Dirac Hamiltonian (1) is given by [13] :

$$H_w = \alpha \cdot \mathbf{p} + \beta\, m^*(\mathbf{r}) + I\, V(\mathbf{r}). \tag{6}$$

If in (4) the WT of the propagator is explicitely written in a power series of $\hbar$:

$$G_w = \sum_{n=0} \hbar^n G_n, \tag{7}$$

by expanding $\exp\left(i\hbar\,\overleftrightarrow{\Lambda}/2\right)$ and equating the coefficients of the same powers of $\hbar$, a set of coupled differential equations for G_n in the $\hbar$ expansion of the propagator is obtained. These equations can be integrated taking into account the boundary condition $G(\eta = 0) = I$ obtaining the following recurrence algorithm :

$$G_n = -\frac{1}{2}G_0^{1/2}\left[\sum_{m=1}^{n}\frac{1}{m!}\left(\frac{i}{2}\right)^m \int_0^\eta d\eta'\, G_0^{-1/2}\left\{H_w, \left(\overleftrightarrow{\Lambda}\right)^m, G_{n-m}\right\} G_0^{-1/2}\right] G_0^{1/2} \tag{8}$$

that allows to get any desired term G_n in the $\hbar$ expansion (7) starting from the lowest order solution G_0. In our case of the Dirac Hamiltonian, G_0 takes the simple form :

$$G_0 = \exp\left(-\eta H_w\right) = \left[\cosh \eta\varepsilon - \frac{\sinh \eta\varepsilon}{\varepsilon}(\alpha \cdot \mathbf{p} + \beta\, m^*)\right]\exp\left(-\eta V\right), \tag{9}$$

with $\varepsilon = (p^2 + m^{*2})^{1/2}$. Since the hyperbolic forms which appear in (9) can be written in term of $\exp\left(-\varepsilon\eta\right)$ and $\exp\left(\varepsilon\eta\right)$, it is clearly seen that G_0 contains the positive and negative energy solutions separately. In addition, all the contributions coming from the expansion of the exponentials in eq.(2) in powers of η have been consistently resummed. From the structure of (8), it is evident that these features of G_0 are kept to any order G_n in the $\hbar$ expansion of the propagator. These are the two main advantages of this method in comparison with the ones used in the non-relativistic case [9,16-18].

The genaeralized density matrix $\hat{\mathcal{R}}$ is obtained from the propagator by an inverse Laplace transform [16] :

$$\hat{\mathcal{R}}(\lambda) = \mathcal{L}^{-1}_{\eta \to \lambda} \left[\frac{\hat{G}(\eta)}{\eta} \right] \tag{10}$$

Inserting the $\hbar$ expansion of the propagator (7) in eq.(10) one gets the expression of the corresponding WT of the density matrix:

$$\mathcal{R}_w = \mathcal{L}^{-1}_{\eta \to \lambda} \left[\frac{G_w(\eta)}{\eta} \right] = \mathcal{R}_0 + \hbar \mathcal{R}_1 + \hbar^2 \mathcal{R}_2 + \dots . \tag{11}$$

In particular, the lowest order of the semiclassical density matrix obtained from (9) simply reads:

$$\mathcal{R}_0 = \frac{\Theta(\lambda^+ - V - \varepsilon)}{2} \left[I + \frac{1}{\varepsilon}(\alpha \cdot \mathbf{p} + \beta \, m^*) \right]$$
$$+ \frac{\Theta(\lambda^- - V + \varepsilon)}{2} \left[I - \frac{1}{\varepsilon}(\alpha \cdot \mathbf{p} + \beta \, m^*) \right], \tag{12}$$

where the first line corresponds to the positive energy states and the second line to the negative ones.

The semiclassical expectation value of a given single-particle operator is given by :

$$\langle \hat{O} \rangle = \frac{1}{(2\pi)^3} \int d\mathbf{r} \int d\mathbf{p} \, Tr^+ \left[\hat{O}(\mathbf{r}, \hat{\mathbf{p}}) \, \hat{\mathcal{R}}(\mathbf{r}, \hat{\mathbf{p}}) \right]_w , \tag{13}$$

where Tr^+ means that the trace is taken disregarding the negative energy terms.

In particular, we are interested in quantities like the particle, scalar and energy densities which in the Wigner-Kirkwood representation are given by:

$$\rho(\mathbf{r}) = \frac{1}{(2\pi)^3} \int d\mathbf{p} \, Tr^+ \left[\hat{\mathcal{R}}(\mathbf{r}, \hat{\mathbf{p}}) \right]_w$$
$$= \frac{k_F^3}{3\pi^2} + \rho_2(k_F, m^*, \nabla V, \Delta V, \nabla m^*, \Delta m^*) \tag{14}$$

$$\rho_{\rm s}(\mathbf{r}) = \frac{1}{(2\pi)^3} \int d\mathbf{p} \, Tr^+ \left[\beta \, \hat{\mathcal{R}}(\mathbf{r}, \hat{\mathbf{p}}) \right]_{\rm w}$$

$$= \frac{m^*}{2\pi^2} \left[k_{\rm F} \varepsilon_{\rm F} - m^{*2} \ln \frac{k_{\rm F} + \varepsilon_{\rm F}}{m^*} \right]$$

$$+ \rho_{s2}(k_{\rm F}, m^*, \nabla V, \Delta V, \nabla m^*, \Delta m^*) \qquad (15)$$

and

$$e(\mathbf{r}) = \frac{1}{(2\pi)^3} \int d\mathbf{p} \, Tr^+ \left[(\alpha \cdot \hat{\mathbf{p}} + \beta \, m^*(\mathbf{r}) + I \, V(\mathbf{r})) \, \hat{\mathcal{R}}(\mathbf{r}, \hat{\mathbf{p}}) \right]_{\rm w}$$

$$= \frac{1}{8\pi^2} \left[k_{\rm F} \varepsilon_{\rm F}^3 + k_{\rm F}^3 \varepsilon_{\rm F} - m^{*4} \ln \frac{k_{\rm F} + \varepsilon_{\rm F}}{m^*} \right] + e_2(k_{\rm F}, m^*, \nabla V, \Delta V, \nabla m^*, \Delta m^*)$$

$$+ V\rho \qquad (16)$$

where k_F is the local Fermi momentum defined by :

$$k_{\rm F} = \left[(\lambda - V)^2 - m^{*2} \right]^{1/2} \qquad (17)$$

and

$$\varepsilon_{\rm F} = \lambda - V = \left(k_{\rm F}^2 + m^{*2} \right)^{1/2} \qquad (18)$$

In eqs.(14)-(16) the lowest order contributions to the particle, scalar and energy densities are explicitely given. The corresponding $\hbar^2$ contributions are rather cumbersome expressions of the variables between brackets and we refer the reader to ref. [13] for their explicit expressions.

The next step towards a relativistic density functional is to eliminate the field V in the Wigner-Kirkwood expressions (14)-(16) in favour of ρ and m^*. To this end we separate the variable k_F into a term k_0 and a term k_2 of order $\hbar^2$. Replacing in (14) we get the semiclassical expansion of ρ consistently up to $\hbar^2$ order:

$$\rho = \rho_0(k_{\rm F}) + \rho_2(k_{\rm F}) + \ldots = \rho_0(k_0) + \frac{\partial \rho_0(k_0)}{\partial k_0} k_2 + \rho_2(k_0) + \mathcal{O}(\hbar^4). \qquad (19)$$

The new variables k_0 and k_2 are determined by demanding that $\rho_0(k_0)$ reproduces the exact density that gives the right normalization at lowest order because the particle number cannot depend on the parameter $\hbar$.
Consequently :

$$k_0 = (3\pi^2 \rho)^{1/3} \qquad (20)$$

$$k_2 = -\frac{\pi^2}{k_0^2} \rho_2(k_0). \qquad (21)$$

Now ∇V and ΔV within ρ_2 are written in terms of the gradients of ρ and m^* through eqs.(17) and (21). Finally the scalar and energy densities are obtained by performing a similar expansion in (15) and (16) :

$$\rho_s = \rho_{s0}(k_F) + \rho_{s2}(k_F) + \ldots = \rho_{s0}(k_0) + \frac{\partial \rho_{s0}(k_0)}{\partial k_0} k_2 + \rho_{s2}(k_0) + \mathcal{O}(\hbar^4). \quad (22)$$

$$e = e_0(k_F) + e_2(k_F) + \ldots = e_0(k_0) + \frac{\partial e_0(k_0)}{\partial k_0} k_2 + e_2(k_0) + \mathcal{O}(\hbar^4), \quad (23)$$

Again we refer the reader to ref. [13] for the explicit RETF expressions of ρ_{s2} and e_2.

The complete semiclassical energy density for the $\sigma - \omega$ model has a similar structure to the one given in the RMFT [1,2] except that now the new variables are the neutron and proton densities instead of the wave-functions. In the semiclassical approach the energy density reads:

$$e = e_0 + e_2 - m\rho + g_v V_0 \rho + e A_0 \rho_p + \frac{1}{3} b\phi_o^3 + \frac{1}{4} c\phi_o^4 + \frac{1}{2}\left[(\nabla\phi_o)^2 + m_s^2\phi_o^2\right]$$
$$- \frac{1}{2}\left[(\nabla V_0)^2 + m_v^2 V_0^2\right] - \frac{1}{2}\left[(\nabla b_0)^2 + m_\rho^2 b_0^2\right] - \frac{1}{2}(\nabla A_0)^2, \quad (24)$$

where V_0 is the vector field associated with the ω meson, ϕ_0 is the scalar field that represents the σ meson and related with the effective mass through $m^* = m - g_s\phi_0$, b_0 is the isovector field corresponding to the ρ meson and A_0 is the electromagnetic field.

The nucleon, ω-meson and ρ-meson masses take their experimental values $m=939$ MeV, $m_v=783$ MeV and $m_\rho=769$ MeV. The σ-meson mass m_s, the non-linear coefficients b, c and the coupling constants g_s, g_v and g_ρ are the free parameters of the model. They are usually adjusted to reproduce the nuclear matter or finite nuclei properties.

The semiclassical ground-state densities and the meson and photon fields are obtained by solving the Euler-Lagrange equations associated with the semiclassical energy density (24) with the constraint of the conservation of the number of each kind of particles.

$$\varepsilon_{0q} - m + \frac{\pi^2}{k_{0q}^2} \frac{\delta e_{2q}}{\delta k_{0q}} + V_q - \lambda_q \quad (25)$$

where

$$V_q = g_v V_0 + \frac{1}{2} g_\rho \tau_3 b_0 + e\frac{1}{2}(1 + \tau_3)A_0 \quad (26)$$

$$(\Delta - m_v^2)V_0 = -g_v\rho, \quad (27)$$

$$(\Delta - m_s^2)\phi_o = -g_s\rho_s + b\phi_o^2 + c\phi_o^3, \quad (28)$$

$$(\Delta - m_\rho^2)b_0 = -\frac{1}{2}g_\rho(\rho_p - \rho_n), \quad (29)$$

$$\Delta A_0 = -e\rho_p. \quad (30)$$

Once the values of the free parameters of the model are fixed, the semiclassical variational equations must be solved numerically under appropiate boundary conditions. This has been done by using the so-called imaginary time-step method for eq.(25) and Gaussian elimination for eqs.(27)-(30) (See ref. [13] for more details).

Table 1. Total energy (in MeV), proton and neutron r.m.s. radii (in fm) of ^{40}Ca and ^{208}Pb with the parameter set SRK3M7. The energies have not been corrected for centre-of-mass motion.

	^{40}Ca			^{208}Pb		
	E	r_p	r_n	E	r_p	r_n
H	−337.0	3.30	3.25	−1620.1	5.44	5.63
H*	−336.6	3.26	3.24	−1617.2	5.47	5.64
TF$\hbar^2$	−348.1	3.21	3.17	−1679.8	5.45	5.59
TF	−355.2	3.22	3.19	−1697.3	5.47	5.60

3 Finite Nuclei Results

The total energy, proton and neutron r.m.s.radii corresponding to ^{40}Ca and ^{208}Pb nuclei calculated in several approaches are shown in Table 1 for the parameter set SRK3M7 [19] in comparison with the quantal Hartree results (label H). In the approximation denoted by H*, the shell effects have been perturbatively added by means of the expectation value method [10,19,20]. In this case the Dirac-Hartree equations [2] are solved once with the semiclassical RETF fields as input and, without interating, the calculated quantal densities are used to solve the Helmholtz equations for the meson fields (eqs.(27)-(30)). Table 1 shows that this perturbative treatment of the shell effects is a good approximation to the full self-consistent quantum calculation. In the same Table 1 we also display the pure semiclassical calculations performed in the simple Thomas-Fermi approach (label TF) and using the RETF approximation (label TF$\hbar^2$). From this Table it is seen that both semiclassical approaches overbind the Hartree results and give smaller r.m.s. for neutron and proton densities than the quantal calculation. However, no definitive conclusions can be drawn from these results because they are strongly correlated with the values of the parameters of the effective force.

To clarify this fact we show in Figure 1 the difference between the semiclassical and Hartree energies E^{SC}-E^H as a function of the effective mass at saturation (m^*_∞/m) for ^{40}Ca and ^{208}Pb. Results are given for four parameter sets with the following fixed nuclear matter properties [21] : $a_v = $ -15.75 MeV, $\rho = $ 0.16 fm^{-3}, K= 220 MeV and $a_{sym} = $ 30 MeV. The effective masses are $m^*_\infty/m = $

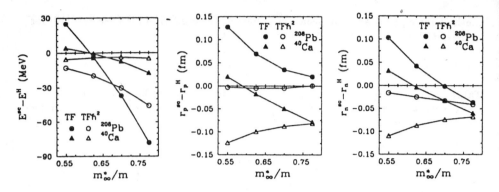

Fig. 1. From the left: (a) Deviation of the semiclassical energies from the Hartree ones as a function of m_∞^*/m, for ^{208}Pb (circles) and ^{40}Ca (triangles). Full symbols: TF, open symbols TF$\hbar^2$. The lines connecting the symbols are only to guide the eye. (b) Same as (a) for the proton r.m.s. radii (c) Same as (a) for the neutron r.m.s. radii

0.55, 0.625, 0.70 and 0.775 covering the typical range of values considered in the literature. A scalar mass $m_s = 400$ MeV that is within the commonly accepted values has been chosen. From Figure 1a it can be seen that the TF solution exhibits a strong dependence on m_∞^*/m. The TF energies are less bound than the Hartree results for small m_∞^*/m values but becomes more bound than the Hartree and TF$\hbar^2$ for larger m_∞^*/m. The discrepancy between TF and Hartree energies varies almost linearly with the effective mass and give rouhgly the same result for $m_\infty^*/m \approx 0.6$. On the contrary, the TF$\hbar^2$ approximation sistematical-ly yields more binding than the Hartree calculation for all the analyzed range of m_∞^*/m. Comparing the TF and TF$\hbar^2$ energies with the Hartree ones, it is observed a better ability of the TF$\hbar^2$ model in reproducing the behaviour of the Hartree results as a function of m_∞^*/m. However in the region $m_\infty^*/m \approx$ 0.6-0.7, the agreement is better for the TF approach. This raises the question wether or not the $\hbar^2$ corrections actually improve the TF results. Semiclassical and quantum calculations must differ in the shell energy which is not taken into account by the semiclassical functionals. Indeed, the shell correction is not esti-mated as well in the TF case as in the TF$\hbar^2$ approach because the semiclassical expansion is less converged in the former than in the latter. Thus, the agreement between the TF and Hartree results around $m_\infty^*/m \approx 0.65$ should be considered only as accidental. Of course for a given m_∞^*/m there is some dependence of $E^{SC} - E^H$ on the remaining parameters of the force, especially on the scalar mass m_s. However the general trends shown in Figure 1a by the TF and TF$\hbar^2$ solutions as a function of m_∞^*/m is not significantly affected by changing the values of other saturation properties or the mass number A [13,19,21,22]. To some extent, a similar phenomenology is found in non-relativistic calculations with Skyrme forces.

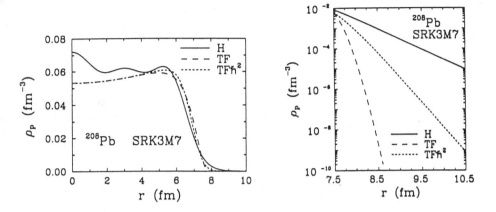

Fig. 2. From the left: (a) proton density of ^{208}Pb obtained with the parameter set SRK3M7 in the Hartree (H), TF and RETF (TF$\hbar^2$) approximations. (b) The same densities of (a) in the outer surface region on a semi-logarithmic scale.

In Figure 1b-1c we display equivalent plots to Figure 1a for the calculated proton and neutron r.m.s. radii of the same nuclei. The dependence of the TF results on m_∞^*/m is again more visible than for the TF$\hbar^2$ approach. As a rule, the semiclassical radii are smaller or larger than the Hartree ones depending on wether the binding energies are, respectively, larger or smaller than the Hartree results. The TF proton radii for ^{208}Pb are an exception and they are always too large even for overbound nuclei. The quality of the semiclassical results for the radii depends on the number of particles. While for ^{40}Ca the TF radii approach to the Hartree ones better than TF$\hbar^2$, for ^{208}Pb the situation is reversed and TF$\hbar^2$ generally improves the TF values.

Concerning the density profiles, the proton density of ^{208}Pb calculated in the Hartree (H), TF and RETF (TF$\hbar^2$) has been plotted in Figure 2a using the relativistic parameter set SRK3M7. The semiclassical densities do not present oscillations due to the absence of shell effects but they average the Hartree densities. In the bulk the TF and TF$\hbar^2$ densities are very similar. However, as expected, the gradient corrections incorporated by the TF$\hbar^2$ functionals improve the densities at the surface which come closer to the Hartree ones and show a better decay than the TF densities despite the fact that the fall-off is still too step in comparison with the Hartree solutions. This behaviour is illustrated in

Figure 2b which shows a semi-logarithmic plot of the densities in the outer surface region.

4 Liquid Drop Model Coefficients

In the liquid droplet model model formulated by Myers and Swiatecki [23] the energy of a spherical nucleus is written as :

$$E = a_v A + 4\pi \int_0^\infty dr\, r^2 \left[e(r) - a_v \rho(r)\right],\tag{31}$$

where a_v is the energy in infinite matter and $e(r)$ and $\rho(r)$ are the energy and particle densities respectively. Starting from (31) and following the method outlined in refs.[23,24], one obtains for the surface and curvature coefficients in semi-infinite nuclear matter the expressions :

$$a_s = 4\pi r_0^2 \int_{-\infty}^\infty dz\, \left[e(z) - a_v \rho(z)\right]\big|_{\kappa=0},\tag{32}$$

$$a_c = 8\pi r_0 \left[\int_{-\infty}^\infty dz(z - z_0)\left[e(z) - a_v \rho(z)\right]\big|_{\kappa=0}\right.$$
$$\left. + \int_{-\infty}^\infty dz\, \frac{\partial\left[e(z) - a_v \rho(z)\right]}{\partial \kappa}\bigg|_{\kappa=0}\right],\tag{33}$$

where $r_0 - (3/4\pi\rho_0)^{1/3}$ is the nuclear matter radius, z_0 is the location of the equivalent sharp surface and κ is the curvature ($2/R$ for a sphere of radius R).

The two contributions to the curvature energy in eq.(33) are called geometrical and dynamical respectively. The geometrical contribution only involves the variation of surface energy density $e(z) - a_v \rho(z)$ across the surface parallel to the z-axis, while the dynamical part corresponds to the variation of the surface energy density when the plane surface is infinitesimally bent. The surface energy density depends on the curvature κ in two different ways: one of them is the explicit dependence of $e(z)$ on the Δ operator (consider for example the non-relativistic definition of the kinetic energy density which in the limit of $R \to \infty$ reads $d^2/dz^2 + \kappa d/dz$). The other one corresponds to the implicit curvature dependence of the density and meson fields.

In the relativistic $\sigma - \omega$ model, the semiclassical energy density is free of an explicit Δ dependence since it can be removed by partial integration. On the other hand, the implicit curvature dependence of the nuclear density and of the meson fields does not contribute to the dynamical part of the curvature energy in a self-consistent calculation because the surface energy in semi-infinite nuclear matter is stationary with respect to changes in the density and meson fields [23,30]. Consequently the curvature energy is only given by its geometrical contribution. (see ref.[30] for a complete discussion).

The nuclear curvature coefficient a_c poses a problem, the so-called curvature energy puzzle, which has been widely discussed in the literature and still remains open. The value of a_c obtained from theoretical non-relativistic calculations [10,23-25] is of the order of 10 MeV. On the contrary, the empirical coefficient determined from the analysis of experimental data on ground-state masses and fission barriers is compatible with a vanishing value [26-28]. Several explanations have been suggested to solve this anomaly. For instance, the considerations of finite range interactions, Friedel oscillations, ground-state correlations or relativistic effects. The influence of Friedel oscillations as well as the effect of the finite range has been analyzed in [29] and it has been found that they are not responsible for the discrpancy. The role of the relativistic effects has been analyzed in [30] and we want to present here the main conclusions reported in this reference.

Table 2 collects the surface and curvature energy coefficients as well as the surface thickness (t) of the semi-infinite density profile (standard 90%-10% distance) calculated with the parameter sets PW1 (that corresponds to the original Walecka model) [2] and the non-linear sets P1 [31], HII [32] and HIV [33]. From this Table it is seen that the quantal and semiclassical calculations of the surface energy and thickness agree fairly well. Comparing the results for the linear and non-linear models, it is seen that in the linear case is not possible to obtain, even qualitatively, a reasonable prediction for a_s and t at the same time due to the high value of the nuclear incompressibility (K) of the model. This shortcoming can be overcome by introducing the non-linear meson coupling [34], which phenomenologically simulates many-body forces and renormalization effects. With this non-linear models that give a more reallistic K is possible to get a_c and t in agreement with the empirical values. However, the prediction for the curvature energy is too large for reasonable predictions of the surface energy and thickness (P1 and HII parameter sets). Values of a_c close to zero can only be found using parameter sets that give unacceptable values for a_s and t (HIV).

Table 2. Surface energy (a_s), surface thickness (t) and curvature energy (a_c) corresponding to some parameterizations of the $\sigma-\omega$ model and to different Thomas–Fermi approximations. Also displayed are some Hartree (H) results (from Ref. [35]), and the values of the effective mass and nuclear incompressibility κ at saturation.

	m^*/m	κ (MeV)	a_s (MeV)			t (fm)			a_c (MeV)	
			H	$TF\hbar^2$	$TF\hbar^0$	H	$TF\hbar^2$	$TF\hbar^0$	$TF\hbar^2$	$TF\hbar^0$
PW1	0.56	546.	33.8	32.8	35.9	2.38	2.27	2.71	18.8	20.1
P1	0.57	206.	15.8	17.4	19.9	2.31	2.09	2.96	12.4	16.0
HII	0.68	345.		16.0	17.3		1.23	1.57	7.2	7.5
HIV	0.80	240.		11.1	10.0		1.07	1.00	4.9	3.7

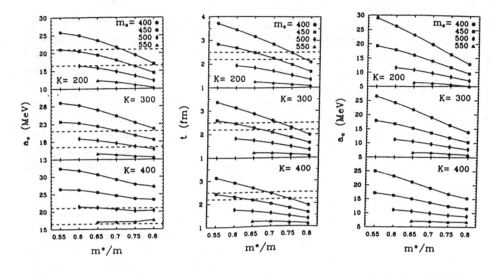

Fig. 3. From the left : (a) Surface energy a_s of the semi-infinite nuclear matter calculated in the TF$\hbar^2$ approximation for the non-linear parameters given in the text. The incompressibilities K and the scalar mass m_s are in MeV. The dashed lines approximately indicates the empirical region for a_s. The solid lines connecting the symbols are only to guide the eye. Note the different values on the vertical axis. (b) Same as (a) for the surface thickness t. (c) Same as (a) for the curvature energy a_c.

The non-linear model gives the opportunity for studing the surface properties as a function of the incompressibility, effective mass and the mass of the sigma meson. We have performed this analysis by using parameter sets with energy per particle and density at saturation of $a_v = -15.75$ MeV and $\rho = 0.16$ fm^{-3}. The values of m_∞^*/m varies between 0.55 and 0.80, K covers the range 200-400 MeV and m_s is taken from 400 up to 550 MeV.

The results for a_s, t and a_c are given if Figure 3a-c where the dashed lines show the region compatible with the empirical data. As it is expected, when m_s increases the three quantities a_s, t and a_c decrease, which is more visible for small m_∞^*/m. For constant m_s and m_∞^*/m, a_s increases with K. The global tendency of t is to decrease with increasing K because high incompressibility prefers a sharp surface [36], but t increases with K if $m_\infty^*/m \geq 0.7$ and $m_s \geq 500$ MeV. The calculated a_c decreases with increasing K if $m_\infty^*/m \leq 0.7$ whereas the opposite behaviour is found if $m_\infty^*/m \geq 0.7$. Aa a general trend a_s, t and a_c go down with increasing m_∞^*/m. In order to understand this result for t it is useful to take into account that the effective mass gives a measure of the non-local effects whose contributions tend to make the surface more diffuse [36,37]. A larger value of the effective mass is associated with less non-locality, and therefore favours a sharper surface.

Looking at the empirical region for a_s and t, one realizes that there exist a number of non-linear parametrisations that are able to give a realistic value separately either for a_s or for t. However, if one demands that both a_s and t (calculated in the TF$\hbar^2$ approach) lie in the empirical region, it restricts K around 200 MeV and m_s=400 -500 MeV. Typically, a small value of m_∞^*/m favours $m_s \approx 500$ MeV while $m_s \approx 400$ MeV is preferred for higher m_∞^*/m.

Concerning the curvature energy, it should be pointed out that in the region where a_s and t have a realistic value, a_c is sistematically around 13 - 15 MeV in agreement with the range of values found in the non-relativistic case [10,25,38,29]. At the price of too low surface energy and too small surface thickness, values for a_c compatible with the empirical coefficient can be found. For instance, with K = 250 MeV, m_∞^*/m=0.8 and m_s = 700 MeV, a_c=0.8 MeV but a_s and t take the unphysical values of 6 MeV and 0.6 fm respectively.

5 Summary

We have presented here some aspects of the semiclassical relativistic theory, which have been complemented with illustrative applications to finite nuclei and semi-infinite nuclear matter in the $\sigma - \omega$ model. A recurrent scheme to obtain the semiclassical $\hbar$ expansion of the propagator associated with a time-independent single-particle Hamiltonian with a matrix structure has been given. The calculation of the Wigner-Kirkwood expressions of the density matrix and of the particle and energy densities as well as the derivation of the corresponding density functionals are also presented.

We have also analyzed the RETF model in numerical calculations by comparing the relativistic semiclassical and Hartree approximations. Results for the ground-state properties of finite nuclei have been presented. The TF solutions show a strong dependence on the adjustable parameters of the relativistic model especially on the effective mass. The second order gradient corrections to the TF approximation generally improve the agreement with quantal calculations in a sensitive way and the results are less dependent on the relativistic interaction. The RETF model takes into account the non-local spin-orbit and effective mass contributions consistently up to $\hbar^2$ order and provides a more realistic description of the surface and nuclear densities.

We have performed a semiclassical study of the semi-infinite nuclear matter by using the RETF approach. In particular the surface and curvasture energies as well as the surface thickness have been analyzed. It has been shown that the empirical values of the surface energy and thickness can be simultaneosly obtained in the model only if the nuclear incompressibility is around 200 MeV and the scalar mass lies in the range 400-500 MeV for effective masses between 0.55 and 0.80. It has been found that the relativistic effects can not explain the difference between the empirical and calculated curvature and energy coefficient within the RMFT.

This work was supported in part by the DGICYT (Spain) under grant PB92-0761 and the Human Capital and Mobility Program (Contract CHRX-CT92-0075).

References

1   J. D. Walecka, Ann. Phys. (N.Y.)**83**, 491 (1974); S. A. Chin, Ann. Phys. (N.Y.)**108**, 301 (1977).

2   B. D. Serot and J. D. Walecka, Adv. Nucl. Phys. **16**, 1 (1986).

3   T. H. R. Skyrme, Philos. Mag. **1**, 1043 (1956); D. Vautherin and D. M. Brink, Phys. Rev. **C5**, 626 (1972).

4   S. Lundqvist and N. H. March (eds.), *Theory of the Inhomogeneous Electron Gas* (Plenum, New York, 1983); R. O. Jones and O. Gunnarsson, Rev. Mod. Phys. **61**, 689 (1989).

5   R. M. Dreizler and J. da Providencia (eds.), *Density Functional Methods in Physics*, Vol. 123 NATO ASI Series B (Plenum, New York, 1985).

6   R. M. Dreizler and E. K. U. Gross, *Density Functional Theory* (Springer, Berlin, 1990).

7   R. W. Hasse, R. Arvieu, and P. Schuck (eds.), *Workshop on Semiclassical Methods in Nuclear Physics*, J. de Phys. Colloque **C6** (1984); I. Zh. Petkov and M. V. Stoitsov, *Nuclear Density Functional Theory* (Clarendon Press, Oxford, 1991).

8   P. Hohenberg and W. Kohn, Phys. Rev. **136**, B864 (1964).

9   B. Grammaticos and A. Voros, Ann. Phys. (N.Y.)**123**, 359 (1979); *ibid.* **129**, 153 (1980).

10   M. Brack, C. Guet, and H. -B. Håkansson, Phys. Rep. **123**, 275 (1985).

11   M. Centelles, M. Pi, X. Viñas, F. Garcias, and M. Barranco, Nucl. Phys **A510**, 397 (1990).

12   M. S. Vallarta and N. Rosen, Phys. Rev. **41**, 708 (1932).

13   M. Centelles, X. Viñas, M. Barranco, and P. Schuck, Nucl. Phys **A519**, 73c (1990); M. Centelles, X. Viñas, M. Barranco and P. Schuck, Ann. Phys. (N.Y.)**221**, 165 (1993).

14   M. K. Weigel, S. Haddad, and F. Weber, J. Phys. **17**, 619 (1991); D. Von-Eiff, S. Haddad, and M. K. Weigel, Phys. Rev. **C46**, 230 (1992).

15   C. Speicher, R. M. Dreizler, and E. Engel, Ann. Phys. (N.Y.)**213**, 312 (1992).

16   P. Ring and P. Schuck, *The Nuclear Many-Body Problem* (Springer, Berlin, 1980).

17   D. A. Kirzhnits, *Field Theoretical Methods in Many-Body Systems* (Pergamon, Oxford, 1967).

18   B. K. Jennings, R. K. Bhaduri, and M. Brack, Nucl. Phys **A253**, 29 (1975).

19   M. Centelles, X. Viñas, M. Barranco, S. Marcos, and R. J. Lombard, Nucl. Phys **A537**, 486 (1992).

20   O. Bohigas, X. Campi, H. Krivine, and J. Treiner, Phys. Lett. **B64**, 381 (1976).

21   M. Centelles, NATO ASI on *Density Functional Theory* 36elvecchio Pascoli,Italy, August 1993 (Plenum Press in print)

22   C. Speicher, E. Engel, and R. M. Dreizler, Nucl. Phys **A562**, 569 (1993).

23   W. D. Myers and W. J. Swiatecki, Ann. Phys. (N.Y.)**55**, 395 (1969); *ibid.* **84**, 186 (1974).

24   W. Stocker and M. Farine, Ann. Phys. (N.Y.)**159**, 255 (1985).

25   J. Treiner and H.Krivine, Ann. Phys. (N.Y.)**170**, 406 (1986).

26   W. D. Myers, *Droplet model of atomic nuclei*, (IFI/Plenum, New York, 1977)

27   H. J. Krappe, J. R. Nix and A. J. Sierk, Nucl. Phys **A489**, 252 (1988).

28   P. Möller, W. D. Myers, W. J. Swiatecki, and J. Treiner, At. Data Nucl. Data Tables **39**, 225 (1988).

30   M. Centelles and X. Viñas, Nucl. Phys **A563**, 173 (1993).

29   M. Durand, P. Schuck, and X. Viñas, Z. Phys. **A346**, 87 (1993).

31   P. -G. Reinhard, M. Rufa, J. Maruhn, W. Greiner, and J. Friedrich, Z. Phys. **A323**, 13 (1986).

32   J. Boguta, Phys. Lett. **B109**, 251 (1982).

33   N. K. Glendenning, Phys. Lett. **B185**, 275 (1987).

34   J. Boguta and A. R. Bodmer, Nucl. Phys **A292**, 413 (1977).

35   D. Hoffer and W. Stocker, Nucl. Phys **A492**, 637 (1989).

36   X. Campi and S. Stringari, Nucl. Phys **A337** 313 (1980)

37   J. L. Friar and J. W. Negele , Adv. Nucl. Phys. **8** 219 (1975)

38   W. Stocker, J. Bartel, J. R. Nix and A. J. Sierk, Nucl. Phys **A489** 252 (1988)

Photonuclear Reactions

E. Oset

Departamento de Física Teórica and IFIC,
Centro Mixto Universidad de Valencia-CSIC,
46100 Burjassot (Valencia) Spain.

Abstract. A successful many body approach to evaluate inclusive photonuclear cross sections is reviewed. The lectures begin with an exposition of the state of the art in the elementary reactions $\gamma N \to \pi N$ and $\gamma N \to \pi\pi N$, which are the basic building blocks for the many body theory. Through the skilful use of Field Theoretical Techniques one can disentangle the different reaction channels in the total photonuclear cross section. The method makes use of genuine reaction probabilities, defined as one step processes and calculated microscopically, and a Monte Carlo simulation procedure by means of which one generates all possible multistep processes which occur in the actual physical reactions. This allows one to compare directly the theoretical results with the experimental cross sections like $(\gamma, \pi), (\gamma, N), (\gamma, NN), (\gamma, N\pi)$ etc. in nuclei.

1 Introduction

Because of the relative small strength of the electromagnetic interaction and its adequate knowledge from the theoretical point of view, electrons and photons have been taditionally excellent probes of nuclear structure and continue to be so. With increasing experimental hability to perform high resolution as well as coincidence experiments, the wealth of experimental information has reached a point which allows one to learn about details of nuclear structure as well as to dig into the complicated or subtle excitation mechanisms of the nucleus. One theoretical approach which has proved particularly rewarding to put in a unified framework all the different photonuclear reactions is the one based on a many body approach to the general problem using Quantum Field Theoretical Techniques, which allows the separation of all the reaction channels in the (e, e') or γ-nucleus interaction and which we will expose here. For this purpose we shall begin with the study of the $\gamma N \to \pi N$ and $\gamma N \to \pi\pi N$ reactions. Then they serve as the starting point in a many body theory which generates two and three nucleon mechanisms for the γ nuclear interaction. A skilful classification of the diagrams allows the separation of genuine one step mechanisms, where all meson correlating two nucleons are off shell, from those diagrams where some

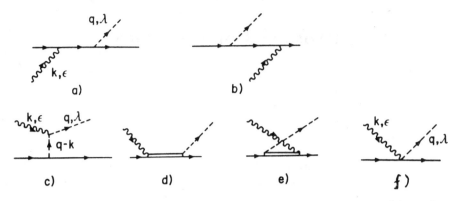

Figure 1: Feynman diagrams considered for the $\gamma N \to \pi N$ process. (a) Nucleon pole. (b) Crossed nucleon pole. (c) Pion pole. (d) Delta pole. (e) Crossed delta pole. (f) Kroll Ruderman term.

meson correlating two nucleons are off shell, from those diagrams where some intermediate meson is mostly on shell and which can be better dealt with by associating the mechanism to a two step process. These two step and multistep processes are considered by means of a Monte Carlo simulation procedure on which one has a certain control and which requires as input only the probabilities associated to the genuine one step procecess (involving one or many nucleons). With this procedure one can evaluate cross sections for all the channels of the inclusive $\gamma - A$ reaction, $(\gamma, \pi), (\gamma, N), (\gamma, 2N), (\gamma, N\pi)$ etc.

A general good agreement with experiment in all the channels is reached. The method allows the investigation of channels so far unmeasured which contain very much information on magnitudes like the pion absorption probabilities or the nucleon mean free path in the nucleus.

2   The $\gamma N \to \pi N$ reaction

From the very beginning we will disregard processes involving $\gamma N \to \gamma N$ which are of order e^2 in the amplitude and hence we shall start from the photoproduction amplitude $\gamma N \to \pi N$, and associated amplitudes, which are of order e in the amplitude.

We shall illustrate the model of ref. [1] which follows closely the one introduced in [2]. The model contains the nucleon and delta pole terms, plus their crossed terms, the pion pole and the Kroll Ruderman terms, as shown in fig. 1.

The amplitudes for these diagrams can be easily obtained from standard

Feynman rules using the basic vertices which we detail below:

$$-i\delta\tilde{H}_{\gamma NN} = -ie\left(\gamma^\mu + i\frac{\chi_p}{2M}\sigma^{\mu\nu}k_\nu\right)\varepsilon_\mu,$$

$$-i\delta\tilde{H}_{\pi NN} = \frac{f}{\mu}\boldsymbol{\sigma}\cdot\boldsymbol{q}\tau^\lambda$$

$$-i\delta\tilde{H}_{\gamma\pi\pi} = \mp ie\,(2q-k)^\mu\varepsilon_\mu = \pm ie2q\varepsilon,$$

$$-i\delta\tilde{H}_{\gamma N\Delta} = -\frac{f_\gamma}{\mu}(\boldsymbol{S}^+\times\boldsymbol{k})\cdot\boldsymbol{\varepsilon}\,T_3^+ + \text{h.c.},$$

$$-i\delta\tilde{H}_{\pi N\Delta} = \frac{f^*}{\mu}(\boldsymbol{S}^+\cdot\boldsymbol{q})\,T_\lambda^+ + \text{h.c.},$$

The Kroll Ruderman term is generated from the pseudovector πNN Lagrangian,

$$\delta\mathcal{L}_{\pi NN} = -\frac{g}{2M}\bar{\Psi}\gamma^\mu\gamma_5\vec{\tau}\Psi\partial^\mu\Phi \tag{1}$$

by minimal coupling, $\partial_\mu \to \partial_\mu + ieA_\mu$. However, it is customary to separate the nucleon propagator into the positive and negative energy state contributions, as done in eq. (2) and include the contribution of the nucleon direct and crossed terms with negative energy states in the Kroll Ruderman term.

$$\frac{\not{p}+M}{p^2-M^2+i\epsilon} = \frac{M}{E(\boldsymbol{p})}\left(\frac{\sum_r u(\boldsymbol{p},r)\bar{u}(\boldsymbol{p},r)}{p^0-E(\boldsymbol{p})+i\epsilon} + \frac{\sum_r v(-\boldsymbol{p},r)\bar{v}(-\boldsymbol{p},r)}{p^0+E(\boldsymbol{p})-i\epsilon}\right),$$

$$\tag{2}$$

Then only the positive energy intermediate states are considered in the nucleon pole term and the Kroll Ruderman term becomes

$$-iT_\alpha = e\frac{f}{\mu} C(\alpha)\boldsymbol{\sigma}\cdot\boldsymbol{\varepsilon}\,,$$

(3)

with $C(\alpha)$ given by

$$C(\alpha) = \begin{cases} \sqrt{2}\left(1-\dfrac{\omega(q)}{\sqrt{s}+M}\right) & \text{for } \gamma p \to \pi^+ n \\[3mm] -\sqrt{2}\left(1+\dfrac{\omega(q)}{E(k)-\omega(q)+E(k+q)}\right) & \text{for } \gamma n \to \pi^- p \\[3mm] -\omega(q)\left(\dfrac{1}{\sqrt{s}+M}+\dfrac{1}{E(k)-\omega(q)+E(k+q)}\right) & \text{for } \gamma p \to \pi^0 p \\[3mm] 0 & \text{for } \gamma n \to \pi^0 n \end{cases}$$

(4)

in agreement up to $0(k/M)$ terms with the results obtained from current algebra and PCAC [3]

In the vertices of the Feynman rules M is the nucleon mass, μ the pion mass, $f = 1$, $f_\gamma = 0.12$, $\vec{\sigma}$ the nucleon spin, $\vec{q}, \vec{k}$ the pion and photon momenta and ε^μ the photon polarization vector. We work usually in the Coulomb gauge, $\varepsilon^0 = 0$, $\vec{\varepsilon}.\vec{k} = 0$.

In the vertex γNN , χ_p is the proton anomalous magnetic moment. For the coupling of the photon to the neutron the γ^μ term does not appear and χ_p is replaced by χ_n, the neutron anomalous magnetic moment. $E(p)$, $\omega(q)$ are the relativistic energies of the nucleon and the pion and s in eqs. (4) the πN Mandelstam variable.

The operators $\vec{S}, \vec{T}$ appearing in the $\pi N\Delta$ and $\gamma N\Delta$ vertices are the spin and isospin transition operators from $1/2$ to $3/2$.

The model exposed is not exactly unitary and different prescriptions are taken to unitarize it and force it to satisfy Watson's theorem. The practical changes are small. In [1] the method of Olson [4] is used to implement unitarity in the model by multiplying the delta pole term by a small phase such that the $\gamma N \to \pi N$ amplitude has the same phase as the $\pi N \to \pi N$ scattering amplitude in every partial wave and isospin channel (Watson's theorem). The agreement of the model with experiment is generally good as can be seen in fig. 1 where we show two examples of cross sections obtained with this model.

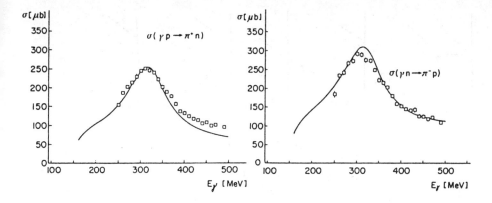

Figure 2: Cross sections for the $\gamma p \to \pi^+ n$ and $\gamma n \to \pi^- p$ reactions as a function of the photon energy. Experiment from [5].

3 The $\gamma N \to \pi\pi$ reaction

Much experimental information is piling up on this reaction. To the old data in the late 60's [6, 7] one is adding new data from Mainz with an improved precision [8, 9]. The theoretical work has followed a similar trend. There is an early model which reproduces qualitatively the basic features of the reaction [10], but recently the model has been enlarged to include extra terms which prove to be important [8, 11]. The most complete model is the one of ref. [11] which contains 67 Feynman diagrams and accounts for N, $\Delta(1232)$, $N^*(1440)$, $N^*(1520)$ baryon intermediate states plus terms in which the two final pions couple strongly to the ρ-meson. Schematically the diagrams are classified according to the scheme of fig. 3.

The reader is addressed to ref. [11] for details but here I would like to stress one of the most interesting findings of the model. The dominant term in the $\gamma p \to \pi^+ \pi^- p$ reaction is the one shown in fig. 4a, which involves the $\gamma N \Delta \pi$ Kroll Ruderman term and the decay of the Δ into $N\pi$

The term in fig. 4b involves the intermediate $N^*(1520)$ resonance, decaying into $\Delta\pi$ plus the Δ decaying into πN. This term is essential to reproduce the peak of the cross section which we see in fig 5.

Contrary to simple intuition, the experimental peak of fig. 5 is not associated with the Δ excitation in the dominant term of fig. 4a. The reason is that there is no particular γ energy which places the Δ on shell since the π^- already takes away some of the photon energy and its energy is distributed according to phase

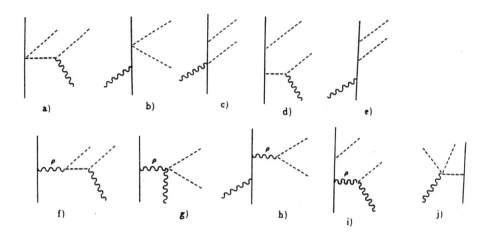

Figure 3: Classification of the Feynman diagrams for $\gamma N \to \pi\pi N$ into one point, two point and three point diagrams.

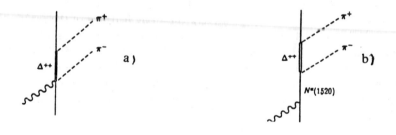

Figure 4: a) Dominant term in the $\gamma p \to \pi^+\pi^- p$ reaction. b) term involving the $N^*(1520)$ resonance and its decay into the $\Delta\pi$ system, which interferes with a) producing the peak of the cross section.

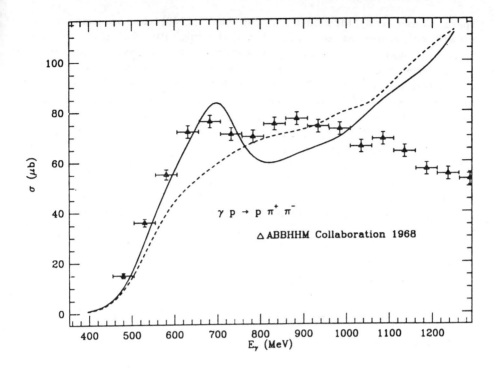

Figure 5: Total $\gamma p \rightarrow \pi^+\pi^- p$ cross section. Dashed line: the model of [11] omitting the $N^*(1520)$ terms; continuous line: complete model.

space. Consequently, the model, omitting the $N^*(1520)$ terms has a monotonous growth with energy, as seen in fig. 5, and the qualitative features of the cross section are missed. The term of fig. 4b happens to have the same spin structure as the term in fig. 4a, and then there is a constructive interference before the on shell energy of the $N^*(1520)$ is reached in the intermediate state and a destructive interference after. In this way the peak of the solid line shown in fig. 5 is reached as an interference phenomenon. The strength of the $N^* \to \Delta\pi$ coupling is taken from experiment. The sign is taken so as to have the best agreement with experiment in fig. 5 (the opposite sign produces unacceptable results) and it agrees (also the strength) with results for that coupling obtained in the constituent quark model for baryons [12].

The model of [11] also reproduces quite well mass distributions for the $(p\pi^-), (p\pi^+)$ and $(\pi^+\pi^-)$ systems. The discrepancies of the model with experiment from $E_\gamma = 1000\ MeV$ on should be expected since at those energies many more resonances than those considered in [11] would play a role in the reaction. Work on the other isospin channels is in progress along the lines of model [11] and experiments are also coming [8, 9].

4   Photonuclear cross section. The many body approach

The many body approach followed in [1] begins by looking at the behaviour of a photon in infinite nuclear matter. Through the interaction of the photon with the matter it acquires a selfenergy $\Pi(k)$, or equivalently the photon of momentum k feels an optical potential $V_{opt}(k)$ defined in terms of $\Pi(k)$ as $2kV_{opt}(k) \equiv \Pi(k)$. Now let us take the wave function of the photon which will have its time dependence modified by the effect of the extra potential.

Hence we have

$$\Psi_\gamma \sim e^{-ikt}e^{-iV_{opt}t}\ldots \sim e^{-iiImV_{opt}t}\ldots$$

$$|\Psi_\gamma| \sim e^{2ImV_{opt}t}\ldots \equiv e^{-\Gamma t}\ldots \tag{5}$$

where the latest result is obtained because the other terms with t dependence have modulus one. Hence we observe that the presence of the optical potential induces a depletion of the photon wave through the medium such that

$$-\frac{1}{N}\frac{dN}{dl} = -\frac{1}{N}\frac{dN}{dt} = \Gamma = -2ImV_{opt} = -\frac{1}{k}Im\Pi(k,\rho) \tag{6}$$

where we have written Γ explicitly as a function of k and ρ, the density of the medium. Thus, $-Im\Pi/k$ is the probability of photon reaction per unit length. Multiplying this probability by dl and ds we obtain a contribution to a cross section. Next, in order to obtain σ_A in a finite nucleus one makes use of the

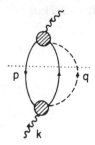

Figure 6: Photon selfenergy diagram obtained by folding the $\gamma N \rightarrow \pi N$ amplitude. The circle indicates any of the terms of this amplitude from fig. 1.

local density approximation, by substituting $\rho \rightarrow \rho(\vec{r})$ and by integrating over the whole nuclear volume one obtains the formula

$$\sigma_A = -\int d^3r \frac{1}{k} Im\Pi(k, \rho(\vec{r})) \tag{7}$$

There is always an approximation involved in the use of the local density approximation. However, in the present case because the contribution to σ_A is a volume contribution, the local density approximating is an excellent tool and the results one obtains with eq. (7) are the same that one obtains if one convolutes the density with a certain range of the interaction [1].

This appreciacion results in an economical and accurate tool to evaluate photonuclear cross sections, since calculations in nuclear matter are much simpler than the equivalent ones in finite nuclei.

The next step consists in the evaluation of the photon selfenergy and we proceed in parts. By starting from the model for $\gamma N \rightarrow \pi N$, folding the diagrams with their complex conjugates and summing over the occupied states we obtain the set of 36 Feynman diagrams represented in fig. 6

The evaluation of $\Pi(k)$ for the diagramas of fig. 6 is easy. Assume for simplicity only one term, the Kroll Ruderman term, in the $\gamma p \rightarrow n\pi^+$ amplitude. We obtain then

$$-i\Pi_{p,n}(k) = \int \frac{d^4q}{(2\pi)^4} i\bar{U}_{p,n}(k-q)iD_0(q)(e\frac{f}{\mu}\sqrt{2}\frac{2M}{2M+k})^2 \tag{8}$$

where $\bar{U}_{p,n}$ is the Lindhard function for ph excitation of fig. 6 of the (p, n) type and $D_0(q)$ the pion propagator. $Im\Pi(k)$ can be easily obtained using Cutkosky rules [13]. The imaginary part of Π is obtained when in the intermediate integrations the states cut by the dotted line in fig. 6 are placed on shell. In practical terms this is carried out by using the following useful rules (Cutkosky rules): Substitute

$$\Pi \to 2iIm\Pi$$

$$\bar{U}(q) \to 2i\theta(q^0)Im\bar{U}(q) \tag{9}$$

$$D_0(q) \to 2i\theta(q^0)ImD_0(q)$$

By making this substitution and summing over the 6 terms of the $\gamma p \to \pi^+ n$ amplitude we obtain

$$Im\Pi(k)_{p,n} = \int \frac{d^3q}{(2\pi)^3}\theta(k^0 - \omega(q))Im\bar{U}_{p,n}(k-q)\frac{1}{2\omega(q)}\bar{\Sigma}_{si}\Sigma_{sf}|T_{\gamma p \to \pi^+ n}|^2 \tag{10}$$

By summing now also over the (p,p), (n,p), (n,n) excitations one would obtain the whole photon selfenergy. It is instructive to see what has been accomplished with this step. For this purpose recall the limit when $\rho_p \to 0$

$$\theta(q^0)Im\bar{U}_{p,n}(q) = -\pi\rho_p\delta(q^0 - \vec{q}^2/2M) \tag{11}$$

By means of this and the standard formulas for the cross section in terms of $|T|^2$, and using eq. (7) we obtain

$$\sigma_A = \sigma_{\gamma p}\int \rho_p(\vec{r})d^3r + \sigma_{\gamma n}\int \rho_n(\vec{r})d^3r$$

$$= \sigma_{\gamma p}Z + \sigma_{\gamma n}N \tag{12}$$

with $\sigma_{\gamma p}, \sigma_{\gamma n}$ the cross section for $\gamma p \to \pi N$ and $\gamma n \to \pi N$ respectively. This result is the impulse approximation. Eq. (11) neglects the effect of Pauli blocking and Fermi motion of the Lindhard function. It is thus clear that the use of eq. (10) improves over the impulse approximation precisely on these two points. However, it only takes into account the $\gamma N \to \pi N$ channel of the photonuclear reactions. The absorption channels are still missing.

The absorption channels can be easily obtained by allowing the pion in fig. 6 to excite a ph as shown in fig. 7. Then, when in the intermediate integrations the states cut by the dotted line in fig. 7 are placed on shell, this will give a contribution to $Im\Pi$ which is tied to the photon disappearing and producing a $2p2h$ excitation.

The evaluation of the absorption diagrams is similar to the one we sketched for the (γ, π) channel, and now two Lindhard functions are involved in the evaluation of Π. There are many technical details which must be skipped here but can be followed from ref. [1], but we would like to mention that there are extra diagrams evaluated where the crossed circle happens in the second ph excitation, there are some symmetric terms which carry a symmetry factor $\frac{1}{2}$,

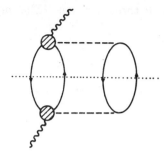

Figure 7: Photon selfenergy corresponding to photon absorption by a pair of nucleons.

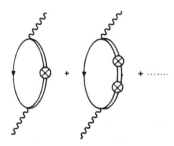

Figure 8: Δh excitation contribution to the photon selfenergy. The crossed circle indicates the Δ selfenergy insertion.

and long range as well as short range correlation corrections are implemented into the scheme.

Around the region of excitation of the Δ resonance, the dominant contribution comes from the diagram of fig. 8

One finds there that

$$\Pi_\Delta(k) = (\frac{f_\gamma}{\mu})^2 \frac{4}{9} \vec{k}_{CM}^2 M \frac{1}{\sqrt{s_\Delta} - M_\Delta + \frac{i\tilde{\Gamma}(k)}{2} - \Sigma_\Delta(k)} \rho \tag{13}$$

where $\tilde{\Gamma}(k)$ is the free Δ width corrected by Pauli blocking and $\Sigma_\Delta(k)$ is the Δ selfenergy. The model for the Δ selfenergy is taken from ref. [14] and has both a real and an imaginary part. The latter part accounts for $\Delta N \to NN$ or $\Delta NN \to NN$ channels, which when implemented in eq. (13) lead to mechanisms of direct photon absorption by two or three nucleons. The important thing to keep in mind is the fact that

$$\frac{\tilde{\Gamma}(k)}{2} - Im\Sigma_\Delta(k) = \frac{\Gamma_{eff}}{2} > \frac{\Gamma(k)}{2} \tag{14}$$

where $\Gamma(k)$ is the free Δ width. Hence, when we are close to the Δ pole, $\sqrt{s} - M_\Delta - Re\Sigma_\Delta(k) \simeq 0$, the Δh propagator in eq. (13) behaves as $2/i\Gamma_{eff}$ and are find that

$$|Im\Pi_\Delta| \sim \frac{1}{\Gamma_{eff}} < \frac{1}{\Gamma} \tag{15}$$

and consequently the photonuclear cross section per nucleon with the renormalized Δ becomes smaller than the free γN cross section.

This feature is clearly visible in the experimental and theoretical cross sections which we show in figs. 9 and 10.

In figs. 9 and 10 we also show the cross section corresponding to direct γ absorption. As we described before, the method we use traces back the γ cross section to the imaginary part of the γ selfenergy, and by looking at the different sources of imaginary part we can associate them to particular channels like (γ, π) or γ absorption (see figs. 6 and 7). But we must be cautious in the interpretation. Eq. (7) gives a contribution to σ_A from each element of volume d^3r, and in each of these elements of volume we made the classification into (γ, π) or γ absorption. We are not saying anything about the fate of the particles emitted, nucleons or pions, in their may out of the nucleus, but are making a semiclassical assumption and this is that the final state interaction of these particles will not change the cross section, it will only redistribute its strength in different channels. This means that in some events that originally were of the (γ, π) type, the pion will be absorbed in its way out and show up as γ absorption, in the sense that only nucleons and no pions will be detected. We shall, thus, make a difference between this way of absorbing photons, which we call indirect photon absorption, and the direct one where originally no real pions were produced and the photon was directly absorbed.

The agreement with the data in figs. 9 and 10 is quite good. It is quite remarkable to see that even the small differences between σ_A/A for ^{12}C and ^{208}Pb that one obtains in the calculation have found experimental support in recent, very precise, measurements [18].

The assumption of the cross section not being modified by final state interaction is accurate for relatively energetic pions or nucleons (100 MeV or 50 MeV kinetic energies respectively). For low energy pions is should be less accurate, but then, for phase space reasons, the contribution to σ_A is small since the cross section is dominated by γ absorption. Thus, for photons with $\omega_\gamma \simeq 100$ MeV on, the assumption leads to accurate total cross sections.

It is also interesting to note that although σ_A/A is nearly constant for nuclei, up to the small differences apreciated in [18], the direct γ absorption per nucleon

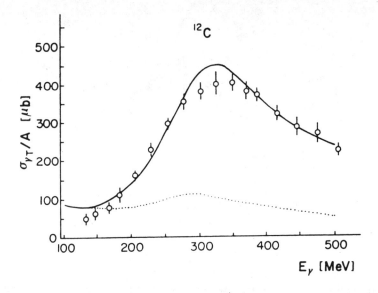

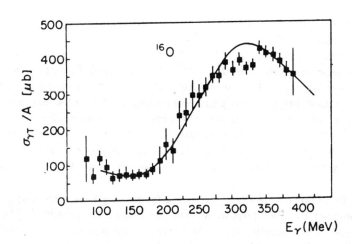

Figure 9: σ_γ/A for ^{12}C and ^{16}O. The experimental data are from [15] and [16] respectively.

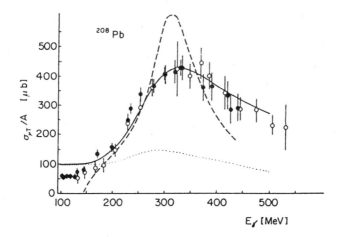

Figure 10: σ_p/A for ^{208}Pb. The dashed line shows the impulse approximation $(Z\sigma_{\gamma p} + N\sigma_{\gamma n})/A$. The dotted line shows the results for direct photon
absorption. The experimental data are from: full circles [17], open circles [15].

is not so constant for different nuclei, as comparison of the results for ^{12}C and
^{208}Pb in figs. 9 and 10 shows.

It is also interesting to note that for γ energies below $\omega_\gamma = 150 \ MeV$, the
absorption cross section is dominated by the Kroll Ruderman and pion pole
terms (see fig. 1). These terms are only relevant for charged pions and this
implies γ absorption by pn pairs, something corroborated experimentally and
which has been the foundation of the quasideuteron model for γ absorption [19].
Here we find a microscopical picture for γ absorption which supports some of
the features of the quasideuteron (empirical) model and extends it to higher
energies where the presence of the Δ allows for γ absorption or pp, pn or nn
pairs.

We should also note that by starting from the $\gamma N \rightarrow \pi\pi N$ amplitude, taking
the dominant Kroll Ruderman Δ term and folding the amplitude like in the case
of the $\gamma N \rightarrow \pi N$ amplitude, we find terms contributing to the γ selfenergy like
those depicted in fig. 11 and which contribute about 20% to the γ total cross
section at energies around $450 \ MeV$.

5   Inclusive (γ, π) cross section

In the former section we calculated σ_A and separated from it the cross section
corresponding to direct photon absorption. The difference between σ_A and σ

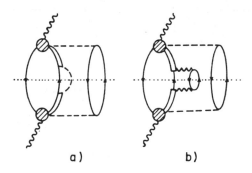

Figure 11: Channels tied to the $\gamma N \rightarrow \pi\pi N$ amplitude. a) corresponds to $2ph1\pi$ excitation; b) corresponds to $3ph$ excitation.

of direct absorption corresponds to events which originally were (γ, π), this is, where a real pion was produced in the first step. However, this pion is produced at a point $\vec{r}$ and before it goes out and can be detected it has to sort out many obstacles. Indeed, the pion can scatter with nucleons in its way out of the nucleus changing direction and energy; it can also be absorbed by pairs or trios of nucleons and then it will not be detected, instead, some nucleons will be ejected from the nucleus and the event will correspond to γ absorption of the indirect type as we discussed.

In order to be able to get cross sections for (γ, π) which can be compared to experiment it is imperative to follow the evolution of the pions in the nucleus once they are produced. This is done in ref. [20] and we outline the procedure here.

In ref. [21] a thorough study of all the inclusive pionic reactions was done: quasielastic, single charge exchange, double charge exchange and absorption. The procedure used there was the following: through a many body theory similar to the one described here for photons, one evaluated the pion reaction probabilities per unit length for the different reaction channnels. Then all this information was used in a Monte Carlo simulation procedure which, according to the calculated probabilities, decides the different steps that the pion follows in its attempt to leave the nucleus. The procedure proves to be very accurate and effective to calculate all the pion reaction channels for pions above $E_\pi = 100\ MeV$ [21]. One could even prove there that the results for the total reaction cross section calculated with the Monte Carlo procedure were in very good agreement ($\sim 5\%$ differences) with the full quantum mechanical calculation of this channel. This latter calculation was done by solving the Klein Gordon equation for the pion with the pion nucleus optical potential calculated microscopically, evaluating then the total cross section from the forward scatter-

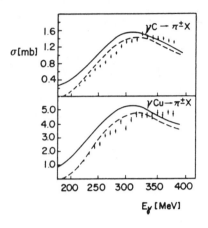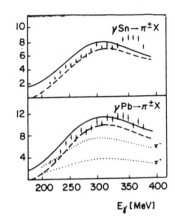

Figure 12: Integrated cross sections for $(\gamma, \pi^{\pm})$ for several nuclei as a function of the photon energy. Continuous line: integrated cross section; dashed line: calculation omitting the pions with $T_{\pi} < 40 \ MeV$ to compare with experiment. Experimental data from ref. [22]. Dotted line: theoretical contribution from (γ, π^{+}) and (γ, π^{-}), both of them without pion threshold.

ing amplitude using the optical theorem, and subtracting the integrated elastic cross section.

The same procedure used in [21] was followed in ref. [20] to follow the fate of the pion produced in a (γ, π) step. Some of the pions are absorbed and then we get indirect photon absorption. Others scatter quasielastically with nucleons and eventually change their charge. The simulation then tells us in which direction, and with which charge and energy the pions come out, allowing direct comparison with experiment.

The agreement of the results of [20] with experiment is rather good in all charge pion channels, double differential cross sections, angular or energy distributions and for different nuclei. We show in fig. 12 results for $(\gamma, \pi^{\pm})$ for different nuclei, where one observes that, once the proper cuts considered in the experiment are taken into the account, the agreement of theory and experiment is rather good.

Since the (γ, π) cross sections come out fine from our calculations and so do the total cross sections, then we are also making accurate predictions for the total absorption cross section. However, our calculations allow us to distinguish between direct and indirect photon absorption. In fig. 13 we show the direct absorption, total absorption and total cross sections per nucleon for photons interacting with ^{12}C and ^{208}Pb.

It is very interesting to note that the amount of indirect photon absorption per nucleon in ^{208}Pb is about three times larger than in ^{12}C. The result is intuitive because in a heavy nucleus the pions have less chances to leave the nucleus without being absorbed. However, one should not overlook this finding which shows that the amount of γ indirect absorption in nuclei provides more information on pion absorption than the experiments of pion absorption with real pions. The reason is that for pions close to the resonance the pion absorption probability per unit length is so big that all pions which come into the geometrical cross section of the nucleus get absorbed and the cross section goes roughly as πR^2 (R, nuclear radius). In this case this cross section is rather insensitive to the intrinsic absorption probability per unit length. Indeed, if we increase this probability in a calculation, the pions are absorbed sooner but are absorbed anyway. In simple words, a black disk can not become blacker. On the other hand the probability that a pion created in a point leaves the nucleus is proportional to

$$e^{-\int P_{abs} dl} \tag{16}$$

and this exponential is very sensitive to P_{abs}. The discussion above can be summarized in simple words by saying that the magnitude of indirect photon absorption in nuclei can provide more information about pion absorption probabilities than the pion absorption cross sections obtained from the scattering of real pions on nuclei. In view of this finding it looks quite useful to invest efforts into the experimental separation of the two sources of photon absorption throughout the periodic table. The techniques to be used would ressemble those used in the separation of genuine three nucleon pion absorption from quasielastic steps followed by two body absorption [23].

6   Inclusive $(\gamma, N), (\gamma, NN), (\gamma, \pi N) \ldots$ reactions

The detailed work on these channels has been done in ref. [24]. The work is a natural continuation of the study in the (γ, π) channel in ref. [20]. Here, in addition to the final state interaction of the pions, one also pays attention to the final state interaction of the nucleons. The nucleons coming from (γ, π) , γ absorption or pion absorption can collide with other nucleons changing energy, direction and charge. A calculation which attempts to compare with experiment necessarily has to address this problem.

The key ingredient in the final state interaction of the nucleons is the nucleon mean free path through matter, which is known to be larger than the semiclassical estimate, $\lambda^{-1} = \sigma_{NN}\rho$ [25]. A more suited magnitude than the mean free path is the imaginary part of the N-nucleus optical potential. The analysis of [24] uses a semiphenomenological model, developed in [26], which reproduces fairly well the microscopic results of ref. [27] obtained with a substantially larger computational time.

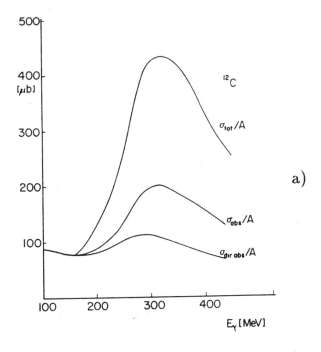

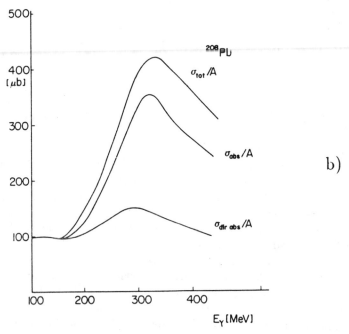

Figure 13: a) Photonuclear cross section for ^{12}C. Upper curve: total photonuclear cross section; middle curve: total absorbtion; lower curve: direct photon absorption. b) Same for ^{208}Pb.

The sources of nucleons in photonuclear reactions are varied:

i) direct γ absorption

ii) (γ, π) knock out

iii) (π, π') knock-out

iv) π absorption

v) NN collisions

In [24] once again a Monte Carlo computer simulation was performed to account for final state interaction of pions and nucleons. It was also proved in [24] that the reaction cross sections for N-nucleus scattering calculated with the simulation or quantum mechanically were also very similar, with about $5 - 8\%$ differences for different nuclei and nucleons with kinetic energy bigger than $40\ MeV$. This observation gives one confidence in the Monte Carlo simulation to investigate the different reaction channels, which would be technically forbidden otherwise.

Less ambitious studies of NN emission, with the consideration of direct photon absorption as the only source, have been carried out elsewhere [28, 29].

The results of [24] for the (γ, N) channel agree only semiquantitatively with experiment. The agreement is good in some nuclei while in other nuclei and certain energies there can be a disagreament of about a factor of two.

In the (γ, NN) and $(\gamma, \pi N)$ channels there is much work going on experimentally [30] and comparisons are being made with the predictions of ref. [24]. Work is still in progress but we expect to be able to extract much information about the dynamics of basic processes like pion absorption, photon direct absorption, nucleon mean free path in nuclei, isospin dependence of basic mechanisms, etc.

7 Conclusions

From the discussions above one can draw the following conclusions, which I classify into some pertaining to the theoretical side and others to the experimental side.

A) Theoretical side:

i) A microscopic many body picture for the photonuclear processs is possible combining elements of pion physics with photonuclear dynamics.

ii) All inclusive channels $(\gamma, \pi), (\gamma, N), (\gamma, 2N), (\gamma, \pi N)$, etc., can be addressed simultaneously.

iii) The γ absorption cross section depends strongly on A and has two sources:

a) Direct absorption.

b) Indirect absorption $((\gamma, \pi)$ followed by π absorption).

In heavy nuclei the indirect absorption dominates the absorption cross section. The indirect absorption cross section goes roughly as A^α with $\alpha > 1$ (in π absorption $\alpha \simeq 2/3$). The study of indirect photon absorption will teach us much about pion absorption in nuclei.

B) Experimental side:

i) Work advisable to separate $(\gamma, pp), (\gamma, pn), (\gamma, nn)$ as a function of the energy to pin down the basic dynamics.

ii) Work necessary to disentangle direct from indirect photon absorption to avoid misleading messages about γ absorption.

iii) Information from pion physics is essential to interprete correctly the results.

iv) The amount of experimental information in coincidence experiments is so large that real efforts are needed to classify it into a practical and useful way, which allows one to learn about the basic dynamics of the physical processes. In view of the fairly large amount of channels present in these reactions, the extraction of the basic dynamics is not easy without a strong theoretical guidance. It is most available that the experimental analysis goes side by side with theoretical calculations, as complete as possible, in order to establish solid facts about the basic dynamics of the processes by learning both from the agreement and disagreement of the predictions with the data. The fact that many experimental groups have reached this conclusion [30] is an encouraging sign that much progress is ahead.

Acknowledgements are given to all my collaborators in different parts of the work reported here: R.C. Carrasco, L.L. Salcedo, M.J. Vicente Vacas and J.A. Gómez Tejedor.

This work is partly supported by CICYT, contract number AEN 93-1205.

References

[1] R. C. Carrasco and E. Oset, Nucl. Phys. A536(1992)445.

[2] I. Blomqvist and J.M. Laget, Nucl. Phys. A280(1977)405.

[3] S. Fubini, G. Furlan and C. Rossetti, Nuovo Cim. 40 (1965)1171.

[4] M. G. Olson, Nucl. Phys. B78(1974)55.

[5] T. Fujii et al., Nucl. Phys. B120(1977)395.

[6] Aacher-Berlin-Bonn-Hamburg-Heidelberg-München collaboration, Phys. Rev. 175(1968)1669.

[7] G. Gianella et al., Nuovo Cimento 63A(1969)892.

[8] L. Murphy, PhD Thesis, Saclay, 1993.

[9] H. Ströher et al., invited talk at the Int. Conf. on Mesons in nuclei, Dubna, May 1994.

[10] L. Lüke and P. Söding, Springer tracts in modern physics, vol. 59 (Springer, Berlin, 1971) p. 39.

[11] J. A. Gómez Tejedor and E. Oset, Nucl. Phys. A571(1994)667.

[12] F. Cano, Tesina de Licenciatura, Universidad de Valencia, 1994.

[13] C. Itzykson and J.B. Zuber, Quantum Field Theory (Mc Graw Hill, New York, 1980).

[14] E. Oset and L.L. Salcedo, Nucl. Phys. A468(1987)631.

[15] L. Guedira, PhD Thesis, University of Paris Sud, 1984.

[16] J. Ahrens et al., Nucl. Phys. A490(1988)655.

[17] P. Carlos et al., Nucl. Phys. A431(1984)573.

[18] N. Bianchi, private communication.

[19] J. S. Levinger, Phys. Rev. 84(1951)43.

[20] R. C. Carrasco, E. Oset and L.L. Salcedo, Nucl. Phys. A541(1992)585.

[21] L.L. Salcedo, E. Oset, M.J. Vicente-Vacas and C. García-Recio, Nucl. Phys. A484(1988)557.

[22] H. Rost, PhD Thesis, Universität Bonn, 1980.

[23] H. Weyer, in the Int. Workshop on pions in nuclei, Peniscola, June 1991, E. Oset et al. Edts., World Scientific, pag. 441.

[24] R. C. Carrasco, M.J. Vicente-Vacas and E. Oset, Nucl. Phys. A570(1994)701.

[25] D. F. Geesaman et al., Phys. Rev. Lett. 63(1989)734.

[26] P. Fernández de Córdoba and E. Oset, Phys. Rev. C46(1992)1697.

[27] S. Fantoni, B. L. Friman and V. R. Pandharipande, Nucl. Phys. A399(1983)51; S. Fantoni and V. R. Pandharipande, Nucl. Phys. A427(1984)473.

[28] C. Giusti, F. D. Pacati and M. Radici, Nucl. Phys. A546(1992)607.

[29] J. Ryckebush, M. Vanderhaeghen, L. Machenil and M. Waroquier, Nucl. Phys. A568(1994)828.

[30] N. Rodning et al.; P. Grabmayer et al.; A. Sandorfi et al.; M. Ling et al., in preparation.

Notes on Scaling and Critical Behaviour in Nuclear Fragmentation

X. Campi and H. Krivine

Division de Physique Théorique[1]
Institut de Physique Nucléaire
91406- Orsay Cedex, France

Abstract. We discuss the relevance of the concepts of scaling and critical behaviour in nuclear fragmentation. Experimental results are reviewed to check whether the signals of a percolation or liquid-gas phase transition manifest themselves in the data.

1   Introduction

Excited nuclei decay according to various deexcitation modes. At energies up to a few MeV/nucleon, the main mode is light particle emission. At energies of the order of five to ten MeV/nucleon, this mode is relayed by the emission of medium mass fragments (see Fig. 1). The aim of these lectures is to discuss this nuclear fragmentation in the general framework of modern theories of fragmentation. The study of this new and fast developing field of statistical mechanics, concerns many physical systems of the microscopic world, including polymers, gels, atomic clusters, nuclei and elementary particles. One of the goals of this research is to investigate what are the features that are common to various fragmentation processes, regardless the nature of the elementary constituents or the binding interactions of the fragmenting objects. A particularly signifi-cant outcome of this research is the realization that in many circumstances the fragment size distribution (FSD) is scale invariant. Another realization is that quantities associated to the FSD can show a critical behaviour, i.e. can diverge at some critical points. These realizations clear the way to a model independent classification of the various fragmentation mechanisms in terms of the scaling properties and critical behaviour. Applying these concepts to nuclear fragmen-tation would be very fruitful. It would clarify what features are generic, (for example common to all finite size systems with short range attractive forces) and what features are specific of nuclei (for example, what is the influence of quantum effects, of the combination of a short range attraction with a long range coulomb repulsion, of the presence of two types of particles, etc.).

---

[1] Unité de Recherche des Universités de Paris 11 et Paris 6 associée au CNRS

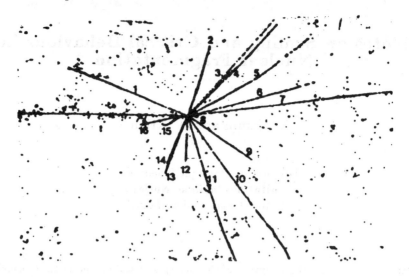

Figure 1: *The collision of a 850 MeV ^{12}C projectile arriving from the left with an Ag (Br) target in a nuclear emulsion. The charge $1 \leq z_i \leq 4$ of 16 fragments has been identified. From Ref. [1].*

The presence of this critical behaviour in the FSD can be the manifestation of a phase transition in nuclear matter. One of the present goals of heavy ion physics is the determination of the equation of state of infinite nuclear matter. The knowledge of this object in a wide range of physical conditions is fundamental for future developments of nuclear theory as well as a basic input for astrophysics. Particularly interesting is the behaviour of the equation of state in the vicinity of phase transitions. For example, one expects [2] that at a temperature of about 6 to 10 MeV takes place in infinite nuclear matter a sort of liquid-gas phase transition. We ignore if it is possible to study this phase transition in the laboratory. By energetic heavy ions collisions we are able to produce during short times very small pieces of nuclear matter at the required conditions, but we dont know what are the signals of a "phase transition" in a transient state of a so small system. In this notes we will discuss how to approach this problem. The reader may find it useful to consult review papers that emphasize different and complementary aspects of nuclear fragmentation [3, 4, 5].

These notes are organized as follows. In Section 2 we give a brief report on models of fragmentation, with emphasis on practical realizations in nuclear physics. Section 3 is devoted to develop the concepts of scaling and critical behaviour. In Section 4 we discuss how this concepts would apply to nuclear fragmentation. Some final remarks are made in Section 5.

m	n_s
6	6 0 0 0 0 0
5	4 1 0 0 0 0
4	3 0 1 0 0 0
4	2 2 0 0 0 0
3	2 0 0 1 0 0
3	1 1 1 0 0 0
3	0 3 0 0 0 0
2	1 0 0 0 1 0
2	0 1 0 1 0 0
2	0 0 2 0 0 0
1	0 0 0 0 0 1

Table 1: *Partitions of $S = 6$ according to the multiplicity m. The rows indicate the values of n_s.*

2   Elements of Theory

A theory *ab initio* of fragmentation, i.e. starting for instance from the Schrödinger equation, is of course not available. In fact such theory would be probably useless. On the other hand, it is dubious that a unique theory explaining the fragmentation of all kinds of clusters (nuclear, atomic, molecular, macroscopic...) can be constructed. The statistics obeying the constituents (fermions or bosons), their size and their interaction (shape, intensity and range) may play crucial roles. In this section we will describe a few very simple fragmentation models that make abstraction of all these (important ?) details. In contrast, we put the emphasis on global properties, like scaling, that would be universal for certain classes of fragmentation.

2.1   Partitions

A partition by definition is a collection of integers (with given sum) without regard to order. The integers collected to form a partition are called its parts, and the number which is the sum of these parts is the partitioned number. It is conventional to abbreviate repeated parts by use of a multiplicity vector.

We will use the following notations : S will be the size (mass or charge) of the nucleus that is partitioned, s the size of its parts (fragments), which appear with multiplicity n_s. For example the 11 partitions of number 6 are displayed in Table 1. Let also be s_{max} the size of the largest fragment. Constituents number conservation is expressed through the multiplicities n_s by

$$S = \sum_{s=1,s_{max}} s n_s. \tag{1}$$

With any given partition one can associate its parts (fragments) multiplicity m, defined as

$$m = \sum_{s=1,s_{max}} n_s. \tag{2}$$

The number $N(S, m)$ of partitions having multiplicity m is calculated by the recurrent relation

$$N(S, m) = \sum_{0 \leq k \leq (S/m)-1} N(S - km - 1, m - 1).$$

The total number of partitions

$$N(S) = \sum_{m=1,S} N(S, m)$$

grows very rapidly with S. For large S this number can be estimated by[6]

$$N(S) \simeq \frac{e^{\pi\sqrt{2/3S}}}{4\sqrt{3}S}. \tag{3}$$

For example, $N(100) = 190569292$ while formula 3 gives $19.9.10^7$.

The basic quantity in a fragmentation process is the probability $P\{n_s\}$ to observe a partition $\{n_1, \cdots, n_S\}$. This probability may be governed by conservation laws, by geometrical constraints, by equilibrium (thermal, chemical, statistical...) conditions, etc.

Assuming that the probability p_s of emitting a fragment of mass s is independent of how the rest of the nucleus disintegrates, the probability of observing a partition of multiplicity m is given by the multinomial distribution

$$P(n_1.....n_S) = \frac{p_1^{n_1} p_2^{n_2} \cdots p_S^{n_S}}{n_1! n_2! \cdots n_s!} m! \tag{4}$$

restricted by $\sum_s s n_s = S$. Because of the very large number of partitions, this hypothesis can only be checked in detail for small systems. Cole and collaborators [7] have studied the fragmentation of carbon nuclei ($S = 12$, 57 different partitions). Taking p_s from the experimental inclusive mass distribution σ_s (all fragmentation events)

$$p_s = \frac{\sigma_s}{\sum_s \sigma_s},$$

they found that Eq. 4 accounts for data better than a factor of two for most partitions. The largest deviations are observed in partitions for which typical nuclear physics effects are strongest (for example the decay in three α particles).

To begin with a study of fragmentation we do not need all the information carried by $P\{n_s\}$. The average fragment size distribution of partitions of a given type suffices, in a first step, to characterize the models and the data. In what follows, we denote this average by n_s. We turn now to the description of various fragmentation models of increasing complexity.

2.1.1 Maximum Entropy

A natural starting point, when no a priori information on the fragmentation mechanism is known, is to make the hypothesis that all partitions of a given type are equally probable. This is called the "Maximum Entropy" principle. Assuming that only the mass (or charge) S is conserved, Aichelin et al. [8] derived the asymptotic expression

$$n_s \simeq \frac{1}{e^{(1.28s/\sqrt{S})} - 1}.$$

One of the shortcomings of this approach is the lack of a control parameter associated to the "violence" of the fragmentation. Moretto and Bowman [9] have added the constraint of a given total surface of the fragments. The idea is that the excitation energy is proportional to the extra surface created by the formation of fragments. The FSD is then

$$n_s = \frac{1}{e^{Ds+As^{2/3}} - 1}. \tag{5}$$

The coefficients A and D are obtained by solving the equations

$$S = \sum_s \frac{a}{e^{Ds+As^{2/3}} - 1},$$

$$Surf = k \sum_s \frac{a^{2/3}}{e^{Ds+As^{2/3}} - 1},$$

where $Surf$ is the surface produced and $k = 3^{2/3}$ for spherical shapes. In order to confront Eq. 5 with data, it remains to convert this surface production in excitation energy.

Notice that, in all circumstances, these FSD are exponentially decaying. We will see below that for a critical phenomenon one expects a power law decay.

2.1.2 Geometrical Models

These models consist in ensembles of points in a *space*, which are *linked* by some mechanisms. The points represent the positions of the constituents in the space (coordinate, phase space...). The positions of the   points are either fixed (static models) or change with time (kinetic models). The linkage mechanisms are in general very simple *random* mechanisms based on *proximity* rules. Remark that "geometrical" is used here in a very loose sense. We will describe percolation models in some details with the purpose to introduce critical phenomena and finite size scaling manifestations in cluster production.

A percolation model is a collection of *static* points (sites) distributed in space, certain pairs of which are said to be adjacent or linked [11]. Whether or not two sites are linked is governed by a *random* mechanism the details of which depend on the context in which the model is used. There are two main classes of percolation known as bond and site percolation. In the site problem, the sites are occupied at random with a probability p. All bonds between occupied sites located at less than a certain distance are active. In the bond problem, all sites are occupied and bonds are active with probability p. The sites may be partitioned into *clusters* such that pairs of occupied sites in the same cluster are connected by active bonds. There is no path between sites in different clusters.

When p is close to zero, most sites will be isolated or form very small clusters. In the opposite, if p is close to one then nearly all sites are connected forming *one* large cluster, which occupies most of the available sites (or joins the borders of the system). This is called a "percolation" cluster. It is observed that in sufficiently large systems (see below), there is either one or none, but never two or more such percolation clusters. For infinite systems a sharply defined percolation threshold p_c exists such that for $p < p_c$ no percolating cluster exists and for $p > p_c$ *one* percolating (infinite) cluster exists. The transition from a non-percolating state to a percolating state is a kind of phase transition. The main difference between percolation and other phase transitions is the absence of a Hamiltonian. The percolation transition is a purely *geometrical* phenomenon in which the clusters are clearly defined *static* objects.

For theoretical purposes it is simpler to consider percolation models on a regular lattice. This reduction is done without any loss of generality because the behaviour of the percolation model near the critical point is independent on the details of the lattice. This behaviour is characterized by the *critical exponents,* which are "universal" to all short range linkage models of the same euclidean dimension d. This means, for example, that the exponents for triangular and square lattices ($d = 2$), site or bond percolation, are strictly the same although the percolation thresholds p_c may vary by a factor of two. In percolation theory, the critical exponents are associated to the FSD. The average number of clusters *per lattice site* has been shown [11] to follow approximately a scaling relation near p_c for large cluster sizes s :

$$n_s(\epsilon) \sim s^{-\tau} f(s/s_\xi(\epsilon)) + \cdots \tag{6}$$

where $\epsilon \sim p - p_c$ and $s_\xi(\epsilon)$ is the characteristic cluster size, that diverges at $p \to p_c$ as

$$s_\xi \sim |\epsilon|^{-1/\sigma} . \tag{7}$$

Here τ and σ are two critical exponents and f a scaling function satisfying $f(0) = 1$, i.e at the critical point $n_s \sim s^{-\tau}$. Remark that τ and σ are "universal" once d is fixed, but f is model dependent.

The size s_{max} of the percolation cluster, that exists only for $p > p_c$, goes to zero as

$$s_{max}/S \sim (p - p_c)^\beta \qquad (8)$$

where β is another critical exponent. This quantity plays the role of the *order parameter* in percolation theory (it is null in the most symmetric phase and takes a finite value in the other).

For percolation in $d = 1$ and $d = 2$ dimensions, the exact values of these exponents are known. For $d \geq 6$, the mean field values become exact. For $d = 3$ and 4, only approximate values obtained from Monte Carlo simulations are known.

In Fig. 2 we show the behaviour of n_s for the bond percolation model in a cubic lattice with 4^3 sites. Fig. 1a is for $p \ll p_c$ and Fig. 1d for $p \gg p_c$. One remarks that in both cases the distribution of light fragments is an exponential-like decreasing function. One also sees in Fig. 1d the broad distribution of s_{max} (centered around $s = 55$). At the critical "point" (Fig. 1b), the distribution is a power law (remark the change to a log-log representation). For p lightly above this point (Fig. 1c), one sees the rise of the percolation cluster.

The site and bond percolation models considered so far can be combined into a site-bond percolation model. Then the sites are occupied with probability q and the bonds between nearest neighbour occupied sites are active with probability p. As shown in Fig. 3, for any probability q larger than the site percolation threshold, there is a percolation threshold $p_c(q)$ for the bond probability. For $p > p_c(q)$, a percolation cluster exists. There is strong numerical evidence that the whole percolation line is described by the usual percolation exponents.

For pedagogical purposes, we will derive Eq. 6 in one dimension. Consider a linear chain of sites containing L sites. Each of these sites is randomly occupied with probability p (empty with probability $(1 - p)$) and all the bonds between nearest neighbours are active (site percolation model). The probability that s arbitrary sites are occupied is p^s. The probability of one end having an empty neighbour is $(1 - p)$ and therefore the total probability to find an $s -$ cluster is $p^s(1 - p)^2$. When L is large and $p < 1$, boundary effects are negligible. Then $Lp^s(1 - p)^2$ is the total number of $s -$ clusters, and the number of $s -$ cluster per lattice site is

$$n_s^{sit.}(p) = p^s(1 - p)^2. \qquad (9)$$

Analogous reasoning gives for the bond percolation

$$n_s^b(p) = p^{s-1}(1 - p)^2. \qquad (10)$$

We now discuss the consequences of Eq. 9.

When $p = 1$ all sites are occupied and the chain contains a single infinite cluster called *percolation cluster*. For every value $p < 1$ the chain will have on

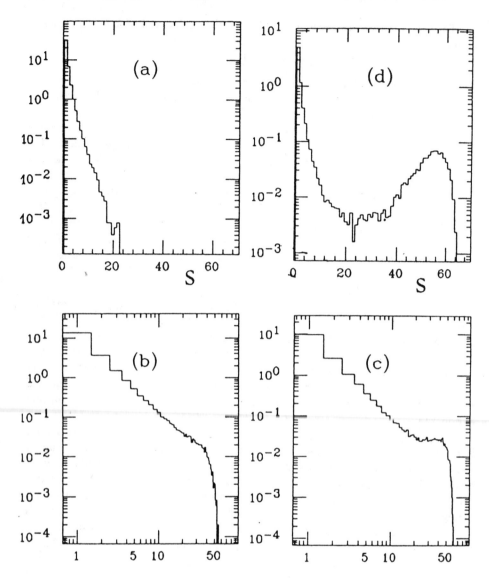

Figure 2: *Fragment size distribution n_s in a percolation model with a cubic lattice containing 4^3 sites : a) $p \ll p_c$, b) $p = p_c$, c) $p \geq p_c$, d) $p \gg p_c$.*

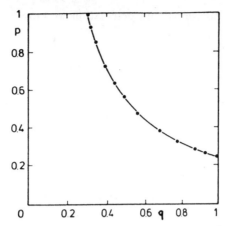

Figure 3: *Phase diagram of site-bond percolation in the cubic lattice. In this lattice the bond percolation threshold is at 0.248 and the site percolation threshold at 0.312. From ref. [10].*

average $(1-p)L$ empty sites and there is no percolation cluster. The percolation threshold is at $p = 1$. Because $p \leq 1$, there is no phase transition in this one dimensional model, as usual [12]. Nevertheless we will see that the system approaches a critical behaviour as given by Eq. 6 when $p \to p_c = 1$. We can write as in Ref. [13]

$$p = e^{\log p} \simeq e^{p-1} = e^{p-p_c},$$

therefore

$$n_s^{sit.}(p) \simeq s^{-2}[(p_c - p)s]^2 e^{-(p_c-p)s} = s^{-\tau} f(s/s_\xi)$$

with $\tau = 2$, $\xi = (p_c - p)^{-\sigma}$, $\sigma = 1$ and $f(z) = z^2 e^{-z}$. Remark that when $p \ll p_c$, n_s decreases exponentially with s (and as a power law when $p \to p_c$, see Eq. 9).

Cluster size distributions given by Eqs. 9 or 10 may be seen as signatures of fragmentation of 1-dimensional objects. In most practical application is better to work with the moments of n_s,

$$m_k = \sum_s s^k n_s.$$

or by the combination

$$\gamma_2 = \frac{m_2 m_0}{m_1^2} \tag{11}$$

which characterizes the width of the FSD n_s.

For example for 1-dimensional site percolation we have

$$\left.\begin{array}{rcl} m_o & = & 1-p \\ m_1 & = & p \\ m_2 & = & p(1+p)/(1-p) \\ & \vdots & \end{array}\right\}$$

The second moment of the FSD can be rewritten in the form

$$m_2 = p(1+p)(1-p)^{-\gamma}$$

that diverges at threshold with the critical exponent $\gamma = 1$.

The *correlation function* $g(r)$ is defined as the probability that a site a distance r apart from an occupied site belongs to the same cluster. Obviously $g(0) = 1$, $g(1) = p$ and in one dimension,

$$g(r) = p^r = e^{r \log p}$$

and

$$\log p = \log\left[1 - (1 - p)\right] \simeq p - 1.$$

Therefore

$$g(r) \simeq e^{-\frac{r}{\xi}} \simeq \frac{1-r}{\xi}$$

where

$$\xi = \frac{1}{1-p} \simeq (p_c - p)^{-\nu}$$

is the *correlation length* and $\nu = 1$ another critical exponent.

In summary, in 1-dimension percolation we have the set of critical exponents : $\tau = 2$, $\gamma = 1$, $\nu = 1$ and $\sigma = 1$.

2.1.3   Nuclear Percolation

Percolation ideas have been implemented in various nuclear fragmentation models. The nucleus is idealized by an ensemble of nucleons (sites), linked by the short range nuclear force (bonds). In a cold nucleus, all sites are occupied and all bonds between nearest neighbours are active. During a nuclear collision, the number of active sites decreases (because nucleons are ejected out of the nuclear volume or because the system expands the density decreases), and the number of active bonds also decreases (because the expansion breaks the short range bonds). The bond activation parameter plays the role of an excitation energy and the site parameter the role of the density.

In ref. [14], one uses a one-dimensional bond percolation model in combination with a standard particle evaporation model. Three-dimensional bond percolation in a cubic lattice was examined by Bauer et al. [15], continuous percolation (percolation without a lattice) in ref. [16] and site-bond percolation by Desbois in ref. [17]. Sometimes this model is coupled to a dynamical theory that gives information about the distribution of p and q parameters [18, 19, 20].

2.1.4 Statistical Equilibrium

In the theories of cluster fragmentation we have considered so far the weight of the various partitions as given only by combinatorics and geometry. Although these theories are very successful in predicting many global features of data (see Section 4), in some physical situations it is necessary to take into account explicitly energy and momentum conservation. This is the case for example in low bombarding energy fragmentation, where the binding energies of the individual clusters play a dominant role in defining the fragmentation pathway. The simplest way to incorporate this information is to adopt a thermodynamic point of view and say that after a complicated break-up phenomenon the available phase-space is homogeneously filled by the events. Let's be more precise [21]. For a complicated process the transition probability $P_{i\alpha}$ from the initial i to any find state α is mainly constrained by particle number S, energy E and momentum $\vec{P}$ conservation, i.e. $P_{i\alpha}$ is essentially the same for all allowed final states and zero otherwise.

We have

$$P_{i\alpha} = 1/\xi \quad or \quad P_{i\alpha} = 0$$

where

$$\xi = \sum_{\{n_i^\alpha\}} \delta(E - E_\alpha)\delta(S - \Sigma\, n_i^\alpha s_i)\delta^3(\vec{P} - \Sigma\, n_i^\alpha \vec{P_i}) \tag{12}$$

is the available phase space (the partition function in the microcanonical ensemble) of decay channels. The various decay channels are well defined by specifying the fragments in each channel α, i.e. their number n_i^α their mass s_i, position $\vec{r_i}$, and momentum $\vec{P_i}$.

The channel energy E_α is defined as

$$E_\alpha = \sum_i n_i^\alpha \left(B_i + \frac{P_i^2}{2m_i} + \varepsilon_i^* \right) + C_\alpha$$

where B_i is the binding energy of fragment i, ε_i^* its internal excitation energy and C_α is the Coulomb interaction energy between the fragments. The partition function (Eq. 12) has been calculated in a nuclear context using an infinite matter approach [22], solving a selfconsistent system of many (up to 3000) non linear coupled equations or using a Monte-Carlo simulation of possible decay channels putting at random fragments inside a sphere of a fixed volume [23]. Work on similar lines has also been done by the Copenhagen group[24].

This model explicitly incorporates quantum shell effects through the binding B_i and internal excitation energies ε_i^* and the effects of the long range repulsive interactions between the clusters. It shows characteristic phase transitions, one at a temperature of about 5 MeV, which is of fission type and specific of nuclear systems and another at $T \sim 6 - 7$ MeV of droplets condensation type [25]. The canonical heat capacity, which is the classical observable of a phase transition, clearly shows two peaks at these temperatures (see Fig. 4).

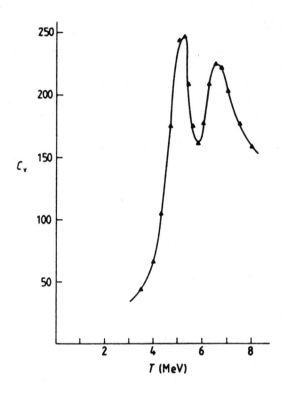

Figure 4: *Canonical heat capacity for* ^{131}Xe *as a function of temperature, calculated with the statistical equilibrium model. From ref. [4]*

2.1.5   Rate Equations Theory

So far we have considered *static* approaches of cluster fragmentation. However in some cases it is important to take into account the time evolution of the fragmentation process.

Assuming that the breakup process is driven by an external source, the time evolution of the fragment-size distributions for systems undergoing fragmentation may be described by a system of *linear* rate equations,

$$\frac{\partial}{\partial t} n_x(t) = -a(x)n_x(t) + \int_x^\infty n_y(t)a(y)f(x \mid y)dy. \tag{13}$$

Here $n_x(y)$ is the number of fragments of size x per unit volume at time t. The first term in the right-hand side represents the loss of particles of size x due to the breaking into smaller ones at a rate $a(x)$. The integral term represents the increase of particles of size x because of the breakup of larger ones. The rate at which x is produced from y is denoted by $f(x \mid y)$. This quantity must be normalized so that mass is conserved

$$\int_0^y x\, f(x \mid y)dx = y.$$

The number of fragments is given by

$$\bar{N}(y) = \int_0^y f(x \mid y)dx.$$

For the particular case of binary breakup, $f(x \mid y) = f(y - x \mid y)$ and $\bar{N} = 2$. Cheng and Redner [26] developed a scaling theory for homogeneous breakup kernels for which

$$a(x) = x^\lambda,$$

where λ is the homogeneity index. Homogeneity also implies that,

$$f(x \mid y) = y^{-1}b(x \mid y). \tag{14}$$

The analysis of Eq. 13 is done with the scaling cluster size distribution given by 6

$$n_y(t) \sim s^{-2}\phi(y/s(t)).$$

Here the typical cluster mass decreases according to Eq. 13 as

$$s(t) \sim t^{-1/\lambda},$$

for $\lambda > 0$. The main results are summarized as follows. For large x, $\phi(x)$ has the universal asymptotic form

$$\phi(x) \sim x^{b(1)-2}e^{-ax^\lambda}, \qquad x \to \infty$$

with $b(1) \geq 0$.

For small x, $\phi(x)$ may decay as

$$\phi(x) \sim \exp\left[-\frac{\lambda}{2\ell n x_0}(\ell n^2 x)\right] \qquad x \to 0,$$

when it exists a sharp cut off in the kernel at small fragment sizes, i.e. $b(x) = 0$ for $x < x_0$ and $0 < x_0 < 1$. When there is no small size cut off the decays of $\phi(x)$ approaches a power law

$$\phi(x) \sim x^\nu \qquad x \to 0.$$

A "shattering" transition takes place when $\lambda < 0$, in which clusters of indefinitely decreasing size are produced. This process is analogous to, but opposite from, the gelation found in aggregation systems.

Exact solutions of Eqs. 13 have also been obtained by Mc Grady and Ziff for analogous rate kernels[27] and for discrete binary breakup models[28].

Some attempt has been made to describe nuclear fragmentation in terms of a sequential-binary decay process. We will mention the realistic calculations of the statistical decay code GEMINI [40] and the semi-analytical model of Richert and collaborators [41].

2.2   Comparison with Experimental Data

Here we will compare with experimental data some predictions of the previously described models. The models are the following.

a) The sequential-binary decay model, in the GEMINI version [40].

b) The statistical equilibrium model, in the Copenhagen version [38].

c) The site-bond percolation model.

The first and second models, need as input parameters the distributions of excitation energies and the sizes of the breaking systems. These informations have been taken from a BUU calculation. For the percolation model [34], the bond probability has been adjusted at $p = 0.45$, while the site occupation probability varies randomly $0 < q < 1$, to mimic the variation of the size of the system with the violence of the collision. The experimental data is from the ALADIN experiment [34]. It concerns the fragmentation of Au projectiles bombarding different targets at energies of 600 MeV/nucleon. Event by event, all charged fragments are detected (except those with $z = 1$ and some with $z = 2$). The events are classed according to the variable Z_{bound}, that is the sum of all the detected charges of the event. There should be some anti-correlation between this variable with excitation energy, since small Z_{bound} means large number of (undetected) protons and large excitation energy.

As a function of Z_{bound}, we show in Fig. 5 the following quantities. a) The average charge $< Z_{max} >$ of the largest fragment. b) The average multiplicity of intermediate mass fragments (IMF). c) The average value of the relative

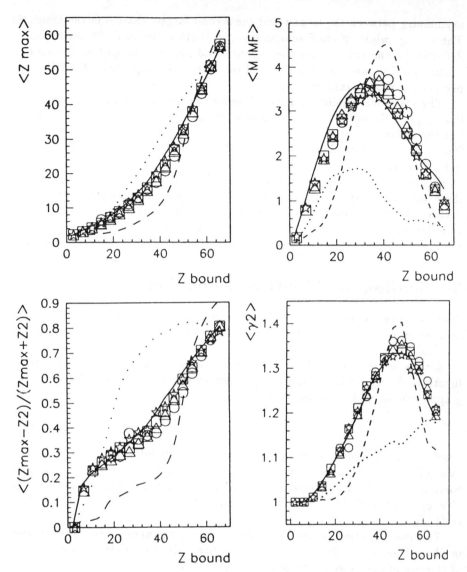

Figure 5: *The average charge* $< Z_{max} >$ *of the largest fragment (a), the mean number of intermediate mass fragments* $< MIMF >$ *(b), the average value of the relative asymmetry for the two largest fragments (c) and the average reduced width of the fragment charge distribution* $< \gamma_2 >$ *as a function of* Z_{bound} *(see text). Experimental data is from ALADIN experiment on Au 600 Mev/nucleon on C (circles), Al (triangles), Cu (squares) and Pb (stars). The lines are the statistical equilibrium (dashed), GEMINI (dotted) and percolation (continuous) predictions.*

asymmetry between the largest and the second largest charges in the event. d) The average width of the fragment size distribution (see Ref. [34] for the exact definition). The symbols are the experimental points. The dashed, dotted and continuous lines are the predictions of the statistical equilibrium, GEMINI and percolation models respectively.

The statistical equilibrium model under the size of the largest fragment at low Z_{bound}. The position of the peak of $< M_{IMF} >$ and $< \gamma_2 >$ is almost correct, but the width of the curves is too small. The relative asymmetry of the largest fragments is underpredicted[2]. The GEMINI model fails in all predictions. In particular, it does show no clear maximum for $< \gamma_2 >$. This is because this model does not have a critical behaviour. The percolation model can describe these data with surprising accuracy, despite their simplicity and their absence of nuclear physics content. This probably means that the observables a)-c) reveal an universal behaviour of a finite size fragmenting system interacting with short range forces and that this behaviour is well represented by percolation theory.

3   Scaling and Critical Behaviour

First we recall some basic definitions concerning critical phenomena. For a detailed discussion, the book of Stanley[29] remains the standard reference in the field.

Let's start with a mathematical introduction of scaling. In physics, many functions $F(x, y)$ of two variables approach to the leading order, if both x and y approach zero, the simpler form

$$F(x, y) = x^A f(y/x^B) \tag{15}$$

where f is a continuous function. (If a variable goes to infinity, use the reciprocal value of this variable). For example, $F(x, y) = (x+\sin(x/y))/(y+xy)$ looks very complicated, but if x and y go to zero with fixed ratio than we get asymptotically $F(x, y) = (x/y) \sin(x/y)/x$. Thus in Eq. 15, $A = -1$ $B = 1$ and $f(z) = \sin(1/z)/z$.

This mathematical property applies in many domains of physics, particularly in phase transitions. For example, in the Fisher droplet model[30] the number of clusters of size A at temperature T is given by

$$n(s, T) = s^{-\tau} e^{-K(T-T_c)s^\sigma} \tag{16}$$

i.e. by a scaling invariant part $s^{-\tau}$ and a scaling function $f(z)$. Here K is a constant, σ and τ are two critical exponents and T is the critical temperature. We recall that critical phenomena are the phenomena that manifest in the vicinity

---

[2]By adjusting as free functions the energy and source size distributions, it is possible to reproduce well the data with a statistical equilibrium model [39], but the adjusted energies are as much as a factor of two smaller than the BUU predictions.

of the transition point of a second order phase transition. A critical exponent is a number that describes the behaviour of a physical quantity near that critical point. Consider a real and non-negative function $f(x)$ defined in the interval $]0, x_0]$. If the limit

$$\lambda = \lim_{x \to 0^+} \frac{\log f(x)}{\log x} \tag{17}$$

exists when x goes to zero on the positive side, then λ is called the critical exponent to be associated with the function $f(x)$. (The definition is extended straightforwardly to the interval $[x_0, 0[$ when $x \to 0^-$). It is important to stress that the shorthanded notation that will be frequently used

$$f(x) \sim x^\lambda,$$

does not imply that

$$f(x) = Ax^\lambda,$$

where A is a constant. (Take for example the function $f(x) = x^\lambda \mid \log x \mid$, that has λ as critical exponent).

3.1 Signals of Scaling in Fragmentation

We have seen that the scaling *ansatz* for the cluster size distribution

$$n_s \sim s^{-\tau} f(s/s_\xi) \tag{18}$$

apply to geometrical fragmentation (or aggregation) models as well as to rate equations models. In this section we discuss how to test the existence of such scaling property using experimental data from a fragmentation process.

3.1.1 Critical Behaviour and Critical Exponents

First we concentrate on the possible manifestation of a critical behaviour. In infinite systems, near critical points there are clusters of all sizes and the characteristic cluster size diverges in leading order usually with a power law behaviour

$$s_\xi(x) \sim \mid x - x_c \mid^{-1/\sigma}$$

Then $n_s(x_c) \sim s^{-\tau}$ and $f(o) = 1$. Here x_c is the critical value of the variable x that defines the physical state of the system. For example, in percolation theory $x \equiv p$ (see Eq. 7. For this value x_c the cluster size distribution n_s is singular because all moments m_k with $k > \tau - 1$ diverge. This is shown [11] by

$$m_k(\epsilon) = \sum s^k n_s(\epsilon) \simeq \int_0^\infty s^{k-\tau} F(\epsilon s^\sigma) ds.$$

Or

$$m_k(\epsilon) = |\epsilon|^{(\tau-k-1)/\sigma} / \sigma \int_0^{\pm\infty} |z|^{(1+k-\tau)/\sigma} z^{-1} F(z) dz = C^{\pm} |\epsilon|^{(\tau-k-1)/\sigma}$$
(19)

where we have introduced the variables $\epsilon \equiv x - x_c$, $z = \epsilon s^{\sigma}$, made the change of function $F(z) = f(z^{1/\sigma})$ and replaced the sum by an integral. Notice that for $p > p_c$ the integral runs from 0 to ∞ and from 0 to $-\infty$ for $p < p_c$. $C^{\pm}$ are the corresponding values of these integrals. The exponents of the moments $k = 0, 1$ and 2 are called $2 - \alpha, \beta$ and $-\gamma$ respectively, in analogy with thermal critical exponents. Then exponent relations like

$$\gamma + 2\beta = 2 - \alpha = (\tau - 1)/\sigma$$
(20)

are automatically fulfilled.

3.1.2   Finite Size Scaling

The above considerations strictly apply to infinite systems. In finite systems, Monte-Carlo simulations show that a fingerprint of a transition remains, but spread out in a finite region around p_c.

For example, the behaviour of Eq. 18 is illustrated in Fig. 2 for the bond percolation model in a cubic lattice. Figures 1a and 1d show the cluster size distribution far below and above the critical point respectively. We remark that in both situations the distribution of light fragments is a very fast (exponential-like) decreasing function. In addition, above the critical point it exists a distribution of large clusters (centered in this example around $s = 55$) which is not accounted by Eq. 18. Right at the critical point (Fig. 1b) the distribution is a power law (remark the change to a log-log plot). Slightly above the critical point (Fig. 1c) one sees the rise of the large cluster at $s = 35$. This behaviour is quite general. Analogous distributions appear in thermal phase transitions and in kinetic processes of cluster formation .

How do the fragment size distribution $n(s)$ and related quantities behave near the critical point as a function of the size S of the fragmenting system ? One predicts[11] a finite size scaling of the type

$$n(s, S) \sim S^{-x} g(s/S^x),$$
(21)

where x is a new exponent and g a scaling function. The size s_{max} of the largest fragment produced at the critical point scales as

$$s_{max} \sim S^x$$
(22)

In the framework of a geometric picture of cluster production the exponent x is the ratio

$$x = D_f/d$$

of the "fractal dimension" D_f of the clusters to the euclidean dimension d of the physical space. Here the "fractal dimension" is defined through the relation between the mass s and the radius R^3 of the clusters,

$$s \sim R^{D_f}$$

which does not necessarily imply selfsimilarity. A "fractal" is an object with $D \neq d$. In dimension $d = 3$ percolation theory predicts $D_f = 2.5$ at the critical point. Far below the critical point, where only very small fragments are present $D_f = 2$. Far above, the largest fragment is obviously a compact object with $D_f = 3$. Different values of D_f are obtained in other fragment production mechanisms. The fractal dimension can be used to sign the fragmentation process. Following these lines, analysis of experimental data can be found in Refs[37] and [36].

4   Critical Behaviour in Nuclear Fragmentation

The first worry one encounters when applying the concepts of critical phenomena to atomic nuclei is how to specify the control parameter ϵ. In principle, looking for a thermal phenomenon one would like to choose $\epsilon = T - T_c$, but unfortunately this information is not directly available from experiment. When looking for a "geometrical" phenomenon, like a in percolation fragmentation, one would took $\epsilon = p - p_c$, where p is the bond activation probability and p_c its critical value, but this quantity is even more inaccessible experimentally. One possible solution is to substitute the temperature (or p) by another quantity that is measurable and that is strongly correlated with it. Furthermore, if this correlation is *strictly linear* in the critical region, (this is the case of the control parameter p and the moment m_0 in percolation theory[33]) then taking $\epsilon = m_0 - m_{0crit}$ one can determine directly the correct values of the critical exponents. In thermal phase transitions, the relation between the temperature T and the multiplicity m_0 (or other control parameter) has to be carefully studied. We will discuss below a method that overcomes this difficulty.

The second and more serious difficulty concerns the finite size of atomic nuclei. Strictly speaking, we cannot talk about critical behaviour in a finite size system because none of the moments m_k can diverge. Nevertheless, some aspects of this behaviour remain valid. The critical point is replaced by a "critical region", the width of which increases when decreasing the system size. In the middle of this region, finite and infinite systems behave very differently, but on both sides, these differences are much smaller. Hence it is in principle still possible to extract some information on the exponents by looking at these two regions, but avoiding the central one. This is shown in Fig. 6, where the quantity m_2/S is plotted as a function of the control parameter m_0/S for a large percolation cubic lattice and for a small one of the typical size of a nucleus. The

---

[3] R is the typical size of any compact object enclosing the clusters.

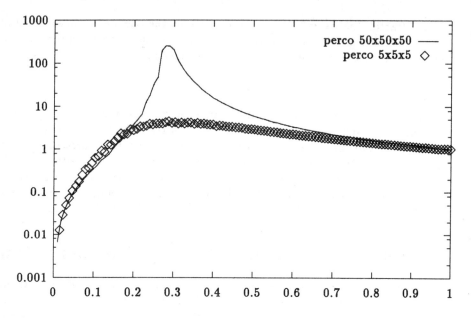

Figure 6: *Mean value of m_2/S as function of m_0/S for small $(S = 125)$ and large $(S = 125000)$ percolation systems.*

Table 2: *Critical exponents for various systems. From refs : Percolation [11],Lattice-gas[35], Statistical Equilibrium Model[36], experimental data on 1 GeV/u Au fragmentation in emulsion [31] and in $Au + C$ reactions[42].*

	τ	β	γ	$1 + \beta/\gamma$
Percolation	2.20	0.45	1.76	1.25
Lattice-gas	2.21	0.33	1.24	1.27
Statis. Equil.	~ 2.2	-	-	2.63 ± 0.07
Au + emul.	2.17 ± 0.1	-	-	1.2 ± 0.1
Au + C	2.14 ± 0.06	0.29 ± 0.2	1.4 ± 0.1	1.21 ± 0.1

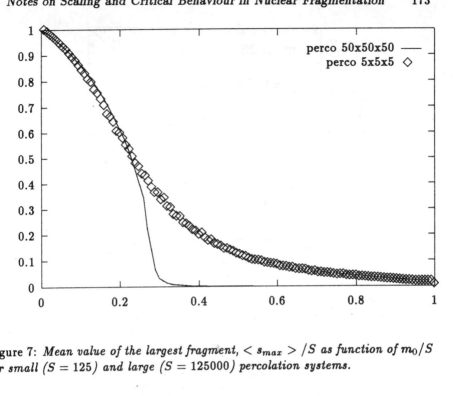

Figure 7: *Mean value of the largest fragment,* $< s_{max} > /S$ *as function of* m_0/S *for small (S = 125) and large (S = 125000) percolation systems.*

similarity is even stronger for the quantity $(S - m_1)/S =< s_{max} > /S$ which plays the role of the "order parameter" in percolation-like theories. In infinite systems, this quantity is finite in the "percolating" phase and zero in the other. We see in Fig. 7 that for $m_0 \ll m_{0crit}$ small and large systems behave similarly but very differently elsewhere.

These two examples give an idea of the difficulties to extract accurate values for the exponents β and γ in nuclear fragmentation. The members of the EOS Collaboration[42] have tried to extract these exponents from their data on 1 GeV/nucleon Au projectiles fragmentation. The control parameter is the multiplicity m_0. The critical multiplicity is determined by looking at the best linearity of the curve $\ln < s_{max} >$ for $m_0 < m_{0crit}$ (see Fig. 8) and at the best linearity and best parallelism of the two branches of m_2 (see Fig. 9). These curves are drawn as a function of $ln| m_0 - m_{0crit} |$. The slopes β and γ are rather sensitive to the choice of m_{0crit}.

In an earlier work[31], the ratio of the exponents β/γ was determined more directly. Representing $\ln < s_{max} >$ versus $ln(m_2/m_1)$ for events of the same type (say, same m_0) one obtains a two-branch curve, the crossing point corresponding to the "critical point" (see Fig. 10). The slope of the lower branch is $1+\beta/\gamma$. The advantage of this method is that there is no need to fix m_{0crit} and no need for a linear relation between T or p and m_0. The price one pays is that

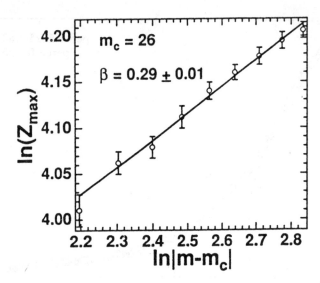

Figure 8: *Example of the determination of the critical exponent β for a particular choice of m_{0crit}. From Ref. [42].*

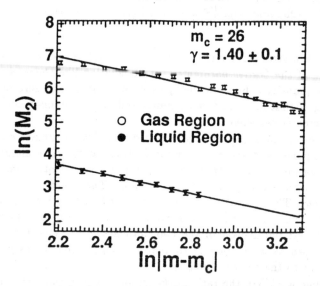

Figure 9: *Example of the determination of the critical exponent γ for a particular choice of m_{0crit}. Liquid and gas regions indicate $m_0 < m_{0crit}$ and $m_0 > m_{0crit}$ respectively. From Ref. [42].*

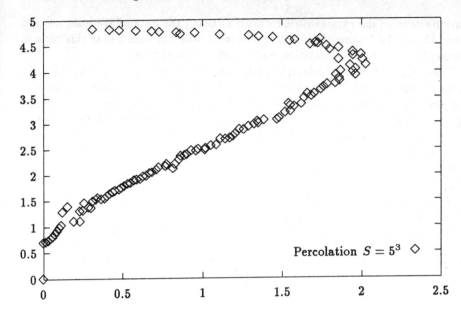

Figure 10: *Size of the largest fragment averaged at fixed multiplicity versus m_2/m_1 on a Log-Log plot.*

only the ratio of the exponents is measured. Using the data of Waddington and Freier[43] on Au fragmentation in emulsion, it was concluded that this ratio is compatible with both percolation and liquid-gas predictions.

The exponent τ is in principle easier to determine, by using Eq. 18 right at the "critical point". The problem is again to define this "point". Here it is also possible to avoid this difficulty, by looking at the slope of $\log m_3/m_1$ versus $\log m_2/m_1$ [31]. Unfortunately, most theories predict very similar values for τ. In any case, one always gets $\tau > 2$ from experiments in nuclear fragmentation.

In Table 2 we synthesize our present knowledge of critical exponents. We see that the two experimental determinations of β/γ [31, 42] are in good agreement with both liquid-gas and percolation predictions. When determined separately by the EOS collaboration [42] β and γ are in better agreement with the former prediction.

Also shown in Table 2 is the value of β/γ calculated with the statistical equilibrium model[36]. It differs from the two independent *experimental* determinations by a factor of two. This discrepancy deserves further theoretical considerations.

5   Concluding Remarks

This Contribution is a short review of the methods to study phase transitions in the fragmentation of finite nuclei. The determination of a set of critical exponents associated to the mean fragment size distribution appears to be the unique way to achieve a definite classification of the transition. This program seems in principle realizable with the new generation of 4π charged particles detectors. However we will temperate this optimism with a few warnings concerning the analysis of experimental data.

We assumed implicitly in the preceeding discussions that the size S of the fragmenting system was invariant. This is not always the case in nuclear fragmentation. At high bombarding energies intermediate mass fragments (the ones that determine mostly the critical behaviour) come from a "spectator" source, which size may change drastically with impact parameter. For example, in ALADIN experiments[34] on 600 MeV/nucleon Au projectiles bombarding Cu targets, for the most violent collisions that have been detected (lowest Z_{bound}) the emitting "spectator" source has on average the size of a Fe nucleus[45]. The influence of this variation of S on the critical exponents should be carefully studied.

At lower bombarding energies (less then 100 MeV/nucleon) we have the problems of the number and the size of emitting sources. As a function of the impact parameter, the reaction mechanism varies from deep inelastic, quasi-fusion and (maybe) total fusion. Even with a complete identification of the fragment momenta it is not possible to determine, event by event, the source of each fragment. This raises various difficulties. What is the multiplicity (or another control parameter) of each source ? What is the largest fragment s_{max} of each source ? (we recall that the subtraction of the largest fragment, at least in the "liquid" or "percolating" phase, is essential to calculate correctly the critical exponents). All these questions deserve a close examination.

References

[1] B. Jakobson et al., *Z. Phys.* **A307** (1982) 293.

[2] L.P. Csernai and J.I. Kapusta, *Phys. Rep.* **131** (1986) 223.

[3] J. Huefner, *Phys. Rep.* **125** (1985) 129.

[4] D.H.E. Gross, *Rep. Prog. Phys.* **53** (1990) 605.

[5] L.G. Moretto and G.J. Wozniak, *Annu. Rev. Nucl. Part. Sci.* **43** (1993) 379.

[6] A. Abramowitz and I.A. Stegun *Handbook of Mathematical Functions* (Dover Inc., New York, N.Y.,1965).

[7] A.J. Cole et al., *Phys. Rev.* **C39** (1989) 891.

[8] J. Aichelin, J. Hüfner znd R. Ibarra, *Phys. Rev.* **C30** (1984) 107.

[9] L.G. Moretto and D.R. Bowman, *XXIV Inter. Winter Meeting of Nucl. Phys.* Bormio. Italy. Ricerca Scientifica cd Educazione Permanente (1986) p. 126.

[10] D. Stauffer, A. Coniglio and M. Adam, *Adv. in Pol. Science* **44**(1982) 103.

[11] D. Stauffer, *Introduction to Percolation Theory* (Taylor and Francis Ed., London and Philadelphia, 1985).

[12] L. Landau and E.M. Lifshitz, *Course of theoretical Physics* Vol. 5, Pergamon Press, 1986.

[13] D. Stauffer and C. Jayaprakash, *Phys. Lett.* **A64** (1977) 433.

[14] X. Campi, J. Desbois and E. Lipparini, *Phys. Lett.* **B142** (1984) 8.

[15] W. Bauer et al., *Phys. Lett.***B150** (1985) 536.

[16] X. Campi and J. Desbois, *GSI Report 85-10; XXIII Int. Winter Conf.* , Bormio, Italy. Ricerca Scientifica ed. Educazione Permanente (1985) p. 498.

[17] J. Desbois, *Nucl. Phys.* **A466** (1987) 724.

[18] C. Ngo et al., *Nucl. Phys.* **A499** (1989) 148.

[19] J. Cugnon and C. Volant, *Z. Phys.* **A334** (1989) 435.

[20] S. Leray et al., *Nucl. Phys.* **A531** (1991) 177.

[21] X.Z. Zhang et al., *Nucl. Phys.* **A461** (1987) 641 and references therein.

[22] J. Randrup and S.E. Koonin *Nucl. Phys.* **A356** (1981) 223.

[23] D.H.E. Gross, *Phys. Scr.* **T5** (1983) 213.

[24] A.S. Botvina et al., *Nucl. Phys.* **A475** (1987) 663.

[25] D.H.E. Gross, Yu-Ming Zheng and H. Massmann, *Phys. Lett.* **B200** (1988) 397.

[26] Z. Cheng and S. Redner, *Phys. Rev. Lett.* **58** (1988) 2450.

[27] E.D. McGrady and R.M. Ziff, *Phys. Rev. Lett.* **58** (1987) 892.

[28] R.M. Ziff and E.D. McGrady, *Macromolecules* **19** (1986) 2513.

[29] H.E. Stanley, *Introduction to Phase Transitions and Critical Phenomena* (Clarendon Press, Oxford, 1971).

[30] M.E. Fisher, *Physics* (N.Y.) **3** (1967) 255; *Rep. Prog. Phys* **30** (1967) 615.

[31] X. Campi, *J. Phys A: Math. Gen.* **19** (1986) L917.

[32] X. Campi, *Phys. Let.* **B208** (1988) 351.

[33] X. Campi and H. Krivine, *Z. Phys.* **A344** (1992) 81.

[34] P. Kreutz et al., *Nucl. Phys.* **A556** (1993) 672.

[35] J.J. Binney, N.J. Dowtick, A.J. Fisher and M.E.J. Newman, *The Theory of Critical Phenomena*, Clarendon Press, Oxford (1992).

[36] H.R. Jaqaman and D.H.E. Gross, *Nucl. Phys.* **A524** (1991) 321.

[37] X. Campi, *Nucl. Phys.* **A495** (1989) 259c.

[38] J.P. Bondorf et al., *Nucl. Phys.* **A443** (1985)321.

[39] Bao-An Li, A.R. DeAngelis and D.H.E. Gross, *Phys. Lett* **B303** (1993) 225.

[40] R.J. Charity et al., *Nucl. Phys.* **A483** (1988) 371.

[41] C. Barbagallo, J. Richert and P. Wagner, *Z. Phys.* **A324** (1986) 97.

[42] A.S. Hirsch et al., To be published in *Phys. Rev. Lett.*

[43] C.J. Waddington and P.S. Freier, *Phys. Rev.* **C31** (1985) 888.

[44] A.S. Hirsch et al., Contribution to the Heavy Ion Conference at LBL, October 1993.

[45] X. Campi, H. Krivine and E. Plagnol, *Phys. Rev. C*, in press.

The Continuum in Nuclei

R.J. Liotta

Royal Institute of Technology at Frescati, S–10405 Stockholm

Abstract: The Green function formalism is used to extend the standard (shell-model) treatment of bound states to processes that occur in the continuum part of nuclear spectra. The Berggren and Mittag-Leffler expansions are introduced and analysed. Applications to single-particle and particle-hole resonances are performed. Giant resonances are studied within the framework of the continuum RPA. In all cases it is found that the expansions agree well with the exact calculation. The mechanisms that induce the clustering of nucleons in nuclei are analysed and the corresponding decay processes are discussed in detail.

1 The Continuum in Nuclei

The study of processes occurring in the continuum part of a quantum spectrum has a long tradition in nuclear physics. In fact, one may say that the first application of proper quantum mechanics, that is by using the probabilistic interpretation of the wave function, was performed by Gamow in 1928 studying the penetrability of the alpha-particle through the nuclear barrier in the nuclear continuum. Perhaps surprisingly this very original problem is still a source of searching and understanding of the behaviour of nucleons in nuclei, as will be seen in the course of these lectures. There are a number of other processes influenced by the continuum that are at present being studied. Among them one may mention the formation and decay of giant resonances, the excitation of nuclei close to the drip line and the formation and subsequently decay of heavy clusters in nuclei.

We will start the study of excitations in the continuum by analysing the spectrum of a particle in a central field V, including the Coulomb interaction. The corresponding Schrödinger equation is

$$[-\frac{\hbar^2}{2\mu}\nabla^2 + V(r)]u_n(r) = \mathcal{E}_n u_n(r) \tag{1}$$

where μ is the reduced mass of the particle. One can generalize the definition of "eigenvectors" of the single particle Schrödinger equation above by requiring the boundary conditions[1,2]

$$\lim_{r \to 0} u_n(r, k_n) = 0$$

$$\lim_{r \to \infty} u_n(r, k_n) = N_n e^{ik_n r}$$

where k_n is the asymptotic momentum of the state with energy eigenvalue $\mathcal{E}_n$, i. e.

$$\mathcal{E}_n = \frac{\hbar^2}{2\mu} k_n^2$$

That is, one assumes that the particle can leave the region of the potential V, as it is in the case of the decay of the system by emitting that particle. This is a time dependent problem, but by using the Schrödinger equation (1) we have transformed it in a stationary process. The price we have to pay for this commodity is that the eigenvalues $\mathcal{E}_n$ can now be complex. We actually assume in these lectures that all energies can be complex. With

$$E_n = Re\mathcal{E}_n \tag{2a}$$

$$\Gamma_n = -Im\mathcal{E}_n/2 \tag{2b}$$

one notices that the time dependency of the wave function is given by

$$e^{-i\mathcal{E}_n t/\hbar} = e^{-iE_n t/\hbar} e^{-\frac{1}{2}\Gamma_n t/\hbar}$$

and, therefore, Γ_n would be the decay width of the system. To verify whether this is the case one can evaluate the decay width as done in elementary quantum mechanics, i. e. to calculate the transmition coefficient through the barrier (centrifugal, Coulomb, etc) induced by the potential V. One then finds that the value of Γ above is indeed the width but to order E_n/Γ_n, that is only for narrow resonances one can use the outgoing boundary solutions to evaluate the width. This is a feature that will appear often in the course of these lectures. For instance, we will see that the partial decay widths of giant resonances calculated as the residues of the S-matrix in the comples energy plane have physical meaning only if the giant resonance is narrow, that is only if its *escape* width is small.

Observable resonances are often narrow. In particular in the α-decay of heavy nuclei it is $10^{-10} \leq \Gamma_n/E_n \leq 10^{-31}$ and the width defined by eqn. (2b) virtually coincides with the physical one. Actually these resonances are so narrow that a numerical calculation of the width (2b) becomes difficult even for the most powerful computers of today because the complex energy would need to be calculated with great precision in order to distinguish the vanishing small imaginary part. But there are processes, like neutron decay of giant resonances, where the calculation of the widths by means of complex energies have the limitation of being too broad, as will be seen below.

The complex eigenvalues of the Schrödinger equation provides a valuable tool to analyse the continuum, but care has to be taken of the metric of the space spanned by the corresponding eigenvectors. Writing

$$k_n = \kappa_n - i\gamma_n$$

one sees that the wave functions diverge except for $\gamma_n > 0$, i. e. for bound states which have real and negative energies. Actually one finds that the eigenvectors corresponding to the complex eigenvalues can be classified in four classes, namely: (a) bound states, for which $\kappa_n=0$ and $\gamma_n < 0$; (b) antibound states with $\kappa_n=0$, $\gamma_n > 0$; (c) decay resonant states (Gamow resonances) with $\kappa_n > 0$, $\gamma_n > 0$ and (d) capture resonant states with $\kappa_n < 0$, $\gamma_n > 0$.

With the standard definition of scalar product only the bound states can be normalized in an infinite interval. Therefore this definition has to be generalized in order to be able to use the generalized "eigenvectors". This can only be done if one uses a bi-orthogonal basis and apply some regularization method for calculating the resulting integrals. As regularization method the complex rotation suggested in ref.[3] which became later known as exterior complex scaling[4] will be used in the application presented in these lectures.

A convenient way of introducing the continuum and the complex states discussed above is by studying the Green function, that is the function that solves the Scrödinger equation with a kernel that is a delta function. In the case of equation (1) the Green function is defined by the equation

$$[E - (-\frac{\hbar^2}{2\mu}\nabla^2 + V(r))]G(rr'; E) = \delta(r - r') \tag{3}$$

The Green function has many important properties. As will be seen below it contains the same information about the system as the Schrödinger equation itself. Originally, it was introduced as a powerfull tool in perturbation theory. Thus, assume that one adds a perturbation $v(r)$ to the interaction V in eq. (1). It is straightforward to find that the solution of the new Schrödinger equation (with the interaction $v + V$) is

$$\psi(r) = \psi^{(0)}(r) + \int dr' G(rr'; E)v(r')\psi(r') \tag{4}$$

where $\psi^{(0)}$ is the general solution of the Schrödinger equation (1) corresponding to the energy E. Notice that E is a parameter in this contex, it could be any scattering (positive) energy or, in general, any complex number. Eq. (4) is known as the Lippman-Schwinger equation.

The Green funtion can be written in terms of the bound states of the system, the so-called spectral representation, as[5]

$$g(r, r'; k) = \sum_n \frac{w_n(r, k_n)w_n(r', k_n)}{k^2 - k_n^2} + \frac{2}{\pi} \int_0^\infty dq \frac{u^{(+)}(r, q)u^{(+)}(r', q)}{k^2 + i\epsilon - q^2} \tag{5}$$

where w_n are the wave functions of the bound single-particle states and $u^{(+)}(r, q)$ are scattering states. That is, the poles of the Green function are the physical energies and the corresponding residues the wave functions. Therefore the Green function contains the same information as the Schrödinger equation itself.

The integration path in eq. (5) is along the real axis in the complex k−plane. Notice that it is the wave function times itself, and not times its complex conjugate, that appears in eq. (5). This will have the important effect of providing complex transition probabilities in processes that occur in the continuum.

For bound states the contribution from the continuum integral in (5) is usually neglected.

The Green function can also be shown to have the form

$$g(r, r'; k) = -u(r_<)v(r_>)/W \tag{6}$$

where $u(r)$ and $v(r)$ are the regular and the irregular solutions of the Schrödinger equation (i. e. the single particle Hamiltonian) and W is their Wronskian. The smaller and the larger between r and r' are denoted by $r_<$ and $r_>$, respectively.

We are now in a position to see the importance of the complex eigenvalues. For more details consult refs. [6-9]. The representation (5) of the single-particle Green function can be generalized by changing the path of integration. That is, instead of integrating over the real k-axis one uses a path over the complex plane. According to the path one chooses one can write the Green function in different forms by applying the Cauchy theorem[8]. Thus, with a path on the fourth quadrant in the complex k−plane Berggren obtained[9]

$$g(r, r'; k) = \sum_n \frac{\tilde{w}_n^*(r, \tilde{k}_n)w_n(r', k_n)}{k^2 - k_n^2} + \frac{2}{\pi} \int_{L+} dq \frac{\tilde{u}^*(r, q)u(r', q)}{k^2 + i\epsilon - q^2} \tag{7}$$

where $\tilde{w}_n(r, \tilde{k}_n)$ is the mirror state of $w_n(r, k_n)$, i. e. the solution with $\tilde{k}_n = -k_n^*$. The sum in eq. (7) runs over the states of type (a) and (c) in the classification done above, that is over bound states plus the decaying resonant states which lie between the real k-axis and the integration path. The advantage of introducing this integration path is that from a physical point of view the resonant states represent the most important process occurring in the continuum. The numerical evaluation of the integral requires the discretization of the wave number k along the integration contour and the solution of the radial equation for each of these complex k-values. The idea of the approximate procedure is to neglect the integral and thus to avoid that formidable task. In the applications we will perform a similar task by computing the exact Green function (6) to compare with the approximate expansions.

Another representation of the single-particle Green function, which can be obtained by choosing another path of integration, is the so-called Mittag-Leffler expansion. If the potential is of a finite range, i. e. it vanishes beyond a certain distance (as it is always the case in practical applications), one can write

$$g(r, r'; k) = \sum_n \frac{\tilde{w}_n^*(r, \tilde{k}_n)w_n(r', k_n)}{2k_n(k - k_n)} \tag{8}$$

where the sum runs now over all classes of poles $(a) - (d)$ mentioned above.

The partial wave expansion of the single-particle Green function is

$$< r \, | \frac{1}{H_0 - \epsilon_p} | r \, ' > = g(r, r \, '; \epsilon_p)$$

$$= \frac{2\mu}{\hbar^2 rr'} \sum_{ljm} \mathbf{Y}_{l,j}^m(\hat{r}) \mathbf{Y}_{l,j}^{m*}(\hat{r}') g_{lj}(r, r \, '; k)$$

The $\mathbf{Y}_{l,j}^m$ are the vector spherical harmonics and μ is the reduced mass of the particle.

Once the Green function is known one can evaluate the response function corresponding to an external field f, i. e.

$$R(E) = \int dr dr \, ' f(r \,)^* G(rr \, '; E) f(r \, ') \tag{9}$$

The response function is easier to analyse than the Green function because it depends only on the energy. But its importance lies in the fact that the cross section corresponding to inelastic processes induced by the field f is proportional to the reponse function. Even the energy weighted sum rule and the strength corresponding to giant resonances are related to the reponse function, as will be seen below.

By using for the Green function the exact expression (6) and the corresponding Berggren (eq. (7) without the integral) and Mittag-Leffler (eq. (8)) expansions one can probe the validity of the approximations. A detail analysis of this was performed in ref. [6], but here we will only present a case for illustration. As seen in figure 1, the exact and the approximated values of the response function agree well with each other.

An important question is whether one can use the Berggren and Mittag-Leffler set of states as representations to expand function in the Hilbert space. This was discussed in detail in ref.[10], where the unity (projection) operator is just assumed to be

$$1 = \eta \sum_n |u_n >< \tilde{u}_n|$$

where u are the states in the four categories mentioned above and $\eta=1$ in Berggren and $1/2$ in Mittag-Leffler. Notice that in the metric of the complex functions it is $< \tilde{u}_n | f >=< f | u_n >$. If the unity as written above is indeed valid then the function

$$U(N) = \frac{\eta \sum_{n=1}^{N} < f | n >^2}{< f | f >} \tag{10}$$

where f is a regular function, should approach unity as the number of states N included in the expansion is increased. This indeed happens, as seen in figure 2.

It thus seems that one can explore the possibility of using the Berggren and Mittag-Leffler expansions as representations in the continuum. Moreover, studies performed in particle-hole excitations also indicate the validity of those expansions. The particle-hole excitations can best be analysed by using the continuum RPA (CRPA). The corresponding Green function is[11]

$$G_{ph}(r, r \, '; E) = G_{ph}^{(0)}(r, r \, '; E) + \int dr_1 dr_2 G_{ph}^{(0)}(r, r_1; E) V_{ph}(r_1 r_2) G_{ph}(r_2, r \, '; E)$$

V_{ph} is the residual (particle-hole) interaction. $G_{ph}^{(0)}$ is the bare particle-hole Green function corresponding to the Hamiltonian H_0 with occupied states ϕ_h and corresponding eigenenergies ϵ_h, i. e.

$$G_{ph}^{(0)}(r, r'; E) = \sum_h \phi_h^*(r')(g(r, r'; \epsilon_h + E) - g(r, r'; \epsilon_h - E))\phi_h(r') \quad (11)$$

where g is the single-particle Green-function discussed above.

The use a separable interaction greatly simplifies the response function. Thus with

$$V_{ph}^\lambda(r_1 r_2) = -\kappa_\lambda Q_\lambda(r_1).Q_\lambda(r_2)$$

where

$$Q_{\lambda\mu}(r) = f_\lambda(r)Y_{\lambda\mu}(\hat{r})$$

The radial dependence of the multipole operator Q coincides with that of the external field f. It is then easy to obtain the expression of the CRPA response function as

$$R_\lambda(E) = \frac{R_\lambda^{(0)}(E)}{1 + \kappa R_\lambda^{(0)}(E)} \quad (12)$$

where $R^{(0)}$ is the bare response function , i. e. from eq. (11). The exact form of the response function is obtained by using for g in (11) the expression (6). The Berggren form is obtained by using the sum in (7) which is the same approximation that leads to the RRPA[12]. The Mittag-Leffler form is obtained by using g as in eq. (8).

The poles ω_n of the response function are then given by the complex roots of the equation $\kappa = -1/R_0(\omega_n)$. The CRPA energies can be complex quantities. The imaginary part of these energies are, as for the single-particle case, related to the total escape widths as

$$\Gamma_n = -2Im(\omega_n) \quad (13)$$

The partial decay widths are usually defined[13] as the residues of the S-matrix, which in our case is

$$S_{cc'} = \delta_{cc'}e^{2i\delta_c} - 2i\pi < \chi_E^{c'(-)}|V_{ph}|\chi_E^{c(+)}> -2i\pi < \chi_E^{c'(-)}|V_{ph}G_{ph}(E)V_{ph}|\chi_E^{c(+)}> \quad (14)$$

where G_{ph} is the CRPA Green function and the unperturbed eigenfunctions of H_0 has the form

$$\chi_E^{c(+)}(r_p, r_h) = e^{i\delta_c}u_\epsilon(r_p)\phi(r_h)$$

where $\phi(r_h)$ is the wave function of the daughter nucleus (one of the hole states), ϵ is the kinetic energy of the decaying particle, i. e. $E = \epsilon + \epsilon_h$, and $u_\epsilon(r_p)$ is the corresponding scattering wave function. We assume that the resonances that we are interested in are not degenerate (no two complex eigenvalues are equal) therefore the S-matrix can be parametrized in a many-level Breit-Wigner form[13]

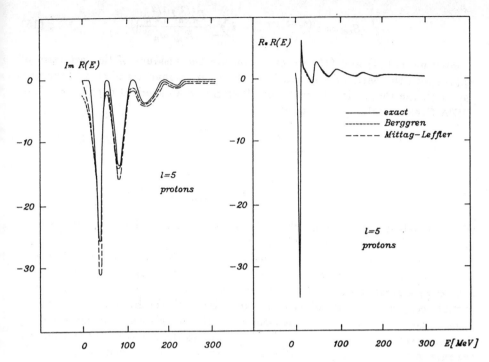

Fig. 1. Imaginary and real part of the single-particle response function corresponding to the partial wave l=5 for protons in a square well potential. This figure is from ref. [6].

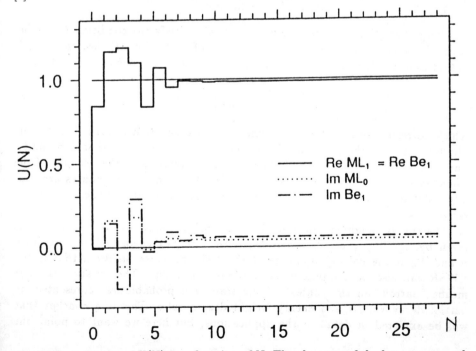

Fig. 2. Convergence of U(N) as a function of N. The elements of the base correspond to $\ell = 9, \jmath = 19/2$. For more details see ref. [10], from where this figure was taken.

$$S_{cc'} = e^{i(\delta_c + \delta'_c)}[\delta_{cc'} - i\sum_n \frac{\Gamma_{n,ch}^{1/2}\Gamma_{n,c'h'}^{1/2}}{E - \omega_n}] \tag{15}$$

Comparing eq.(14) and (15) we can associate the residues of the last term of eq.(14) with the partial decay widths of the giant resonance and that of the first part with the shape elastic scattering. Performing an spectral expansion of the CRPA Green function one obtains

$$\Gamma_{n,ch} = \frac{\mu}{\hbar^2 k_n}\frac{\kappa_\lambda^2}{2j_h + 1}R_n^2 \times \sum_{lj} \frac{1 + (-1)^{l+l_h+\lambda}}{2}(2j+1) < j1/2\lambda 0|j_h 1/2 >^2$$

$$\times [\int dr u_{\epsilon lj}(r)f_\lambda(r)\phi_h(r)]^2 \tag{16}$$

where ch labels the outgoing channel (emitted particle plus residual nucleus) and R_n is the residues of the response function at the pole ω_n.

Generally the partial decay widths are complex numbers and therefore any interpretation of physical properties in terms of those quantities has to be performed with great caution. The interpretation of the poles of the S-matrix as the partial decay widths was induced by comparing with the Breit-Wigner formula in processes where a resonance is isolated[14,15]. In this case one can show[13] that the residues of the S-matrix is real and due to the unitarity of the S-matrix the sum of the partial decay widths is the total width Γ_n defined in eq. (13). To analyse this we define (for details see ref. [7])

$$U_n = -\frac{\sum_{ch}\Gamma_{n,ch}}{2Im(\omega_n)} \tag{17}$$

and conjecture that physical resonances should satisfy the condition $U_n=1$ because this condition implies the conservation of probability. If the resonance is indeed isolated then the partial decay widths are real and one can as well define

$$U_{n,abs} = -\frac{\sum_{ch}|\Gamma_{n,ch}|}{2Im(\omega_n)} \tag{18}$$

which corresponds to the usual definition of partial decay width, i. e. to the absolute value of the residues of the S-matrix. U_n and $U_{n,abs}$ should be unity, but if these conditions are not fulfilled the interpretation of the residues of the S-matrix as partial decay widths is doubtful. Since narrow resonances are usually isolated one may assume that the conditions above are satisfied by narrow resonances.

The energy weighted sum rule (EWSR) corresponding to the resonance n is given by $EWSR_n = \omega_n R_n$, where ω_n is the complex energy of the resonance n and R_n is the corresponding residues of the response function, that is the EWSR can also be complex is the resonance is broad. The question is what means "narrow" in the context of the transition probabilities represented by partial decay widths and energy weighted sum rules. This is a concept that will be explored in detail in the applications, but here we want to point out that the total escape with of giant resonances is typically a few 100 keV. For comparison, one may notice that in the decay of alpha particles from heavy nuclei the values of $\Gamma_n/Re(\omega_n)$ range between $10^{-10} - 10^{-30}$, i. e. the decaying state may well be described by including only one term in the expression (15), as mentioned in the previous lecture. One may expect that in the more dubious case of the decay of giant resonances the analysis of the experimental data is not straightforward. Thus, when the giant resonance is fractioned in nearby lying pieces one cannot apply eq. (16) directly to analyse the experimental data since it is not the decay width that one measures but rather the branching ratio between the decay and the excitation cross section averaged within the energy range of the giant resonance[16]. To probe whether a resonance is isolated one can also calculate the branching ratio and see its dependence on the energy interval. This formalism was recently applied in the lead region and the exact result was compared with the approximated Berggren and Mittag-Leffler expansions[7,16]. It was thus found that most of the giant resonances are not isolated and therefore the calculation of the partial decay width is not meaningful. The dipole giant resonance is one of the few meaningful resonances, as seen in table 1.

Table 1. Dipole resonances that exhaust more than 5% of the isoscalar (SEWSR) or isovector (VEWSR) energy weighted sum rule. The numbers U are as in eqs. (17) and (18). This table was taken from ref. [16].

Energy	SEWSR	VEWSR	U_{abs}	U
11.194-i0.005	5.5-i1.2	3.4+i1.3	0.93	0.93+i0.03
13.651-i0.107	6.9+i0.8	23.2+i5.6	0.96	0.96-i0.05
22.686-i0.002	5.7-i0.2	1.1+i0.5	1.03	1.02+i0.09
23.234-i0.030	6.8-i0.9	0.9+i0.2	0.27	0.21-i0.10
23.992-i0.008	5.9+i1.1	0.3+i0.1	1.03	0.92+i0.34
26.718-i0.057	6.6+i3.9	6.8-i0.5	1.10	0.99-i0.13
27.406-i0.634	9.7+i1.9	7.5-i3.1	0.16	0.14-i0.05
27.638-i0.082	7.3+i3.8	6.7-i1.0	0.96	0.78-i0.15
28.317-i0.446	8.6-i2.1	4.5-i5.4	0.21	0.14-i0.12

One sees that the giant dipole resonance, lying at 13.651 MeV, has values of both U_{abs} and U nearly unity. The corresponding values of the partial decay widths are given in table 2.

Table 2. Neutron partial decay widths in MeV corresponding to the decay of the dipole giant resonance lying at 13.651 MeV in table 2. Only widths larger than 10 keV (in absolute value) are given. The proton contributions are below this value. This table was taken from ref. [16].

Hole state	$\Gamma_{n=V,ch}$
$h_{9/2}$	0.013 - i 0.00
$f_{7/2}$	0.035 - i 0.00
$i_{13/2}$	0.061 - i 0.00
$p_{3/2}$	0.027 - i 0.00
$f_{5/2}$	0.050 - i 0.00
$p_{1/2}$	0.019 - i 0.00

The partial decay widths are practically real and one even finds that the branching ratios corresponding to this giant resonace show that it is isolated.

The quality of the approximated results can be seen in table 3.

Table 3. Exact (i. e. CRPA), Mittag-Leffler and Berggren energies (in MeV) corresponding to the main pieces of the dipole (isovector), quadrupole and octupole (isoscalars) giant resonances in ^{208}Pb. This table was taken from ref. [16].

λ	CRPA	Mittag-Leffler	Berggren
1	13.651 -i0.107	13.630 -i0.104	13.615 -i0.204
2	10.361 -i0.073	10.368 -i0.064	10.553 -i0.036
3	18.484 -i0.083	18.483 -i0.074	18.547 -i0.024

One can say that the approximated results are excellent. The exact results obtained in ref. [16] are even in good agreement with those obtained in ref. [12], where the Berggren approximation was used to derive the resonant RPA. This formalism was also applied in ref. [17] to analyse the temperature dependence of energies and escape widths of giant resonances.

The conclusion of this lecture on giant resonances is that the ratio U between the sum of the partial escape widths and the total width is unity only in cases where the imaginary part of the partial decay widths are (in absolute value) small. In this case even the transition strengths are practically real. This is an important property that can be used to select physically meaningful resonances.

In our CRPA calculations the continuum is included exactly and, therefore, the influence of all possible particle-hole excitations are taken into account. The drawback of this method is that it is very time consuming. To avoid this we used two different pole expansions of the Green function, namely the Mittag-Leffler and Berggren expansions. We found that the Mittag-Leffler expansion reproduces very well the exact results while in the Berggren case the agreement is within 10-15%. These methods are faster than the exact CRPA by several orders of magnitude. This feature, as well as the reliability of the results provided by the expansions, make them a powerful tool to evaluate processes that take part in the continuum part of the spectrum, in particular partial decay widths.

The other subject where the continuum plays an important role and which will be explored in these lectures is alpha decay. The process of clustering of nucleons and subsequent decay of the already formed cluster has a long and rich history in nuclear physics. To go through this development in detail is well beyond the scope of the one-hour lecture that I will give today. Instead I plan to present my own experience in the field as I went through different etages in the study first of alpha decay and then the continuum in nuclei.

This story starts in 1977 when experimentalists at my institution in Stockholm asked me to calculate the alpha decay of ^{212}At. I performed this calculation using the nuclear field theory, which was at the time a popular theory. For the alpha decay process I used the theory of alpha-decay developed in refs. [14,15], which assumes that the decay width is the residues of the S-matrix. As we have seen in the previous lectures, in this theory it is assumed that the decay process is stationary and, as a result, both the energies, i. e. the poles of the S-matrix, and the decay width are in general complex quantities. In the case of alpha-decay the fact that a quantity proportional to a probability (the decay width) is complex is not a practical problem because usually the width is very small and the residues of the S-matrix is practicaly real. In fact, in the one-level case (when one can neglect any overlap among resonances) one can show that the residues of the S-matrix is strictly real. But in the neutron decay of giant resonances this is a serious drawback which usually is ignored, as discussed above.

In the Wigner-Thomas theory the alpha decay width is given by

$$\Gamma_L(R) = 2P_L(R)\frac{\hbar^2 R}{2M}F_L^2(R),$$

where M is the reduced mass, $P_L(R)$ is the Coulomb penetration factor with angular momentum L and $F_L(R)$ is the formation amplitude of the α-particle at the point R.

In my calculations I used for R the sum of the radii of the α-particle and the daughter nucleus while to describe the formation amplitude I used only one shell-model configuration, as it was standard at the time, although mostly to evaluate relative decay widths[18]. I was dismayed to find that my calculated absolute decay width was wrong by several orders of magnitude. I first thought that it was the nuclear field theory which was deficient. Although indeed the NFT was eventually shown to be plagued by divergencies, in the case of alpha-decay the theory was actually equivalent to the shell-model[19].

The weakest point of this calculation was not its total disagreement with experiment but that the penetration P was so strongly dependent on R that one could find a rather acceptable value of the distance for which that agreement improved drastically. It was then obvious that something fundamental was missing since one should expect that Γ_L is independent of R for distances outside the daughter nucleus, where the alpha particle is already formed. The calculated formation amplitude was vanishing small just outside the nuclear surface, indicating that the only configuration included in my calculation was not enough to describe the alpha decay process in the important region where the alpha cluster penetrates the Coulomb barrier. That is, high lying configurations, which are relevant at large distances, should be included in the calculations. This was rather easy to do within the NFT as well as with the multi-step shell-model, as later calculations showed[20]. We found that indeed the effect of high lying configurations was both to make the alpha decay width independent of R in a region close to the nuclear surface and, at the same time, to increase the value of Γ by many orders of magnitude, as seen in figure 3. Notice that the total width is the *product* of the neutron and proton contributions in figure 3 and therefore with 50 configurations the width is about 5 orders of magnitude larger than with 1 configuration.

We also found that the physical reason of this increase is that, through the high lying configurations, the nucleons that eventually become the alpha particle are clustered. Yet, the calculated alpha decay width was more than one order of magnitude too small. We then thought that the neutron-proton interaction, which had been excluded in our and other previous calculations[21,22], was important to cluster nucleons of different isospin. This we introduced through a "giant pairing resonance", that is a high lying collective pairing state scalar in isospin[23]. But within reasonable limits for the mixing of this giant resonance in the mother nucleus wave function the alpha decay width could not be increased more than a few per cent. A proper treatment of the neutron-proton interaction would have required a full shell model calculation, as was later performed within a cluster shell-model configuration mixing model[24]. But the original reason to introduce the approximate treatment of the continuum was to study the alpha-particle formation amplitudes. Although most of the work done so far with the Berggren and Mittag-Leffler expansions is related to giant resonances, the Berggren "representation" was also used to describe the wave function of ^{212}Po and the corresponding alpha formation amplitude[25]. It was found that the $^{212}Po \rightarrow \alpha +^{208}Pb$ decay width is independent of the distance R up to distances well beyond the nuclear surface. This result was just what we expected to obtain by introducing the influence of the continuum in the formalism.

The other attemp to solve the problem of the decay of the alpha particle was to describe clustering microscopically but within a basis which is a combination of shell- and cluster-model set of elements[24]. As mentioned before, within this basis the neutron-proton excitation could be taken into account properly and a good account of the experimental data could be given. This method present the inconvenience that the calculation is rather involved. To circumvent this

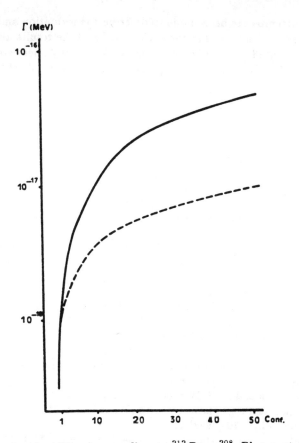

Fig. 3. Absolute decay width corresponding to $^{212}Po \rightarrow ^{208}Pb + \alpha$ at $R = 8fm$ as a function of the number of configurations included in the calculation. Solid (dashed) lines represents neutron (proton) contributions. Taken from ref. [21].

problem, and to study processes in the continuum in general, a new shell-model representation was recently introduced[26]. It consists of a gaussian set of functions for a given value of l and j. In this way one can perform the integrals in the formation amplitude analitycally. But perhaps the main advantage of this representation is that it describes as well all regions of space, even well beyond the nuclear surface.

Also the description of alpha decay from deformed nuclei shows the importance of the continuum part of the spectrum to induce clustering[27-29]. Surprisingly enough (and in contrast to the spherical case) good agreement with experimental data is obtained by including a large shell-model representation. The importance of alpha decay from deformed nuclei is that one in this way expects to obtain information about nuclear shapes[29].

Finally, attemps are been made to describe the formation and decay of heavy clusters microscopically within the shell model and the results are promising[30].

I have presented in these lectures the work of many scientists, as seen in the list of references. I would like to express my gratitude to all of them.

References

1   Y. B. Zel'dovich, JETP (Sov. Phys.)12 (1961) 542.

2   W. J. Romo, Nucl. Phys. A419 (1984) 333.

3   B. Gyarmati and T. Vertse, Nucl. Phys. A160 (1971) 573.

4   B. Simon, Phys. Lett. 73A (1979) 211.

5   R. G. Newton, Scattering theory of waves and particles (McGraw-Hill, NY, 1966) p368.

6   T. Vertse, P. Curutchet and R. J. Liotta, Phys. Rev. C42 (1990) 2605.

7   T. Vertse, R. J. Liotta and E. Maglione, Nucl Phys A, in press.

8   T. Berggren and P. Lind, Phys. Rev. C47 (1993) 768.

9   T. Berggren, Nucl. Phys. A109 (1968) 265.

10   P. Lind, R. J. Liotta, E. Maglione and T. Vertse, Z. Phys. A347 (1994) 231.

11   S. Shlomo and G. Bertsch, Nucl. Phys. A243 (1975) 507; K. F. Liu and N. Van Giai, Phys. Lett. B65 (1976) 23.

12   P. Curutchet, T. Vertse and R. J. Liotta, Phys. Rev. C39 (1989) 1020.

13   C. Mahaux and H. A. Weidenmüller, Shell model approach to nuclear reactions (North-Holland, Amsterdam, 1969).

14   T. Teichmann and E. P. Wigner, Phys. Rev. 87 (1952) 123.

15   R. G. Thomas, Prog. Theor. Phys. 12 (1954) 253; A. M. Lane and R. G. Thomas, Rev. Mod. Phys. 30 (1958) 257.

16   E. Maglione, R. J. Liotta and T. Vertse, Phys. Lett. 298B (1993) 1.

17   G. G. Dussel, H. Sofia, R. J. Liotta and T. Vertse, Phys. Rev. C46 (1992) 558.

18   M. A. Radii, A. A. Shibab-Eldin and J. O. Rasmussen, Phys. Rev. C15 (1977) 1917 .

19   F. A. Janouch and R. J. Liotta, Phys. Lett. 82B (1979) 329.

20   G. Dodig-Crnkovic, F. A. Janouch and R. J. Liotta, Nucl. Phys. A501 (1989) 533, and references therein.

21   J. A. Janouch and R.J. Liotta, Phys. Rev. C25 (1982) 2123.

22   I.Tonozuka and A.Arima, Nucl. Phys. A323 (1979) 45.

23   M. Herzog, O. Civitarese, L. Ferreira, R. J. Liotta, T. Vertse and L. J. Sibanda, Nucl. Phys. A448 (1986) 441.

24   K. Varga, R. G. Lovas and R. J. Liotta, Phys. Rev. Lett. 69 (1992) 37; Nucl. Phys. A550 (1992) 421.

25   S. Lenzi, O. Dragún, E. E. Maqueda, R. J. Liotta and T. Vertse, Phys. Rev. 48C (1993) 1463.

26   K. Varga and R. J. Liotta, Phys. Rev. C, in press.

27   A. Insolia, P. Curutchet, R. J. Liotta and D. S. Delion, Phys. Rev. C44 (1991) 545.

28   D.S.Delion, A.Insolia and R.J.Liotta, Phys. Rev. C46 (1992) 884.

29   D. S. Delion, A. Insolia and R. J. Liotta, Phys. Rev. C46 (1992) 1346.

30   D. S. Delion, A. Insolia and R. J. Liotta, J. Phys. G19 (1993) L189 and in press.

Spherical Shell Model, a Renewed View[1]

Alfredo Poves

Departamento de Física Teórica
Universidad Autónoma de Madrid, 28049 Madrid, (Spain).

Abstract In these lectures the foundations of the spherical shell model description of nuclei are presented with particular emphasis in the relationship between spectroscopic and global properties. Non standard applications of the spherical shell model - to nuclei far from stability and to rotational motion - are explicitly discussed.

1. Basic Concepts

The aim of a nuclear theory is to describe the details of the nuclear phenomenology by means of approximate solutions to the A-nucleon problem. In what follows we shall stress the need of keeping contact with the basic nucleon-nucleon interaction while dealing with the sequence of approximations in such a way that the loss of physical information be minimal.

1.1 The general A nucleon problem

The most conspicuous feature of the A nucleon problem - besides the obvious one i.e. to be a many body problem that cannot be treated statistically - is the fact that the elementary constituents of the system - the nucleons - are composite objects. This lead to a very complicated nucleon-nucleon interaction, derived from the simpler(?) interaction between quarks and gluons. Among these complications, one leads; the very large repulsion at short distances, partly induced by the Pauli principle at the quark level, or in a nucleon representation

---

[1] The content of these lectures is based in the results obtained in collaboration with E. CAURIER and A. ZUKER, Groupe de Physique Théorique, Centre de Recherches Nucleaires, 67037 Strasbourg, (France), and J. RETAMOSA and G. MARTINEZ-PINEDO, Departamento de Física Teórica, Universidad Autónoma de Madrid, 28049 Madrid, (Spain).

due to the fact that nucleons have a size (0.8 fm) and cannot interpenetrate. Only the insistence of the experimental data in demanding it could have justified the description of the nucleus as a system of independent particles giving rise to the nuclear shell model. The work of Brueckner [1] gave theoretical support to the empirical shell model of Goeppert-Mayer. In brief, it can be shown that the strongly correlated nuclear wave function associated to the bare nucleon-nucleon interaction can be substituted by an independent particle wave function provided the bare interaction is changed by a regularized - effective - interaction. The regularized interaction make it possible to attack the nuclear many body problem with techniques that resemble to those used in the study of diluted, weakly interacting systems. Nevertheless the nuclear case is at the borderline of the applicability of these approaches, which brings in enhanced difficulties.

In addition to that, nuclear matter results have shown that no two body nucleon-nucleon interaction can reproduce the saturation point of the nuclear system. This puts an *ab initio* limitation to the shell model description of nuclei in that some phenomenological tampering of the interaction in necessary from the very beginning. This is the case in the broadly used mean field interactions - Skyrme, Gogny - that, even if qualitatively justified in their structure - density dependence- by Brueckner Hartree Fock calculations, contain parameters that must be fitted to the experimental data along the nuclear chart. In any case the parts of an effective interaction responsible for saturation must be fitted.

Because of the regularization procedure, the shell model particles are actually quasiparticles. The work of Pandharipande *et al* [2] has demonstrated that the short range correlations erode the steep Fermi surface. Consequently the bare shell model particles have an overlap of - at best - 0.8 with the real correlated particles. This fact has been experimentally born out by the electron scattering experiments made in ^{206}Pb [3]. These experiments make it possible to extract the charge density distribution of the $3s_{1/2}$ orbit. The shape of the orbit is exactly what the shell model predicts but the charge is depleted. Similar results come out from (d,p) and (e,e'p) experiments [4]

Let's suppose that the nuclear interaction is regularized. Then we can choose a single particle basis $a_i^+|0>$, using some kind of Hartree Fock method, and build the A-particles wavefunctions $a_{i_1}^+|0> ....a_{i_A}^+|0> = |\phi_\alpha>$. The physical states will be expressed as $|\Phi> = \sum_\alpha C_\alpha|\phi_\alpha>$, with $|\Phi>$ satisfying the Schroedinger equation $H|\Phi> = E|\Phi>$. The problem is now cast in matrix form and reduces to the diagonalization of the matrix $<\phi_\alpha|H|\phi_{\alpha'}>$. If the basis $|\phi_\alpha>$ were complete we should get the exact solution to the problem. Obviously, this is not the case, because the complete basis is infinite dimensional. The goal of the spherical shell model is to devise the truncations of the infinite basis that contain the physical degrees of freedom of each particular problem and at the same time preserve the basic symmetries of the Hamiltonian (particle number, total angular momentum, isospin, parity).

1.2 Saturation and the monopole hamiltonian

In second quantification formalism, the effective nuclear interaction can be written as

$$H = \sum_i t_i + \sum_{ijkl,\Gamma} W^{\Gamma}_{ijkl}(a_i^+ a_j^+)^{\Gamma}(a_k a_l)^{\Gamma}$$

with $\Gamma \equiv (J,T)$ and $W^{\Gamma}_{ijkl} \equiv\; < ij(JT)|V_{NN}|kl(JT) >$. If we make a Racah transformation we obtain the hamiltonian in multipole multipole form:

$$H = \sum_i \rho_i + \sum_{ijkl,\lambda} \omega^{\lambda}_{ijkl}(a_i^+ a_j)^{\lambda}(a_k^+ a_l)^{\lambda}$$

with $\omega^{\lambda}_{ijkl} = \sum_{\Gamma} W^{\Gamma}_{ijkl} C6J(i,j,k,l,\Gamma.\lambda)$.

The λ=00,01 terms are proportional to the number and isospin operators $\hat{n}_i$ and $\hat{t}_i$. It has been rigorously proven in [5] that the spherical Hartree Fock energy of any hamiltonian is given by the expectation value of the monopole hamiltonian in the Hartree Fock basis:

$$< H >_{HF} = \sum_i < \rho_i > + \sum_i \frac{1}{2} n_i(n_i - 1)\bar{V}_{ii} + \sum_{ij} n_i n_j \bar{V}_{ij}$$

where $\bar{V}_{ij}$ are the averages of the corresponding diagonal two body matrix elements. Hence, the monopole hamiltonian provides the basic ingredient of a mass formulae. Consequently, the bad saturation properties of the effective interactions obtained from the NN potentials imply that the monopole part of any realistic interaction in incorrect and has to be replaced by a fitted one.

The monopole Hamiltonian is also responsible for most of the unperturbed energy of a given shell model configuration. In this sense, good shell model spectroscopy relies in a good understanding of the global properties of the nucleus (mass, radius etc). The remaining multipoles of the realistic interactions are essentially right.

1.3 Model spaces

Dimensionality considerations force truncations in the full Hilbert space. How to truncate demands a good knowledge of the nuclear phenomenology, the structure of the nuclear force and the renormalization mechanisms. Perturbation theory is often of great help. Classical model spaces in light nuclei are spanned by the subshells of a major oscillator shell. In a given shell the total degeneracy of the space for a given number n_{ν} of neutrons and n_{π} of protons is given by:

$$D = \begin{pmatrix} d \\ n_{\nu} \end{pmatrix} \begin{pmatrix} d \\ n_{\pi} \end{pmatrix}$$

where d is the shell degeneracy, d= 6, 12 and 20 in the p, sd and pf shells respectively. The maximum value of D in the pf shell, $D_{max} = 3.4 \cdot 10^{10}$, gives a clear idea of the extremely rapid increase in the size of the many particles basis. This makes full calculations in higher shells with the diagonalization approach rather unpractical. Nevertheless, more restricted model spaces have been used with fair success [6] in heavier nuclei.

1.4 Effective interactions

There have been two main approaches to the problem of fixing an effective interaction, We shall call them phenomenological and realistic. The phenomenological one was introduced by Cohen and Kurath [7] in their study of the p-shell and consists in a direct fit of the two body matrix elements and the single particle energies to a set of well known energy levels. This approach became later very popular due to the success of the work of Brown and Wildenthal in the sd-shell [8]. In the realistic apprach, the effective interaction is taken to be a G-matrix obtained from a nucleon-nucleon potential. Core polarization corrections (bubbles) are usually added too. Well known examples are the interactions calculated by Kuo and Brown [9]. These realistic interactions have been used by many groups. The status of their use in conexion with the monopole problem can be found in [6].

1.5 Solving the secular problem

There are two main steps in the solution of the secular problem. The first one is the definition of the basis and the calculation of the many particle matrix elements. The choice is between a coupled and an uncoupled or m-scheme. The advantage of working in coupled scheme is that the full basis is splitted in JT-blocks, hence the dimensions of the matrices get reduced. The price to pay is a greater complexity in the calculation of the matrix elements that involves Racah algebra and fractional parentage coefficients. This approach has been implemented by French and collaborators in the Oak-Rigde multishell code [10]. The m-scheme approach makes the calculation of the matrix elements simpler, but the dimension of the basis is larger.

The second step consists in the obtention of the eigenvalues and eigenvectors of the secular problem. In correspondence with the two approaches quoted above, the matrix can be calculated completely and then diagonalized directly (as it is done in the Oak-Rigde code), or the secular problem can be written in tridiagonal form usig the Lanczos algorithm as it is done in the Glasgow code [11]. Modern multishell codes incorporate features belonging to the two basic approaches described above.

Most of the results that will be presented in these lectures have been obtained using the code ANTOINE [12]. It is a m-scheme code that uses the Lanczos method with a few modifications with respect to the Glasgow code. The most important is that the starting vectors ('pivots') have good angular momentum and isospin. They are obtained through diagonalization of the operators J^2 and T^2. This choice does not reduce the basis dimensions but improves very much the convergence of the Lanczos method. Let's give a brief description of the method. First we take a state in the space $|0>$, then we construct

$$|1> = [H - < 0|H|0 >]|0 >$$

orthogonal to $|0>$. The normalized state

$$|\hat{1}> = |1 > / < 1|1 >^{1/2}$$

is used to compute $< \hat{1}|H|\hat{1} >$ and $< \hat{1}|H|0 >$. At this step we have built a 2x2 matrix. It can be diagonalized and yields a first approximation to the energy. The method proceeds defining the next Lanczos vector by the same kind of construction:

$$H|\hat{1}> = H_{01}|0 > + H_{11}|\hat{1} > + H_{21}|\hat{2} >$$

and so on. After N iterations we have a N-dimensional tridiagonal matrix. The procedure is repeated until convergence.

1.6 A new approach; Shell Model Montecarlo

This method is different from the beginning, because the shell model matrix is never constructed, not even in the Lanczos sense. This makes it possible to circunvent dimensionality problems. The approach relies in the use of a finite temperature treatement that, at the end, is extrapolated to T=0. The basic element is the partition function $e^{\beta H}$. This function is written as an integral by using the Hubbard-Stratonovitch decomposition and auxiliary fields introduced in order to convert the two body propagator in a set of one body ones. The final integrals are computed by Montecarlo sampling. In its present status the method has three main shortcomings:

1.– The T=0 extrapolation is very critical, because the convergence of the method deteriorates when T goes to zero. Up to now only results for T=0.5 MeV are presented as the approximate solution for T=0 physics. This excludes the study of most odd nuclei, and all the odd-odds. The rotational nuclei are also out of reach. In the even-even nuclei the results for the ground state are contaminated by components of the first excited states.

2.– The results are affected by systematic errors linked to the auxiliary field method that are not well understood.

3.– The sign problem. Montecarlo sampling collapses if the sign of the function to integrate oscillates. This is the case for realistic nuclear interactions. Consequently one has two possibilities, either to use schematic interactions free from sign problems or to find an approximate way of getting round of the problem. The proposal made by the authors of this method [13] is the following; first the nuclear Hamiltonian is split in two parts, H_g (good) that do not have sign problem, and H_b (bad) that do have, and it is written as: $H = H_g + g H_b$. For g=1 one recovers the real hamiltonian. To avoid the sign problem, they compute for several $g \leq 0$ values and make a blind extrapolation to g=1. In our opinion this step of the method brings in errors whose evaluation is extremely uncertain.

2. pf-Shell Results

We have applid the methods outlined in the first part of these lectures to the study of nuclei of the pf-shell. The valence space is spanned by the four orbits $1f_{7/2}, 2p_{3/2}, 2p_{1/2}$ and $1f_{5/2}$. The single particle energies used in the calculation are extracted from the ^{41}Ca experimental level scheme, $\epsilon(1f_{7/2})=0.0$ MeV, $\epsilon(2p_{3/2})=2.0$ MeV, $\epsilon(2p_{1/2})=4.0$ MeV and $\epsilon(1f_{5/2})=6.5$ MeV. In this model space the $1f_{7/2}$ orbit plays a prominent role. As a matter of fact, the first calculations of nuclei between ^{40}Ca and ^{56}Ni were made taking this single orbit as the full valence space. The results obtained for the energy levels were in reasonable agreement with the experimental data, however, the transition probabilities and electromagnetic moments were mostly wrong. The effective interaction used is based in the G-matrix computed by Kuo and Brown [9]. As explained before the monopole part of a G-matrix has to be corrected in order to give good spectroscopic results. In the pf shell the modifications are minimal and aimed to garantee the right single particle evolution from to ^{41}Ca to $^{55,56,57}Ni$ and read:

$$\delta < f7f7|H|f7f7 > (T = 0) = -350\,keV$$

$$\delta < f7f7|H|f7f7 > (T = 1) = -110\,keV$$

$$\delta < f7(p3, p1, f5)|H|f7(p3, p1, f5) > (T = 0) = -300\,keV$$

$$\delta < f7(p3, p1, f5)|H|f7(p3, p1, f5) > (T = 1) = +300\,keV$$

The importance of these changes can be tested in ^{56}Ni where using the bare interaction the closed shell configuration is excited 5 MeV above the 4p-4h configurations. After the modifications, the closed shell recovers its status and the 4p-4h states start at around 5 MeV excitation energy.

The pf shell description with the modified Kuo-Brown interaction (called KB3) has been applied successfully to many nuclei. We shall refer to published work for details and limit here to give an abstract of the different calculations performed up to now.

- Full pf shell model study of A=48 nuclei [14]. This paper contains a detailed analysis of ^{48}Ca , ^{48}Sc, ^{48}Ti, ^{48}Ti, ^{48}V, ^{48}Cr, ^{48}Mn and ^{48}Fe. Level schemes, electromagnetic moments and transitions fully agree with the experimental results. Gamow Teller properties, including lifetimes, are also well described using the standard quenching factor 0.77. The total Gamow-Teller strength is reduced by the correlations by factors 2-3. The calculations are able to explain the collective features - including backbending - experimentally found in ^{48}Cr and suggest a microscopic description of the onset of rotational motion that will be discussed in the next chapters.

- Spin quenching and orbital enhancement in the Ti isotopes [15]. The spin, orbital and M1 strength functions are computed in ^{46}Ti and ^{48}Ti and compared with (p,p') and (e,e') data. The correlations are shown to produce spin quenching and orbital enhancement in line with what is experimentally found. Total M1 strengths are calculated for Ca and Ti isotopes and found to agree with the experiments if the isovector spin operator is quenched by a factor 0.77.

- A full $0\hbar\omega$ description of the $2\nu\beta\beta$ decay of ^{48}Ca [16]. The lifetime of the 2ν double beta decay of ^{48}Ca is computed using the exact wave functions of ^{48}Ca , ^{48}Sc and ^{48}Ti. The resulting value, using the renormalized axial vector coupling, is $T_{1/2} = 3.7 \cdot 10^{19}$ yr, very close to the experimental bound $T_{1/2} = 3.6 \cdot 10^{19}$ yr.

- Missing and quenched Gamow Teller strength [17]. The Gamow Teller strength function in the resonace region is calculated for the ^{48}Ca (p,n) ^{48}Sc reaction. The standard quenching factor is shown to have the same origin as the depletion factor observed in single particle transfer and (e,e'p) reactions. The observed profile is reproduced almost perfectly up to 10 MeV excitation energy. It is very sensitive to the level density and may become so diluted as to be confused with background.

- Results for A≤50 nuclei without truncations yield excellent spectroscopy and will be published soon. For A>50 minor truncations are necessary and a program in this direction is being implemented.

3. Spherical Shell Model Far from Stability

A very transparent illustration of some of our considerations about the importance of the monopole field in shell model calculations is provided by the nuclear behaviour far from stability. We shall concentrate in the very neutron rich nuclei with N=20 (Ne, Na, Mg). N=20 is a magic number, and in the standard shell model view, semimagic nuclei are spherical. Experimentally the opposite hapens and these nuclei are deformed. The only way to interpret this fact is to accept the vanishing of the N=20 neutron shell closure. In spherical shell model language this is equivalent to say that the ground states of these nuclei are dominated by intruder configurations. The leading role of intruders has a twofold origin. First, the gap between - usually - empty and - usually - occupied *configurations* is reduced far from stability. This gap depends on the isovector monopole part of the effective interaction. Second, the quadrupole-quadrupole part of the nuclear hamiltonian favours energetically the intruder configurations, because they have neutrons and protons outside closed shells. The combined action of these two physical mechanisms make it possible to understand the onset of deformation in heavy Ne, Na and Mg nuclei. A complete study can be found in [18]. A similar situation is found in ^{80}Zr. In a naive shell model view one could expect a N=Z=40 double closure. However, experimentally this nucleus is deformed. Most probably 8-particles 8-holes rotational intruders, that will be discussed in the last of these lectures, are responsible for this anomaly.

The same physics may well be at the origin of a shell model description of the nuclear halo, experimentally found in ^{11}Li. In this nucleus we face again a neutron shell closure, N=8. Nevertheless, the ground state of its nearest isobar, ^{11}Be, is an intruder state $1/2^+$. It is tempting to propose that the ground state of ^{11}Li is also an intruder with two neutrons in the $2s_{1/2}$ orbit. This would solve at least three problems; the nearly unbound $2s_{1/2}$ orbit, which has no centrifugal barrier, can have a very large spatial extension and explain the halo [19]; the relative long lifetime of ^{11}Li can be understood, and a very low $3/2^+$ state - the soft dipole mode - comes in at about the right energy [20].

4. Rotational Motion in a Spherical Basis

It is a well known - but often forgotten - fact that the spherical shell model can cope with the collective aspects of the nuclear dynamics provided the valence space can accomodate the relevant degrees of freedom. Nevertheless, there is only one practical realization of the preceding statement; Elliott's Model [21]. This model describes the region of deformed nuclei at the beginnig of the sd shell (^{20}Ne, ^{24}Mg etc). Its ingredients are two; the valence space (one major oscillator shell) and the effective interaction (a pure quadrupole-quadrupole force).

Under these conditions the problem explicitly exhibits an SU(3) symmetry and rotational bands are described naturally. What is more important, the notion of intrinsic state comes out of the model in an extremely clear way. In spite of it's great conceptual importance, it's limited applicability has somehow obscured it's role in modellig the nuclear collective behaviour in a spherical shell model framework.

The reason for the non-applicability of Elliott's Model to other regions of the nuclear chart has to do with the strong spin orbit term in the nuclear mean field. While Elliott's geometry demands an LS-like mean field, nature does not. The lowest part of the sd shell is the only case where due to the degeneracy of the $2s_{1/2}$ and $1d_{5/2}$ orbits and to the minor role played by the $1d_{3/2}$ orbit, the LS coupling scheme is successfully mocked up.

In what follows we call good rotor to a nucleus whose yrast band follows the J(J+1) law and keeps a well defined intrinsic structure (for low spins). In order to gauge the existence of a well defined intrinsic structure we rely in the constance of the intrinsic quadrupole moment that we extract from the calculated -spectroscopic- one using the well known formula

$$Q_0(J) = \frac{(J+1)(2J+3)}{3K^2 - J(J+1)} Q_{spec}(J) \qquad (1)$$

Besides the sd shell SU(3) examples, there are cases in which rather good rotors come out of spherical shell model calculations with realistic interactions. The first one is -again- ^{20}Ne which rotates also in the $(2s_{1/2}, 1d_{5/2})$ valence space with any current realistic interaction [22]. Another is provided by the configurations involved in the onset of deformation in the very neutron rich sodium and magnesium isotopes [18]. In ^{32}Mg, the configuration $(2s_{1/2}, 1d_{5/2})^{4\pi} (2p_{3/2}, 1f_{7/2})^{2\nu}$ (π protons, ν neutrons) produces a very clear rotational spectrum. In figure 1 the experimental yrast band of ^{48}Cr is compared to the predictions of a full $(pf)^8$ calculation. The collective behaviour is well accounted by this model space. Yet, these collective aspects are already developed in a $(1f_{7/2}, 2p_{3/2})^8$ calculation. This last calculation definitely provides the picture of a good rotor.

If rotational motion does set in under the influence of the quadrupole $(q.q)$ component of the force, to determine the intrinsic states we have to diagonalize $\chi q_{20} + S$, where χ is a constant selfconsistently fixed and S a central field. The matrix elements of interest are:
$< jm|q_{20}|j'm > = (-)^{j'-m} < jmj' - m|20 > q_{jj'} = C(j - j', m)q'_{jj'}$,
where $q_{jj'}$ is a reduced matrix element and $q'_{jj'}$ regroups factors that depend only on jj'. We have: $C(0, m) \approx 3m^2 - j^2$, $C(1, m) \approx m\sqrt{3(j^2 - m^2)}$, $C(2, m) \approx \sqrt{\frac{3}{2}(j^2 - m^2)}$, from which, $\Delta j = j - j' = 0$ favours oblate orbits ($m \sim j$) over prolate ones ($m \sim 0$), $\Delta j = j - j' = 1$ mixes neither prolate nor oblate ones, $\Delta j = j - j' = 2$ mixes, and therefore favours, prolate orbits only. Examples of $\Delta j = 2$ sequences are:

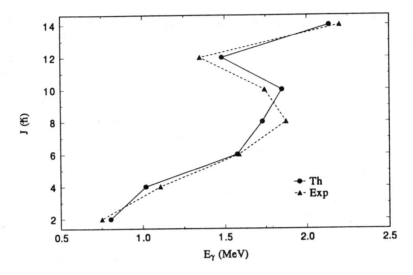

Figure 1: ^{48}Cr yrast band

$(1d_{5/2}, 2s_{1/2})$, $(1f_{7/2}, 2p_{3/2})$, $(1g_{9/2}, 2d_{5/2}, 3s_{1/2})$, $(1h_{11/2}, 2f_{7/2}, 3p_{3/2})$.

In order to get rotational motion it is necessary to have at least one $\Delta j = 2$ block for protons and another for neutrons to allow the action of the strong proton-neutron quadrupole interaction. We shall study the following set of configurations:

$$(1f_{7/2}, 2p_{3/2})^{4\pi} \ (1f_{7/2}, 2p_{3/2})^{4\nu}$$
$$(1f_{7/2}, 2p_{3/2})^{4\pi} \ (1g_{9/2}, 2d_{5/2}, 3s_{1/2})^{4\nu}$$
$$(1g_{9/2}, 2d_{5/2}, 3s_{1/2})^{4\pi} \ (1g_{9/2}, 2d_{5/2}, 3s_{1/2})^{4\nu}$$
$$(1g_{9/2}, 2d_{5/2}, 3s_{1/2})^{4\pi} \ (1h_{11/2}, 2f_{7/2}, 3p_{3/2})^{4\nu}$$

We use the interaction of Lee, Kahana and Scott [23] (KLS), calculated at $\hbar\omega = 9$ MeV. The single particle energies are given by $\epsilon(j) = -\frac{1}{2}j \cdot \epsilon_0$. The calculations are made with $\epsilon_0 = 1$ MeV. Quadrupole effective charges of 0.5 for neutrons and 1.5 for protons are used. The dimensions of the matrices diagonalized by the code ANTOINE [12] reach $2 \cdot 10^6$.

In table 1 we compare the ratios of the excitation energies and intrinsic quadrupole moments of the Yrast states to those of the 2^+, to the rigid rotor values. For quadrupole moments the rigid rotor picture is valid to better than 10% up to J values that go from J=8 in the smallest space to at least J=14 in the largest one. The intrinsic quadrupole moments are 1.33b, 1.40b, 1.89b and 1.93b in the four cases studied. Using the rotational model relationship between

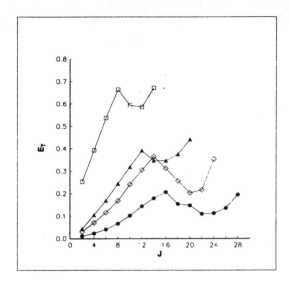

Figure 2: Yrast transition energies $E_\gamma = E(J+2) - E(J)$ for different configurations: $(fp)^8$ (empty boxes), $(fp)^4(gds)^4$ (full triangles), $(gds)^8$ (empty diamonds), $(gds)^4(hfp)^4$ (full circles). KLS interaction.

quadrupole moment and deformation parameter, and a mass value A$\sim$50, the results in the $1g_{9/2}, 2d_{5/2}, 3s_{1/2}$ space would correspond to β=0.7 well into the superdeformed regime.

	$(f_7p_3)^8\ T=0$		$(f_7p_3)^4(g_9d_5s_1)^4$		$(g_9d_5s_1)^8\ T=0$		$(g_9d_5s_1)^4(h_{11}f_7p_3)^4$		rigid rotor	
	$\frac{E(J)}{E(2^+)}$	$\frac{Q_0(J)}{Q_0(2^+)}$	$\frac{E(J)}{E(2^+)}$	$\frac{Q_0(J)}{Q_0(2^+)}$	$\frac{E(J)}{E(2^+)}$	$\frac{Q_0(J)}{Q_0(2^+)}$	$\frac{E(J)}{E(2^+)}$	$\frac{Q_0(J)}{Q_0(2^+)}$	$\frac{E(J)}{E(2^+)}$	$\frac{Q_0(J)}{Q_0(2^+)}$
4	3.26	1	3.44	1	3.32	1	3.29	0.995	3.33	1
6	6.75	0.992	7.37	0.993	7.13	1	6.74	0.990	7	1
8	11.56	0.955	13.04	0.979	12.66	0.995	12.72	0.979	12	1
10	17.49	0.707	20.47	0.921	20.56	0.984	21.81	0.964	18.33	1
12	22.79	0.211	29.59	0.721	30.57	0.947	34.66	0.933	26	1

Table 1: Ratios between the excitation energies E and intrinsic quadrupole moments Q_0 of the Yrast states of several $\Delta j = 2$ configurations and those of the 2^+ Yrast state.

It is natural to wonder about what may happen at higher spins. Our results are extremely suggestive. We have plotted in figure 2 the energies of the gammas emitted along the yrast band against the angular momentum of the

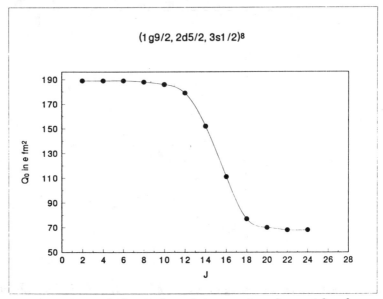

Figure 3: Intrinsic quadrupole moments, gds yrast band

decaying state in the four cases studied. The plot shows clearly the presence of a backbending. It is more pronounced in the spaces involving larger l orbits, and sets in at increasing values of J (J=8,10 12 and 14 for the four spaces studied). The backbending has been extensively described in the framework of models of deformed intrinsic states as due to the alignment of pairs of particles along the rotation axis. It is difficult to translate this interpretation into our laboratory frame description and we will not attempt to do it now. We shall limit ourselves to discuss two sets of results that can be put in correspondence with the onset of backbending.

In figure 3 we show the evolution of the intrinsic quadrupole moment with J for the gds configuration. As in all the cases studied, near the J value at which the backbending takes place, there is an abrupt change in the value of Q_0. Increasing still the angular momentum by a few more units a second region of constant Q_0 (less evident in the smaller spaces) comes in. This can be interpreted as due to the mixing of oblate and prolate shapes in the wave function, much in line with the interpretation of the rotational models. More evidence on this can be extracted from figure 4 where we show the squared amplitudes of the 'oblate' configuration i.e. the one with all the particles in the largest l orbit. We can observe that up to the J value where the band backbends they are almost constant. Afterwards they go on increasing steadily.

Further understanding of rotational motion is obtained by studying the predictions of a pure quadrupole quadrupole interaction in these spaces. Again we particularize to $(gds)^8$. We use the same central field than in the realistic

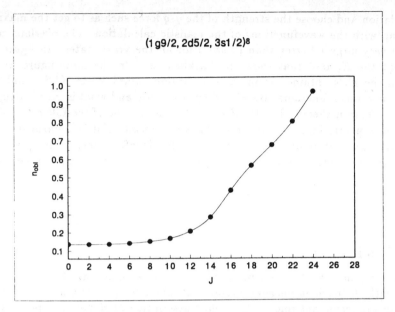

Figure 4: Probability of the $(1g_{9/2})^8$ configuration, gds yrast band

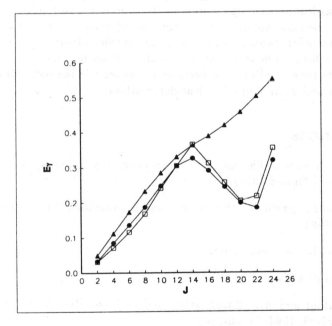

Figure 5: Yrast transition energies $E_\gamma = E(J+2) - E(J)$. Results using a quadrupole quadrupole interaction (full triangles) compared with the KLS result (empty boxes) and with a perturbative calculation of H(KLS) using the quadrupole quadrupole wave fuctions.

calculation and choose the strength of the $q \cdot q$ force such as to get the maximum overlap with the wavefunctions of the realistic calculation. The overlaps can be made very large (better than $(0.95)^2$) for all the yrast states. In figure 5 we present the E_γ plot that shows no backbending. In the same figure we have drawn the E_γ's obtained from the expectation values of the KLS hamiltonian in the $q \cdot q$ wave functions. Backbending reappears and what is more intriguing, in perturbation theory. The third set of E_γ's is the one of the full KLS calculation. It is nearly identical to the perturbative result. If it is confirmed that all the structural information is contained in the eigenfunctions of the quadrupole force, the description of rotors in the laboratory frame will become much easier.

5. Conclusions

A spherical shell model description of nuclei can be made that keeps conceptual links with the nucleon nucleon interaction. The monopole hamiltonian is shown to play a determinant role in the unified description of global and spectroscopic properties of nuclei. Excellent agreement with the experimental results can be found in classical areas of application of the spherical shell model. New fields of interest can be also explored, as it is the case of nuclei far from stability where the role of intruder states may be crucial. Large calculations with a sound choice of the mean field geometry make it possible to describe rotational motion. An approximate form of $SU(3)$ symmetry seems to provide the basis for a spherical shell model understanding of nuclear deformation.

References

[1] K. A. Bruekner, The many body problem, Les Houches, Summer School lectures (Dunod 1958).

[2] V. Pandharipande, C. Papanicolas and J. Wambach, Phys. Rev. Lett. **53**, 1153 (1984).

[3] B. Frois in these proceedings.

[4] P. Vold *et al*, Nuc. Phys. **A302**, 12 (1978)

[5] M. Dufour and A.P. Zuker, submitted to Phys. Rev. Lett., and preprint CRN 93-29, 1993, Strasbourg.

[6] A. Abzouzi, E. Caurier and A.P. Zuker, Phys. Rev. Lett. **66**, 1134 (1991).

[7] S. Cohen and D. Kurath, Nucl. Phys. **73**, 1 (1961).

[8] B. A. Brown and B. H. Wildenthal, Ann. Rev. Nucl. Part. Sci. **38**, 29 (1988).

[9] T.T.S. Kuo and G.E. Brown, Nucl. Phys. **A114**, 235 (1968).

[10] J. B. French, E. C. Halbert, J. B. Mc Grory and S. S.M.Wong in Advances in Nuclear Physics, vol 3,(1969), M. Baranger and E.Vogt eds. Plenum press, New York.

[11] R. R. Whitehead, A. Watt, B. J. Cole and J. Morrison in Advances in Nuclear Physics, vol 9, (1977) 123, M. Baranger and E. Vogt eds. Plenum press, New York.

[12] ANTOINE code, CRN, Strasbourg 1989, to be released

[13] Y. Alhassid, D. J. Dean, S. E. Koonin, G. Lang and W. E. Ormand, Phys. Rev. Lett. **72**, 613 (1994).

[14] E. Caurier, A.P. Zuker, A. Poves and G.Martinez-Pinedo, Phys. Rev. **C** Phys. Rev. july 1994 issue.

[15] E. Caurier, A. Poves and A. Zuker, Phys. Lett. **B256**, 301 (1991).

[16] E. Caurier, A. Poves and A. Zuker, Phys. Lett. **B252**, 13 (1990).

[17] E. Caurier, A. Poves and A. Zuker, submitted to Phys. Rev. C

[18] A. Poves and J. Retamosa, Phys. Lett **B184**, 37 (1987) and Nucl. Phys. **A571**, 221 (1994).

[19] J. M. Gomez, C. Prieto and A. Poves, Phys. Lett. **B 295** 1 (1992)

[20] G. Martinez-Pinedo and A. Poves, work in progress.

[21] J. P. Elliott, Proc. Roy. Soc. London **A245**, 128, 562, (1958) ; J. P. Elliott and M. Harvey, Proc. Roy. Soc. London **A272**, 557 (1963).

[22] A. Arima, S. Cohen, R.D. Lawson and M.H. Mc Farlane, Nuc. Phys. **A108**, 94 (1968).

[23] S. Kahana, H.C. Lee and C.K. Scott, Phys. Rev. **180**, 956 (1969) and code by H.C. Lee.

High Spins and Exotic Shapes[1]

Sven Åberg

Department of Mathematical Physics
Lund Institute of Technology, PO Box 118
S-221 00 Lund, Sweden

Abstract. Some open problems in the fields of high-spin physics and exotic shapes are considered. We discuss pairing properties, octupole effects and the possibility for high-K isomers in superdeformed nuclei, as well as the posible existence of hyperdeformed nuclei. Clustering effects in very light nuclei leading to the formation of α-string nuclei, and effects of heavy di-nuclear systems are also discussed. The deformation degree of freedom for some halo nuclei is stressed.

1 Introduction

The field of high-spin physics is very active, and is today perhaps one of the most exciting areas of nuclear physics. With present experimental detector development of large arrays for γ-rays, such as GAMMASPHERE in the U.S.A. and EUROBALL and GA.SP in Europe, many new and unpredictable things will most certainly be discovered in the close future. In these lectures I shall try to give an overview of the structure of rapidly rotating nuclei, emphasizing exotic shapes such as superdeformations and even more exotic shapes. I remind you about the basic theory of rotating nuclei in section 2, and of superdeformations in section 3. Then I shall approach the front of today in section 4, the way I see it and with no other option than my own preferences and interests. Guided by theoretical calculations I shall discuss some open problems and point at interesting possibilities for new exotic behaviour of the rapidly rotating atomic nucleus.

Since the space in these written notes is limited, and since several good books as well as review articles already exist which well cover the first two parts of my lectures, I only give the list of contents and proper references for sections 2 and 3.

---

[1] This project was supported by the Swedish Natural Science Research Council

2   Basic knowledge about rotating nuclei

This lecture gave an overview on the following basics:

- Deformation – Rotation

- Particle-plus-rotor model

- Rotation aligned and deformation aligned coupling schemes

- Backbending

- Cranking model

- Band termination

- Non-collective rotation

This is mainly text-book material, and I recommend you to study, for example, one of the following rather basic textbooks: Nilsson and Ragnarsson [1], Ring and Schuck [2] or Szymański [3].

3   Superdeformations

Nuclei with abnormally large deformations, with axis ratio of about 2:1, were discussed and discovered many years ago in terms of the so-called fission isomers [4, 5]. Some years later it was suggested that also somewhat lighter nuclei, as e.g. rare-earth nuclei, may posess a similar extreme shape when it is rotating sufficiently fast [6, 7], and 1985 such rapidly rotating *superdeformed* nuclei were discovered by Twin et al [8]. The study of superdeformed nuclei has become a very active and fruitful research field, and much has been learned about nuclear structure at extreme deformations.

With the emphasize on superdeformations in this section I discussed:

- Cranked Nilsson-Strutinsky calculations

- Shell structure at high angular momenta and at large deformations

- The role of high-N intruder configurations

- Electric and magnetic properties

The following references contain rather recent information on mean field calculations [9], and on high-spin physics [1], in particular on superdeformations [10, 11], and may serve as background material and partly cover the contents of this lecture.

4   Some open problems

After this basic introduction of high-spin physics and superdeformations I now will discuss some open problems emphasizing exotic nuclear shapes. This includes the following topics: Superdeformations (subsect. 4.1), Octupole effects in superdeformed nuclei (4.2), Hyperdeformations (4.3), Nuclear sausages – α-string nuclei (4.4), Di-nuclear systems (4.5), Deformed halo nuclei (4.6), and finally a speculation on Superdeformed high-K isomers (4.7).

4.1   Superdeformations

Today several hundreds superdeformed (SD) states in the region around ^{152}Dy as well as around ^{192}Hg have been identified, mainly through γ-ray energies but also by lifetime measurements, see [12] for a collection of SD data. However, in none of these states the spin, parity or excitation energy have been measured. It is certainly of great importance to perform such measurements so that theoretical models can be tested. The excitation energy may be obtained if the discrete linking transitions from the superdeformed to the normal deformed (ND) states are measured. Such measurements would provide important information also on the tunneling process from the SD minimum to the ND minimum, in a situation where the pairing properties probably are rather weak. The decay process may be considered as a transition from an ordered SD state to an ordered ND state over an intermediate region of chaotic ND states. It is an open question if the recently discovered phenomenon of chaos-assisted tunneling [13] will play a role here.

The pairing interaction plays a very important role for the understanding of ground-state properties of nuclei. Furthermore, the way angular momentum is aligned along the rotational axis strongly depends on the pairing properties. In a similar way as the superconducting properties decrease and may eventually dissapear when the superconducter is placed in a strong magnetic field, the pairing properties of the nucleus may dissapear when it is set to rotate fast [14]. The decreasing role of the pairing interaction might be studied along rotational bands in SD nuclei where spin sequences up to angular momentum $I \approx 66\hbar$ have been seen.

In fig.1 I show the drastically different behaviour of the $\mathcal{J}^{(2)}$ moment of inertia $(\mathcal{J}^{(2)} \equiv (\frac{d^2E}{dI^2})^{-1})$ with and without strong pairing correlations. By comparing the calculations with the data one may draw the conclusion that pairing is important for SD ^{194}Hg while not for SD ^{152}Dy. In the case when pairing is strong (Hg-region) the spin alignment of the first pair(s) of nucleons (paired band crossing) can be seen as a bump in the $\mathcal{J}^{(2)}$ moment of inertia (the surface covered by the bump is proportional to the aligned angular momentum), while the corresponding alignment occurs much smoother in the unpaired case, and the corresponding $\mathcal{J}^{(2)}$ moment of inertia is much less affected. In general, paired band crossings are much more rare in the ^{152}Dy region.

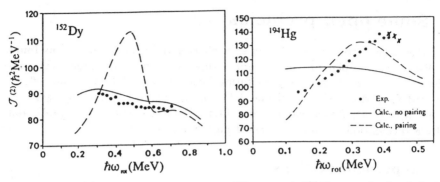

Figure 1: $\mathcal{J}^{(2)}$ moments of inertia of ^{152}Dy and ^{194}Hg as obtained from a cranking calculation neglecting (solid lines) or including the pairing interaction (dashed lines). Experimental values are shown by dots. The most recent data for ^{194}Hg is taken from [16] and shown by crosses. (From [10].)

The lower-frequency part of (presumed) paired band crossings has been seen for several SD bands in the Hg-region as a smooth increase of the $\mathcal{J}^{(2)}$ moment of inertia. Recently, also the beginning of the decreasing part of the $\mathcal{J}^{(2)}$ bump was measured for ^{194}Hg [16], and those new data points have been included in fig.1. This supports the idea that the observed increase of $\mathcal{J}^{(2)}$ with increasing rotational frequency indeed is caused by paired bandcrossings.

The predicted [15] fine competition between deformation changes and the onset of pairing for some SD bands in the ^{152}Dy region may be revealed through $\mathcal{J}^{(2)}$ bumps caused by paired band crossings. All together, the SD nuclei give us a possibility to perform very detailed studies of the pairing phase transition, caused by the alignment of a few nucleon pairs.

Our understanding of nuclear structure at large deformations will certainly be increased if SD bands in new mass regions are observed. Calculations predict SD bands at resonably low angular momenta in nuclei around $^{108}_{44}Ru_{64}$, $^{88}_{44}Ru_{44}$ and $^{78}_{34}Se_{44}$ [17]. In a recent experiment [18] a superdeformed band has been seen in $^{82}_{38}Sr_{44}$.

Based on cranked mean field calculations even more exotic nuclear shapes than superdeformations are predicted to be realized at high angular momenta. Some of these exotic shapes are discussed in the next few subsections.

4.2   Octupole effects in SD nuclei

At superdeformation several single-particle levels emerging from high-lying shells come close to the Fermi level, and the nuclear structure at SD is consequently expected to be quite different than at normal deformations. One example is the high-N orbits which play an important role in understanding [19] observed features of the dynamical moment of inertia ($\mathcal{J}^{(2)}$). Another consequence is that there will be approximately equal amounts of positive- and negative-parity states

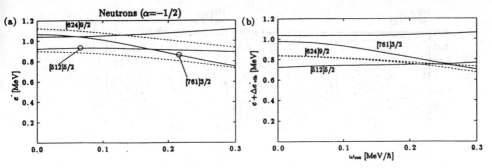

Figure 2: a) Calculated one-quasiparticle Routhians in ^{193}Hg shown vs rotational frequency. In b) an octupole-octupole force including all K-components has been added. (From [21].)

around the Fermi surface. We thus expect several rather low-lying particle-hole states with negative parity. By adding a residual octupole-octupole interaction to the mean field Hamiltonian, low-lying collective octupole vibrational states are likely to appear; or even static octupole deformations. In principle, the octupole states can have any K-value between 0 and 3. From studies in the pure harmonic oscillator potential [20] it turns out that the 0^- and 2^- states come lowest in energy. The 1^- state come at zero energy for closed 2:1 shells, and may consequently be very important, for example, when the pairing force creates partly filled states (quasiparticles) around the Fermi surface (in a similar way as low-lying collective γ-vibrational states, $K=2^+$, emerge at normal deformations).

Example of the role of octupole correlations in a realistic calculation [21] can be seen in fig.2. By including a residual octupole-octupole force in the cranking Hamiltonian, which subsequently is solved in RPA, it is found that some low-lying states become rather collective, mainly through the coupling with $K=2$ octupole phonons. For example, the state labelled [512]5/2 is lowered by about 350 keV, with the consequence that the rotational frequency where this band crosses the [761]3/2 band is increased to the experimental value [22] $\omega_c=0.22$ MeV. Also, the interaction matrix element between the two bands is increased from approximately zero to 5 keV, still being considerably lower than the experimental value [22] of 26 keV.

The nuclei discussed above are calculated to be rather soft in the octupole deformation degree of freedom, and in some cases even a minimum for a reflection asymmetric shape emerges. However, it is rather difficult to find any SD nucleus which has a stable octupole deformation with a reflection symmetric barrier larger than, say 1 MeV. In fig. 3 we show high-spin potential-energy surfaces in the $(\varepsilon_2,\varepsilon_3)$-plane (the nuclear shape is described by Y_{20} and Y_{30} terms) for ^{146}Gd, ^{152}Dy, ^{194}Hg and ^{200}Rn. The case of ^{200}Rn represents more or less the "best" case of strong octupole effects in SD nuclei (in the present

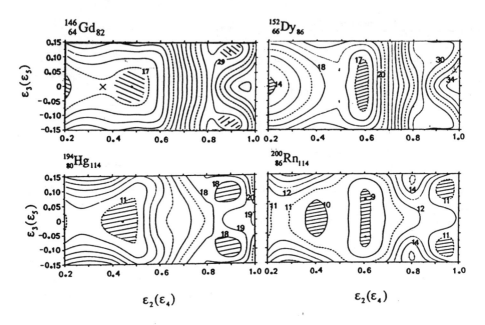

Figure 3: Potential-energy surfaces in the $(\varepsilon_2,\varepsilon_3)$-plane valid at $I^\pi = 40^+$ for ^{146}Gd, ^{152}Dy, ^{194}Hg and ^{200}Rn. The calculations are based on the (diabatic) cranked Nilsson-Strutinsky model utilizing the modified oscillator and exluding the pairing force. Both ε_4 (free) and ε_5 (restricted) deformations are included. (From [23, 11].)

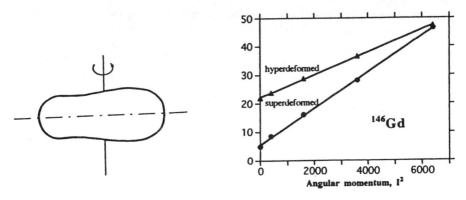

Figure 4: a) Approximate shape of HD ^{146}Gd ($\varepsilon=0.93$, $\varepsilon_3=0.12$, $\varepsilon_4=0.13$ and $\varepsilon_5=-0.056$). b) The energy of lowest SD and HD configurations in ^{146}Gd as functions of angular momentum. (From [11].)

calculations, [23]). If, on the other hand, we move to somewhat larger deformations, $\varepsilon \approx 0.9$, rather deep minima are seen for all four nuclei at $\varepsilon_3 \approx 0.1$. The corresponding nuclear shape can be seen in fig.4a, and corresponds to an approximate axis ratio of 3:1. Such an extended nuclear shape has been called *hyperdeformed* which in this case is reflection asymmetric.

4.3 Hyperdeformations

In the present cranked Nilsson-Strutinsky calculations ^{146}Gd is found to be one of the most favourable cases for hyperdeformation. The hyperdeformed (HD) minimum corresponds to $\varepsilon_2 \approx 0.93$ and $\varepsilon_3 \approx 0.13$, that is considerably larger than the superdeformed minimum of ^{146}Gd that corresponds to $\varepsilon_2 \approx 0.52$. The HD band is calculated to have a moment of inertia of about $140\hbar^2 MeV^{-1}$ ($\approx$ rigid-body value), and crosses the SD band at $I \approx 80\hbar$, see fig.4b. The barrier between the HD and SD minima increases with angular momentum from about 1 MeV at $I=50$ to 3 MeV at $I=90$.

Due to the very large deformation the j-shell character of orbitals emerging from intruders as $i_{13/2}$, $j_{15/2}$, etc (high-N states), is less unique compared to normal deformations, and also compared to superdeformations. Consequently, the special role that high-N states play at superdeformations [19] disappears, and no high-N characterisation of HD rotational bands can be made. Consequently, different HD rotational bands are expected to show quite similar $\mathcal{J}^{(2)}$behaviour [11].

The feeding of Hd states occurs in strong competition with fission. Other conditions are, however, expected to be similar to those favouring the feeding of SD states, namely low level density (one HD band in ^{146}Gd is pushed down by about 1 MeV), and in particular, a very low-lying giant dipole resonance component is expected, corresponding to isovector vibrations along the symmetry axis. Once a HD state is populated there is a strong competition between remaining in the band, fission, and the decay to SD (or ND) states. Due to the expected week pairing the mass parameter connected with quadrupole motion, $B_{\varepsilon\varepsilon}$, is expected to be very large, substantially slowing down the large amplitude collective motion. Furthermore, the low level density of SD states decreases the number of possible daughter states. Finally, an extreme $B(E2)$ value ($\approx 10\ 000$ W.u.) enhances the chance remaining in the HD band. These three effects all favour the decay along the HD band and descrete γ-rays may indeed appear.

Experimental indications exist of a HD band in ^{152}Dy [25]. The evidence is based on γ-γ correlation measurements, where the width of the valley corresponds to a $\mathcal{J}^{(2)}$-value of about 130 $\hbar MeV^{-1}$, i.e. quite close to the calculated value. The rotational frequency varies between $\hbar\omega_{rot} \approx 0.6$ to 0.75 MeV for the observed sequence of ridge structure. In the calculations we find $\mathcal{J}^{(2)} \approx \mathcal{J}^{(1)} \approx \mathcal{J}_{rigid}$ and the observed band would then possess angular momenta, $I = \omega_{rot} \cdot \mathcal{J}^{(1)}$, between 78 and 98 $\hbar$, which are unbelievably large values. Although the calculations are very uncertain it is difficult to think of a process that makes $\mathcal{J}^{(1)} < \mathcal{J}^{(2)}$ over the full interval.

4.4   Nuclear sausages – α-string nuclei

It is well known that many states in even-even $N=Z$ nuclei in the p- and sd-shell region can be understood in terms of α-cluster configurations. This gives a qualitative description of most properties of the states, as well as a nice geometrical picture. But the α-configurations can be understood also in terms of mean field calculations.

The Nilsson diagram shown in fig.5 is valid for protons as well as for neutrons for p-shell and sd-shell nuclei. Each single-particle level may be filled by two protons and two neutrons coupled to spin and isospin zero. Consequently, the Nilsson diagram for light nuclei can be considered effectively as a single-particle diagram valid for α-particles.

Harvey has suggested a phenomenological way to describe the (diabatic) configuration in a fusion reaction between two nuclei [26]. The configuration of the fused system is obtained by fixing one of the nuclei, and filling the particles from the other nucleus into free orbitals under the condition that oscillator quanta may change only in the direction of the reaction axis, e.g. the z-axis. This corresponds to the situation of the double harmonic oscillator potential (see e.g. [27]) where the symmetric state forms the ground state ($N=0$) and the antisymmetric state goes into the first excited harmonic oscillator state ($N=1$), as the distance between the two oscillators decreases, see fig. 6. In the quantum number notation used in fig.5, $[Nn_z\Lambda]\Omega$, this means that N and n_z may change while $n_\perp (= N - n_z)$, Λ and Ω must remain constant.

An α-particle is in an s-state, i.e. the 4 particles fill the orbit $[000]1/2$ coupled to spin and isospin zero. Adding α-particles along the z-axis implies that the second α-particle must occupy $[110]1/2$, the third $[220]1/2$, etc, up to

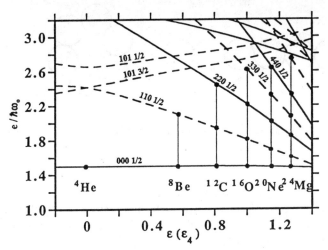

Figure 5: Nilsson diagram valid for protons as well as for neutrons at prolate deformations. The shell filling of n:1 α-string configurations are marked out at relevant deformations for ^4He, ^8Be, ^{12}C, ^{16}O, ^{20}Ne and ^{24}Mg. Single-particle states are denoted by the quantum numbers $[Nn_z\Lambda]\Omega$. (From [24].)

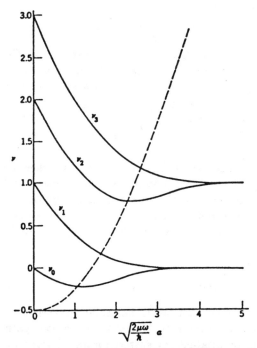

Figure 6: Single-particle levels in the double oscillator as functions of the distance between the oscillators. The dashed line shows the energy where the two oscillators meet. (From [27].)

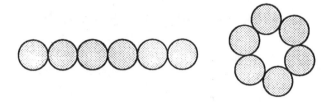

Figure 7: Speculation about possible excited states in ^{24}Mg: α-string nucleus (left) and α-ring nucleus (right).

[n-1 n-1 0]1/2 for n α-particles on a string.

These orbits all carry large single-particle quadrupole moments; at prolate deformations ($\varepsilon > 0$) they are strongly down-sloping in the Nilsson diagram (fig.5). Consequently, the filling of the [NN0]1/2 orbits with α-particles is strongly driving the corresponding nucleus towards prolate shapes. The approximate equilibrium deformations, resulting from a minimization of the shell energy plus the liquid-drop energy (Nilsson-Strutinsky procedure), are marked out in fig.5 for $n=1$, 2, 3, 4, 5 and 6, i.e. for ^4He, ^8Be, ^{12}C, ^{16}O, ^{20}Ne and for ^{24}Mg, respectively [28, 29]. These configurations correspond to excited states in the respective nuclei (except for ^4He and ^8Be). As the intuitive picture of α-strings suggests, the corresponding shapes are prolate with an axis ratio between the z-axis and the perpendicular axes very close to n:1. Since the rigid-body moment of inertia depends on the number of α particles as $\mathcal{J}\sim n^{4/3}$, rotation is a good way to bring down the exotic states of large α-strings.

So, we have obtained a nice relation between the α-cluster description and the deformed shell model, see also [30, 31]. Experimental evidence for these exotic nuclear shapes exist for ^8Be, ^{12}C, ^{16}O [32, 33], and recently also for the 6:1 configuration in ^{24}Mg [34]. Thus, very exotic nuclear shapes do indeed exist for light nuclei.

It is interesting to speculate about the possible existence of even more exotic shapes. By closing the extended α-string a *ring-nucleus* would form, see fig.7. Such an indeed strange shape might have a lower energy than the corresponding string-nucleus due to the gain in surface energy. The decay of an α-ring nucleus would probably occur via a break up into two approximately equal (excited) α-string nuclei.

4.5 Di-nuclear systems

Effects caused by the formation of di-molecular (quasi)states have been seen in very heavy nuclei. In studies by Möller et al [35] two paths to fission were found in the nuclear potential-energy surface for nuclei around ^{264}Fm. One path corresponds to the usual fission path where scission occurs at a quite elongated shape. The other path, "the fusion path", corresponds to shapes where the scission occurs at a much smaller elongation, roughly corresponding to the shape of a double sphere (di-molecule). The latter path was found to be more preferred the closer the neutron number is to $N=164$, which can be understood as an effect of a very favorable formation of two double magic (spherical) ^{132}Sn clusters in the ^{264}Fm compound system. The scenario could give a nice explanation of observed [36] peaks in the total kinetic energy distribution curves [35].

Now let us study di-nuclear systems in lighter nuclei and, in particular, the relation of such shapes to superdeformations. In the anisotropic harmonic oscillator potential, single-particle energy degeneracies occur when the oscillator frequencies, ω_z, and $\omega_\perp$, relate as small integer numbers [37]. Particularly large gaps in the single-particle spectrum show up at ratio 1:1 (spherical shape) and at 1:2 i.e. when the z-axis is double the size of the perpendicular axes. This is the structure underlying the appeerence of superdeformed nuclei.

An interesting observation is that every second SD magic number corresponds to twice a spherical magic number. This may be understood by considering the shape and configuration of the system obtained through the fusion of two equal (double-magic, $N=Z$) spherical nuclei. Through the Harvey prescription [26] it is then found that the resulting configurations exactly correspond to the closed-shell configurations at superdeformation. In the left-hand part of fig.8 we show how the single-particle spectra of two spherical $A/2$ nuclei add up and give rise to a single-particle spectrum of one nucleus with the shape of a double-sphere containing A nucleons. Volume conservation implies that the oscillator frequency of the double-sphere is $2^{1/3}$ times larger than that of a sphere (right-most part of fig. 8).

In the right-hand part of fig. 8 it is shown how the spherical magic numbers change as the sphere is deformed into a superdeformed spheroidal nucleus, with the ratio between the oscillator frequencies 1:2. The same particle numbers for closed shells appear for the spheroidal nucleus as for the double-sphere. Also the single-particle energy distances are the same ($\hbar\omega_\perp=\hbar\omega_{oo}=2^{1/3}\hbar\omega_o$). The double-sphere and the 2:1 deformed oscillator are thus expected to have very similar properties. But although the filled orbitals are the same for the two types of shape, there are also differences. For the spheroidal nucleus additional states appear in the middle of the gaps in the single-particle spectrum of the double-sphere, as shown by dashed lines in fig.8. These states correspond to the excitation of oscillator quanta in the z-direction with the oscillator frequency $\hbar\omega_z$ ($=\frac{1}{2}\omega_\perp$). The degeneracies are thus smaller for spheroidal SD nuclei than for nuclei with a dumb-bell shape which implies a smaller amplitude of the shell energy. (The shell energy of a double-sphere with A particles is expected to be approximately twice the shell energy of one sphere with $A/2$ particles.)

Another general difference between SD spheroids and double-spheres is that the giant dipole resonance (GDR) should look quite different in the two cases. In the superdeformed spheroidal nucleus two peaks are expected to appear, one

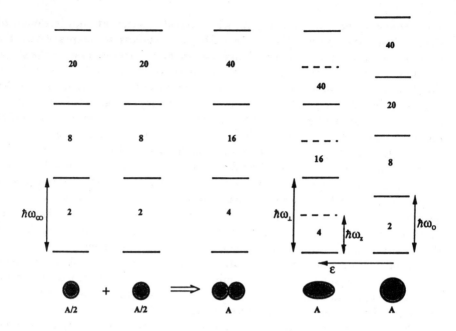

Figure 8. Single-particle spectra for different versions of the pure harmonic oscillator potential. The three spectra on the left-hand side illustrate the fusion into a double-sphere of the two spheres with $A/2$ particles. The two spectra on the right-hand side show how the sphere with A particles changes versus quadrupole deformation, ε, into a 2:1 deformation. (From [24].)

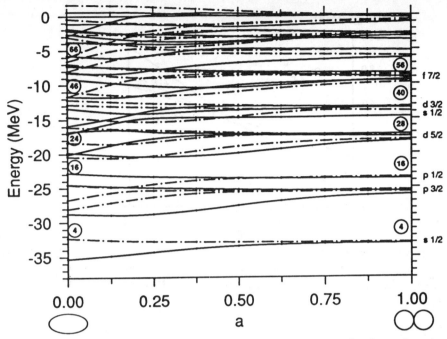

Figure 9: Single-particle energies for Woods-Saxon neutron levels as functions of the shape variable a. The two limits, $a=0$ and $a=1$, correspond to spheroidal 2:1 deformation and double-sphere, respectively, and intermediate a-values correspond to a linear interpolation between these two shapes. Particle numbers at closed shells are encircled. At the double-sphere j-shell configurations are given. Each j-shell has double degeneracy as compared to one sphere. Woods-Saxon parameters from [38] are used. (From [24].)

at about $2\hbar\omega_z$ and one at about $2\hbar\omega_\perp$, while due to the lack of the $2\hbar\omega_z$ states in the system of a double-sphere (see fig.8), the lower GDR-peak disappears and the higher-lying peak becomes correspondingly larger. An increase of the necking degree of freedom is thus expected to decrease the lower component of the GDR peak, and in fact increase the centroid energy, see [39].

Superdeformed nuclei are generally expected to be populated by statistical E1-transitions, where the B(E1) strength becomes strongly enhanced through the low-lying GDR component [40]. In that scenario the population of states which are strongly necked-in should be less favorable than of "normal" SD shapes.

In fig.9 we show the Woods-Saxon neutron single-particle energies valid around ^{96}Zr as functions of the deformation parameter a that describes an interpolation between a spheroid ($a=0$) and a double-sphere ($a=1$). Note the restoration of the spherical quantum numbers at $a=1$, but with a twofold degeneracy as compared to the spherical configuration. Due to the diffuse surface of

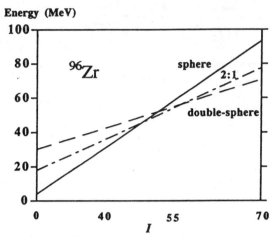

Figure 10: The approximate (rigid-body moments of inertia have been assumed) rotational behaviour of ^{96}Zr at spherical, 2:1 spheroidal and double-spherical shapes. The shell energies were approximated by their spin-zero values. (From [24].)

the Woods-Saxon potential the restoration is only approximative, as compared to two independent spheres, since the two spheres still interact. Nice single-particle gaps are seen at $N=4$, 16, 28, 40, 56,... corresponding to the spherical magic numbers at $N=2$, 8, 14, 20, 28,... . At the spheroidal 2:1 deformation $(a=0)$ closed shells appear for $N=4$, (10), 16, 24, 46, 66, ... , i.e. the same as for the double-sphere for the lowest neutron numbers only. In fig.8 we saw that in the pure harmonic oscillator potential closed shells appear at the same particle numbers for the spheroid and for the double-sphere. Since the pure harmonic oscillator model gives a fairly good approximation of the single-particle spectrum of very light nuclei, this explains the agreement between the two types of shapes for low particle numbers. This also explains the appearence of a semi-closed shell at $N=10$ for the spheroidal shape. For heavier nuclei the two kinds of shapes are favoured at quite different particle numbers, and the occupied orbits are quite different. For example, for $N=66$ and 86 (^{152}Dy) there are (as expected) no favouring at all for double-spherical configurations, while for $N=40$ and 56 the shell structure strongly favours double-spherical shapes (two ^{48}Ca nuclei).

For such exotic shapes as the double-sphere it is quite necessary to include a finite-range surface energy term [41]. For example, for ^{96}Zr the double-sphere thereby gains almost 30 MeV while the gain in the spheroidal 2:1 nucleus is only about 4 MeV. Still, the surface energy disfavors the double-spherical shapes, and for ^{96}Zr the total energy comes about 10 MeV higher for the double-sphere than for the 2:1 deformed spheroidal. However, the moment of inertia of the double-sphere is about 2.2 times larger than the value for the rigid sphere, while the corresponding value for the 2:1 deformed spheroid is about 1.5. Since the rotational energy is inversely proportional to the moment of inertia (which at high spin should take the rigid-body values), this implies that the double-sphere would easily get more favoured at high spins, see fig.10, where the double-spherical shape of ^{96}Zr is seen to become more favoured than the spherical as well as the 2:1 spheroidal shapes for spin values above I$\approx$50.

4.6   Deformed halo nuclei

A few years ago it was found that some very light nuclei close to the neutron drip-line have a very large interaction cross sections [43]. This indicated the existence of a large neutron tail extending far outside the nucleus. Such a "halo" of neutron matter surrounding the nucleus is believed to be an effect mainly of a very low neutron separation energy [44]. Among the berryllium isotopes the halo phenomenon has been observed in ^{11}Be (one-neutron halo) and ^{14}Be (two-neutron halo). It is interesting to note the Borromean character of ^{14}Be (and of ^6He and ^{11}Li): of the three parts, two neutrons and the remaining nucleus, any binary subsystem is unstable [45].

As an unexpected anomali the ground state of ^{11}Be has positive parity $(1/2^+)$. This means that the odd neutron is excited from the p-shell to the sd-shell. In the pure shell model this corresponds to an excitation energy of more than 10 MeV. The excitation energy can be decreased considerably if the neutron is placed in the orbital [220]1/2 and the quadrupole deformation degree of freedom is utilized, see fig.5. In Nilsson-Strutinsky calculations [46] the 1/2$^+$ state is calculated to have a very large (and triaxial) deformation, $\varepsilon_2 \approx 0.6$ and $\gamma \approx 40$, i.e. the ground state of ^{11}Be is superdeformed!

The modified harmonic oscillator potential, which was used in the calculations, has infinite walls. This means that the tail of the wavefunction will always drop off too much, and halo nuclei are in principle impossible to describe, in particular the halo region. If we in spite of this calculate the rms radius of the nuclear density for the beryllium isotopes, a large variation will appear due to the strong variation with particle number in the calculated equilibrium deformations, see fig.11. The measured rms-radii are in fact rather well reproduced solely from the variation in deformation from one isotope to another, completely neglecting the halo effects, i.e. the exterior parts of the wavefunction. The halo region certainly plays a most important role, but anyhow, fig.11 stresses the importance of including deformation degrees of freedom in the study of halo nuclei, at least for ^{11}Be. (For ^{14}Be the measured $< r^2 >$-value is much larger, and the deformation effects of minor importance.)

4.7   Superdeformed high-K isomers

Superdeformed nuclei in the region around ^{194}Hg have much lower angular momenta than in the region around ^{152}Dy. Pairing is much more important,

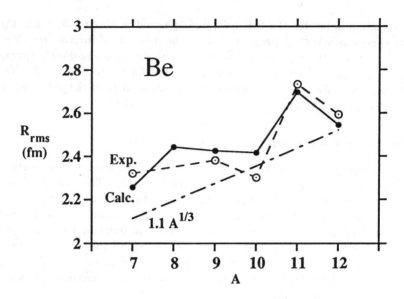

Figure 11: Calculated rms values of the nuclear radius (filled circles) for some Be isotopes are compared to corresponding measured quantities [43] (open circles). The calculated deviation from the smooth $1.1A^{1/3}$ line (dot-dashed line) is due to a strong variation in equilibrium deformation. No "halo effect" has been incorporated.

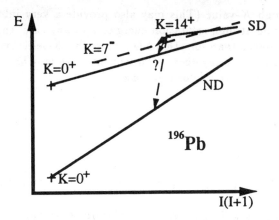

Figure 12: Schematic picture of energy vs angular momentum, $I(I + 1)$, for superdeformed and normal deformed bands with different K-values in ^{196}Pb. Some different decay paths are indicated for the discussed $K=14^+$ isomer.

and it seems as if different rotational bands behave much more similar than bands in the ^{152}Dy region. Another difference, partly connected to the other, is that in the Hg-region there are many more single-particle states with high Ω quantum numbers (see relevant Nilsson diagrams, e.g. in refs. [42, 10]) This is particularly true for the neutrons (as for example, [512]5/2 and [624]9/2 are situated just above neutron number $N=112$), but also proton high-Ω states exist close to the Fermi surface ([642]5/2 and [514]9/2 above proton number $Z=80$). These single-particle states are rather weakly coupled to the rotation (i.e. they are deformation aligned), and give a minor contribution to the $\mathcal{J}^{(2)}$ moment of inertia.

The existence of several high-Ω states suggests the possibility of creating excited superdeformed high-K states. For example, $K=7^-$ in superdeformed ^{196}Pb can be obtained via a favourable two-proton excitation $(\pi\{([512]5/2)^{-1}$ $[624]9/2\}_{7^-})$, as well as via a two-neutron excitation $(\nu\{([642]5/2)^{-1}$ $[514]9/2\}_{7^-})$. Taking blocking into account we may estimate the excitation energy relative the SD ground state to be about 1 MeV for each of them. If the two 7^--states are combined a high-K isomer, $K=14^+$ $((\pi7^-)\otimes(\nu7^-))$, appears at an unperturbed excitation energy of about 2 MeV. This may be compared with the excitation energy, $E_{exc}\approx 1$ MeV, of the $I^\pi=14^+$ state that belongs to the superdeformed ground-state band. In fig. 12 the discussed states are schematically drawn. The decay of the $K=14^+$ state may take place either to the SD $K=7^-$ bands, to the SD $K=0^+$ band, or directly to normaldeformed states. For the latter case the very special configuration may give rise to a strong hindrance, partly because of the presumably small superfluidity, and partly because of the high K-value (This may also provide a very interesting test of the K-selection rule at high excitation energies.) In any case, the lifetime must be considerably longer than for other SD states; the K-selection rule certainly puts strong hindrance for any decay. Consequently, we may have the possibility to identify an isomeric superdeformed state.

References

[1] S.G. Nilsson and I. Ragnarsson, *Shapes and Shells* (Cambridge 1995), to appear.

[2] P. Ring and P. Schuck, *The Nuclear Many-Body Problem* (New York: Springer Verlag 1980).

[3] Z. Szymański, *Fast Nuclear Rotation* (Oxford: Clarendon 1983).

[4] S.M. Polikanov et al, Soviet Phys. JETP 15 (1962) 1016.

[5] S. Bjørnholm and J.E. Lynn, Rev. of Modern Phys. Vol.52, No 4 (1980) 725.

[6] R. Bengtsson et al, Phys. Lett. **57B** (1975) 218.

[7] K. Neergård and V.V. Pashkevich, Phys. Lett. **59B** (1975) 218.

[8] P.J. Twin et al, Phys. Rev. Lett. **57** (1986) 811.

[9] S. Åberg, H. Flocard and W. Nazarewicz, Ann. Rev. Nucl. Part. Sci. **40** (1990) 439.

[10] S. Åberg, Nucl. Phys. **A520** (1990) 35c.

[11] S. Åberg, Nucl. Phys. **A557** (1993) 17c.

[12] R.B. Firestone and B. Singh, preprint 1994.

[13] S. Tomsovic and D. Ullmo, Phys. rev. **E50** (1994) 145.

[14] B.R. Mottelson and J.G. Valatin, Phys. Rev. Lett. **5** (1960) 511.

[15] W. Nazarewicz, R. Wyss and A. Johnson, Nucl. Phys. **A503** (1989) 285.

[16] B. Cedervall et al, Phys. Rev. Lett. **72** (1994) 3150.

[17] S. Åberg, Phys. Scripta **25** (1982) 25.

[18] C. Baktash et al, *Proc. Conf. on Physics from Large γ-ray Detector Arrays*, Berkeley, USA, Aug. 1994.

[19] T. Bengtsson, I. Ragnarsson and S. Åberg, Phys. Lett. **B208** (1988) 39.

[20] R. Nazmitdinov and S. Åberg, Phys. Lett. **B289** (1992) 238.

[21] T. Nakatsukasa, S. Mizutori and K. Matsuyanagi, Progr. Theor. Phys. **89** (1993) 847.

[22] D.M. Cullen et al, Phys. Rev. Lett. **65** (1990) 1547.

[23] J. Höller and S. Åberg, Z. Phys. **336** (1990) 363.

[24] S. Åberg and L.-O. Jönsson, Z. Phys. **A349** (1994) 349.

[25] A. Galindo-Uribarri et al, Phys. Rev. Lett. **71** (1993) 231.

[26] M. Harvey, *Proc. 2nd Int. Conf. on Clustering Phenomena in Nuclei*, College Park 1975, USDERA Report ORO-4856-26, p.549.

[27] E. Merzbacher, *Quantum Mechanics* (John Wiley 1970, 2nd ed.) pp 65.

[28] G. Leander and S.-E. Larsson, Nucl. Phys. **A239** (1975) 93.

[29] I. Ragnarsson, S. Åberg and R.K. Sheline, Phys. Scr. **24** (1981) 215.

[30] W.D.M. Rae, *Proc. 5th Int. Conf. on Clustering Aspects in Nuclear and Subatomic Systems*, Kyoto, Japan, 1988 (eds.: K. Ikeda, K. Katori and Y. Suzuki) p.77.

[31] J. Cseh and W. Scheid, J. Phys. **G18** (1992) 1419.

[32] H. Morinaga, Phys. Rev. **101** (1956) 245.

[33] P. Chevallier et al, Phys. Rev. **160** (1967) 827.

[34] A.H. Wuosmaa et al, Phys. Rev. Lett. **68** (1992) 1295.

[35] P. Möller, J.R. Nix and W.J. Swiatecki, Nucl. Phys. **A469** (1987) 1.

[36] E.K. Hulet et al, Phys. Rev. Lett. **56** (1986) 313.

[37] A. Bohr and B.R. Mottelson, *Nuclear Structure, Vol. 1*, (New York: Benjamin, 1975).

[38] J. Dudek et al, Phys. Rev. **C26** (1982) 1712.

[39] L.-O. Jönsson and S. Åberg, to be publ.

[40] K. Schiffer, B. Herskind and J. Gascon, Z. Phys. **A332** (1989) 17; K. Schiffer and B. Herskind, Nucl. Phys. **A520** (1990) 521c.

[41] H.J. Krappe and J.R. Nix, *Physics and Chemistry of Fission*, Vol.1 (Vienna: IAEA, 1974) p.159.

[42] M.A. Riley et al, Nucl. Phys. **A512** (1990) 178.

[43] I. Tanihata et al, Phys. Lett. **B206** (1988) 592.

[44] P.G. Hansen and B. Jonson, Europhys. Lett. **4** (1987) 409.

[45] M.V. Zhukov et al, Phys. Rep. Vol. 231(4) (1993) 151.

[46] I. Ragnarsson, S. Åberg, H.-B. Håkansson and R.K. Sheline, Nucl. Phys. **A361** (1981) 1; I. Ragnarsson, T. Bengtsson and S. Åberg, Proc. XIX Int. Winter Meeting on Nuclear Physics, Bormio, Italy, 1981, p.48

Heavy Ion Scattering Problems; Regular and Chaotic Regimes

C.H. Dasso, M. Gallardo [a] and M. Saraceno [b]

The Niels Bohr Institute, University of Copenhagen
Blegdamsvej 17, DK-2100 Copenhagen Ø, Denmark

[a] Departamento de Física Atómica, Molecular y Nuclear
Universidad de Sevilla, Apdo. 1065, E-41080 Sevilla, Spain

[b] Departamento de Física, Comisión Nacional de Energía Atómica
Av. del Libertador 8250, Buenos Aires, Argentina

Abstract: We consider a simple model for a nuclear scattering problem involving the coupling of the relative motion between two heavy ions to an intrinsic (harmonic) degree of freedom. At the classical level, the reaction function - relating unobservable incoming quantities to outgoing observables - is the most sensitive probe of the chaotic nature of the scattering. A general procedure to construct the quantum counterpart to these reaction functions is developed and shown to be very sensitive to the presence of chaos even in the regime of very low quantum numbers. In the regular regime it is possible to reproduce quite accurately the quantal results within a semiclassical approximation, just by letting the different trajectories that lead to an identical outcome interfere. We explore ways to extend such a prescription into the chaotic regime.

1 Introduction

The aim of these lectures is not to cover in general the recent developments in the quantum treatment of chaotic systems but rather to provide an interface - which we hope will be smooth - to the current problems related to the understanding of the role and scope of chaotic behaviour in nuclei, in particular in the scattering of heavy ions in the vicinity of the Coulomb barrier. The aim is to remain equidistant between the two conflicting requirements of a realistic description of the nuclear scattering problem and a detailed mathematical treatment of the chaotic behaviour: the first would obscure the pedagogical nature of these lectures while the second would alienate the interested non "chaotic" specialist who would like to understand if the new ideas and methods of non linear dynamics have any relevance in the study of nuclear structure and reactions. The main idea is to reexamine how some quite old model descriptions of heavy ion scattering appear when looked at in this new light. Along the way we will have to develop novel methods of representation of quantum objects in phase space that will allow a close comparison of the classical and quantum descriptions.

We take the point of view that one of the areas where the traces of chaotic behaviour can be looked for in nuclei is in the phenomenological description of collective degrees of freedom. Inasmuch as we are able we emphasize the *dynamical* as opposed to the *statistical* aspects of the motion.

There are several excellent expositions of classical and quantum chaotic behaviour in hamiltonian systems, which is the theory underlying the dynamics of systems with *few* degrees of freedom. Coherent monographies that present a view of the whole picture are among others, those by Ozorio de Almeida [1], Gutzwiller [2], Reichl [3], Eckhardt [4]. Various aspects of the subject have been the topic of several schools and lecture notes [5-8]. The particular problems related to scattering have been the subject of a recent "focus" issue of *CHAOS* [9], where many relevant methods and applications can be found.

The particular problem that we focus upon is the coupling between relative motion and intrinsic surface vibrations in nuclear reactions between heavy ions [10-11]. The presence of the Coulomb barrier and the coupling introduces – at the classical level – a transition between ordered and disordered motion. Chaotic features set in at energies near the Coulomb and centrifugal barriers in situations of weak absorption and are reflected in a very irregular behaviour of scattering observables. Very similar effects have been studied in detail for the coupling to rotational motion [12-13]. In fact the occurrence is very general: whenever a classical separatrix is perturbed one should expect the development of a chaotic layer in its place, whether in bound or in scattering situations.

The model that we take is simple but still quite realistic and it allows us to analyze in detail its classical, quantum and semiclassical aspects, with particular emphasis in developing methods that display the analogies and the differences between them.

2 Formalism

The relative motion of two heavy ions is described in first approximation by a potential model whose more relevant feature is the barrier created by the balance between Coulomb repulsion and nuclear attraction. When the shape degrees of freedom of the nuclear surface are considered, the relative motion is coupled to their dynamics and the scattering is then described by coupled equations with many degrees of freedom. The deformations of the nuclear shape are described by the macroscopic variables $\alpha_{\lambda\mu}$ which characterize deviations from the equilibrium surface according to

$$R(\theta, \varphi) = R_\circ \left[1 + \sum_{\lambda\mu} \alpha_{\lambda\mu} Y_{\lambda\mu}(\theta, \varphi) \right] . \tag{1}$$

The intrinsic degrees of freedom associated with these variables are significantly excited in peripheral collisions [14]. Consequently, they play an important role in the modulation of the ion-ion interaction that affects near-barrier phenomena, such as those leading to fusion [15-16].

As a simplified model we consider the coupling of the relative motion of two ions to an intrinsic harmonic mode, as expressed by an effective hamiltonian of the form

$$H^{(\ell)}(r, p, \alpha, \Pi) = H^{(\ell)}_{rel}(r, p) + H_{int}(\alpha, \Pi) + V_{coup}(\alpha, r) . \tag{2}$$

Here r is the distance between the centers of mass of the colliding systems, α is the dimensionless variable that measures the amplitude of the vibrational motion and p, Π are, respectively, their conjugate momenta. We take

$$H^{(\ell)}_{rel}(r, p) = \frac{p^2}{2m} + \frac{\ell(\ell+1)\hbar^2}{2mr^2} + U(r) , \tag{3}$$

$$H_{int}(\alpha, \Pi) = \frac{C\alpha^2}{2} + \frac{\Pi^2}{2D} . \tag{4}$$

where m is the reduced mass, ℓ is the angular momentum and C and D the restoring force and mass parameters of the collective vibration. These last two quantities are related to the energy $\hbar\omega$ and deformation parameter β of the mode by

$$C = \frac{\hbar\omega}{2\beta^2}, \qquad D = \frac{\hbar}{2\omega\beta^2} . \tag{5}$$

The real potential $U(r)$ represents the combined effects of the nuclear and Coulomb interactions. For concretness the nuclear potential is taken as a Woods-Saxon version of the Christensen-Winther empirical potential of [17]. The Coulomb potential is screened at large distances. The term that couples the intrinsic and relative motion variables, V_{coup}, arises from the Coulomb and surface-surface nuclear interactions between projectile and target. In leading order both these contributions are proportional to the deformation amplitudes. If a multipolarity for the mode is specified, it is possible within the present scheme to incorporate an effective Coulomb formfactor with the proper radial dependence,

$$V_{coup}(r, \alpha) \approx \left[-R_0 \frac{\partial V_N}{\partial r} + \frac{3Z_1 Z_2 e^2}{(2\lambda + 1)} \frac{R_0^\lambda}{r^{\lambda+1}} \right] \alpha = F(r)\alpha . \tag{6}$$

Note that the different μ-components of a mode of multipolarity λ have been combined in an effective "monopole" amplitude. This procedure, justified in a coupled-channels approach because of their degeneracy in energy, is also appropriate for the head-on case we treat below. However the main justification is to make possible a detailed classical analysis that is not encumbered by the need to display many degrees of freedom. The actual multipolarity of the mode is taken into account in the radial dependence of the Coulomb component of the formfactor.

It is more convenient at times to use the action angle variables n, ϕ for the oscillator. They are related to α, Π by

$$\alpha = \sqrt{(2n+1)\hbar\omega C^{-1}} \cos\phi, \qquad \Pi = \sqrt{(2n+1)\hbar\omega D} \sin\phi \qquad (7)$$

At this point n is a continuous variable, which will eventually become discretized at integer values in the quantum treatment.

The equations of motion derived from the hamiltonian (2) in these variables are

$$\dot{p} = -\partial U(r)/\partial r - 2\beta\, \partial F(r)/\partial r \sqrt{n+\tfrac{1}{2}} \cos\phi$$
$$\dot{r} = p/m$$
$$\dot{n} = 2\beta\, F(r) \sqrt{n+\tfrac{1}{2}} \sin\phi \qquad (8)$$
$$\dot{\phi} = \hbar\omega + \beta\, F(r) \frac{\cos\phi}{\sqrt{n+\tfrac{1}{2}}}$$

When supplemented by initial conditions they define entirely the classical trajectories of the system. Except for energy and distance scales and the specific forms of the potential and the coupling - taken here with a heavy-ion collision in mind - the problem that we treat here is very similar to that of inelastic collisions in atomic and molecular physics treated by Miller [18] many years ago. We refer the reader to that very clear exposition for details on the techniques required for the classical solution.

We now address the quantum mechanical aspects of the problem. The standard numerical treatment is through a coupled channel calculation. In the basis of eigenstates $|n\rangle$ associated with the unperturbed hamiltonian,

$$H_{int}|n\rangle = (n+\tfrac{1}{2})\hbar\omega|n\rangle = \epsilon_n|n\rangle \qquad (9)$$

we can expand the total wavefunction $|\Psi\rangle$ as

$$\langle r\alpha|\Psi\rangle = \sum_n \chi_n(r)\langle\alpha|n\rangle . \qquad (10)$$

Projection into the intrinsic states leads to the set of coupled-channel differential equations

$$\left[-\frac{\hbar^2}{2m}\frac{d^2}{dr^2} + \frac{\ell(\ell+1)\hbar^2}{2mr^2} + [U(r) - E + \epsilon_n] \right] \chi_n(r) = \sum_{n'} F_{nn'}(r)\chi_{n'}(r) , \qquad (11)$$

where the coupling formfactors are defined by

$$F_{nn'}(r) = \int d\alpha \, \langle n|\alpha\rangle\, V_{coup}(\alpha, r)\, \langle\alpha|n'\rangle . \qquad (12)$$

These equations are solved numerically – for $\ell = 0$ – subject to the asymptotic boundary conditions

$$\lim_{r\to\infty} \chi_{n,ni}(r) = \delta_{n,n_i} \exp(-ik_{n_i}r) + r_{n,n_i} \exp(+ik_n r) , \qquad (13)$$

where $k_n = \sqrt{2m(E - \epsilon_n)/\hbar^2}$. In this expression $|n_i\rangle$ is the state in which the intrinsic system is prepared to receive the incoming flux and therefore the index n_i labels both the resulting wavefunctions and reflection amplitudes. While in a normal situation the target would be in the ground state, $n_i = 0$, we contemplate other possibilities in order to explore the scattering matrix in its full extent. A numerical solution of the coupled-channel problem requires the truncation of the harmonic ladder to a finite number of states. The size d of the space is chosen so as to ensure that the outermost state, $|n_d\rangle$, is not significantly populated. The output of this calculation is then the $d \times d$ unitary S - matrix

$$\langle n_f|S(E)|n_i\rangle = \sqrt{\frac{k_f}{k_i}} \; r_{n_f, n_i} \tag{14}$$

which provides all the quantum information about the scattering. In particular the excitation probabilities are given by

$$P(n_f, n_i) = |\langle n_f|S(E)|n_i\rangle|^2$$

Our main goal in these lectures is to study the nature of both the classical and the quantum solutions to this simplified problem and establish as many points of contact as possible between them.

3 Overview of the classical results

A scattering trajectory is determined by the initial conditions

$$\phi_i, \quad n_i, \quad r_i = large, \quad p_i = -\sqrt{2m(E_{tot} - (n_i + \tfrac{1}{2})\hbar\omega)} \tag{15}$$

At constant energy the only interesting asymptotic values are then n_i, ϕ_i and n_f, ϕ_f. The reaction function expresses the relationship $n_f = n_f(n_i, \phi_i, E_{tot})$. The dependence on ϕ_i is periodic and a particular origin is fixed by the value of r_i chosen. (It is possible to remove this dependence by considering the angle $\phi = \phi - \hbar\omega r/\dot{r}$ which is the phase angle aquired by the oscillator *in addition* to the free rotation.) In terms of the asymptotic variables the scattering process can be thought of as a *mapping* of the initial variables n_i, ϕ_i to n_f, ϕ_f. The dependence on the radial variables has been eliminated. We will come back to this mapping when we discuss the semiclassical treatment.

We now study the overall features of this mapping. In a scattering situation it is common to start in the ground state $(n_i = 0)$ and to explore the dependence with the bombarding energy thus changing E_{tot}. An overall picture of the function $n_f(\phi_i, E_{tot})$ at $n_i = 0$ looks as follows

$\beta=0.01$ $\beta=0.04$

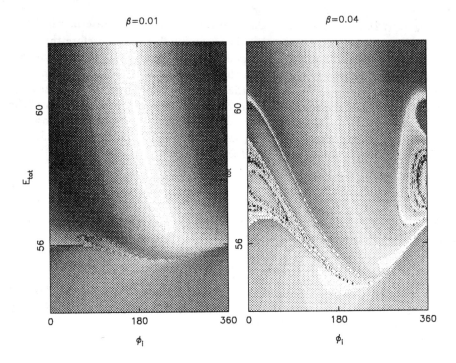

The shades of gray give the value of n_f (white for $n_f = 0$) . The two pictures give an idea of the way in which the barrier region is disrupted by the coupling. In the uncoupled case the picture would be uniformly white ($n_f = n_i = 0$) and there would be a sharp straight line corresponding to the singular trajectory at the top of the barrier ($E_{tot} \approx 56\text{MeV}$). This singularity reflects the fact that the unstable point at the top of the barrier is a *bound* motion and can only be reached asymptotically from the scattering region. The important fact is that this singular trajectory sharply separates the classical motions that occur above and below the barrier. Small values of the coupling ($\beta = 0.01$) bring about the possibility of excitation ($n_f \neq n_i$) along with a mostly *smooth* dependence on ϕ_i. The singular region is slightly distorted but still provides a clear separation for trajectories that scatter below or above the barrier. Larger values of the coupling ($\beta = 0.04$) start to display the full complexity of the barrier region. The "barrier " now extends over several Mev and is characterized by regions of smooth dependence separated by singular regions with structure on all scales. A detail of the previous figure shows clearly this intricate mixture.

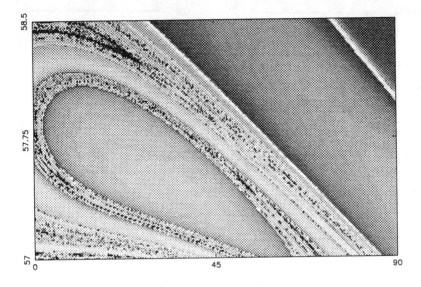

A further blowup would show a very similar structure, with finer and finer details. In spite of the apparent complication, there is a definite organization of the picture, both in energy and angle. Each smooth region is characterized by trajectories that have similar behaviour and which bounce in the interaction region in definite ways. This organization is quite stable when the potential is slightly changed and points to the existence of a general mechanism that is at work in the vicinity of the barrier region.

3.1 From a separatrix to a chaotic layer

We now explain the physical mechanisms that lead to this organization. The clue is the existence in phase space of a *separatrix* which the coupling turns into a chaotic region. Let us explain how this happens starting from the simple example of a one dimensional barrier.

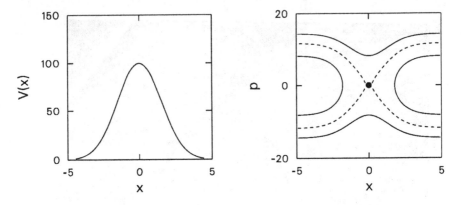

The phase space plot on the right shows the singular trajectory (dotted line) whose energy is exactly at the barrier top: particles with this energy will never reach the top (only in an infinite amount of time) and viceversa a particle sitting at the position of the heavy dot will be in an unstable equilibrium and will be a *bound* – not scattering – solution to the equations of motion. The singularity is manifest in many ways – for example the time delay becomes infinite there – and nearby purely scattering trajectories are also strongly affected. This is the simplest example of how a bound motion can have a strong influence on the scattering.

What happens when this simple scheme is perturbed by some coupling ? This can happen either because we make the problem time dependent – we shake the barrier – or, as for our problem, because we couple in some other degree of freedom. The effect is the same in both cases and it is simpler to think of the barrier as oscillating periodically in height. In this case the trajectories are not obtained any more as $H(p, x) = E$ but have to be obtained by integration of the time dependent equations. The general features of the resulting motion can be studied by recording the phase space position at equal time intervals, *i.e.* by studying the behaviour of the map that generates the sequence $p_0, x_0, p_1, x_1,$ ..., etc. In this map the point at the top of the barrier remains a fixed point, but nearby initial conditions will diverge exponentially from it. The separatrices become the stable and unstable manifolds of this fixed point (*i.e.* the lines of initial conditions that either converge to or diverge from it for infinite times). By itself, the presence of an unstable point of this type will not suffice to create chaotic behaviour. What is needed is that the stable (W^-) and the unstable (W^+) manifolds intersect in such a way that phase space points that are ejected from the hyperbolic region in the vicinity of W^+ are reinjected on W^-. This can happen if the absortion is weak enough so that we no longer face an open problem inside the potential pocket. Then the separatrices bend and eventually cross giving rise to one – and consequently an infinity – of homoclinic points.

One of the standard results of the analysis of the motion close to a hyperbolic point [6] reveals the fact that a transversal crossing of its stable and unstable manifolds at a homoclinic point creates an infinity of new unstable periodic orbits. These orbits in turn have manifolds that extend to infinity and therefore deeply affect the scattering process. One way to think of this process is the following. In the uncoupled regime the separatrix is the only initial condition that is *singular* because motion along it will take an infinite time to reach the barrier and will not exit. But when a homoclinic crossing occurs each one of the infinite new periodic motions will create its own singularity and therefore the set of singular initial conditions is now very complicated (it is in fact a fractal set called a *chaotic repeller* [7]).

In physical terms we can say that the Coulomb barrier provides *hyperbolicity* while the possibility of *folding* is provided by the inner potential barrier. These are the two mechanisms that create the chaotic layer around the separatrix. It is also to be noticed that, because the relative motion slows down near the barrier, the motion is *never* adiabatic there. Thus, when thinking in purely classical terms, the chaotic regime sets in at the point when the adiabatic one ceases to be valid and the two descriptions complement each other.

4 Classical reaction functions

Let us now look in more detail at the reaction functions at a fixed energy. These will be cuts of the previous global pictures that show in detail the possible n_f outcomes. We first choose a regular situation well above the barrier region with parameters $E_{tot} = 75$ MeV, $\hbar\omega = 2$ MeV, $\beta = 0.035$ and $n_i = 3$. All trajectories in this region are "above barrier" with only one turning point inside the potential.

$\beta=0.035$ $E_{tot}=75$MeV

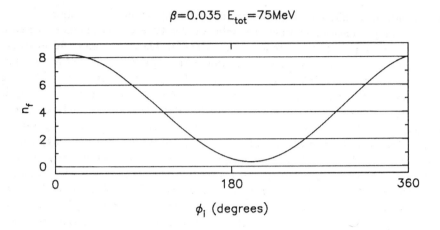

The reaction function is smooth and we observe that all values of n_f in the range $0 < n_f < 8$ can be reached. Moreover for each allowed value of n_f there are two classically allowed trajectories that contribute.

If the energy is lowered towards the barrier the situation changes drastically:

$\beta=0.035$ $E_{tot}=65$MeV

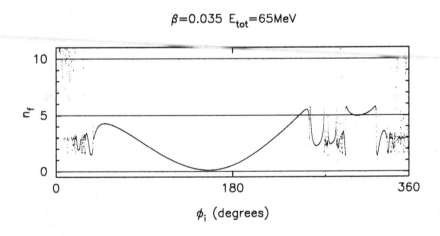

The smooth regions are now interspersed with irregular regions where structure can be found at all scales. The overall range of allowed n_f values has not changed much but an essential difference has appeared: there are now many (in fact an infinite amount) initial conditions leading to a given n_f. A detail of the previous picture shows the appearence of more regular regions separated by finer structure.

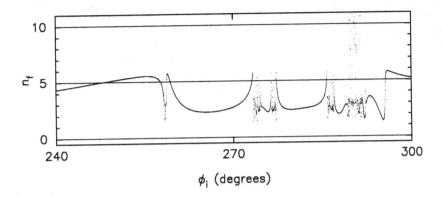

$$\phi_i \text{ (degrees)}$$

Further enlargements would only show similar patterns. These figures give a good idea of the structure of the repeller set. Many other quantities, like the time delay or the angle ϕ_f also show an irregular behaviour at the same places. This behaviour is very general. In the context of nuclear reactions it has also been observed for coupling to rotational motion [12],[13] and our previous arguments show that it will be the norm rather than the exception – whatever the nature of the degree of freedom that is coupled – whenever the inner barrier forces the crossing of the manifolds emanating from the Coulomb barrier.

5 Classical probability distributions

The value of deflection – or in our case reaction – functions is that they provide the basis for the calculation of cross sections. (Recall the elementary relation $\sigma(\vartheta) \sim [d\vartheta/db]^{-1}$.) In cases where the behaviour of the reaction function is chaotic this expression would not seem to be immediately useful, but a slightly different definition brings about its physical meaning and provide a simple way of calculating a cross section for the chaotic case. Consider a uniform sampling of the *unobservable* variable ϕ_i. The probability of reaching a final value of n_f is

$$P_{cl}(n_f) = \frac{1}{2\pi} \int_0^{2\pi} d\phi \, \delta(n_f - n(\phi)) \ . \tag{16}$$

In cases where the reaction function is smooth the integration can be done explicitly and the resulting cross-section is

$$P_{cl}(n_f) = \frac{1}{2\pi} \sum_{\gamma} \left| \frac{dn_f}{d\phi_i} \right|_{\gamma}^{-1} , \tag{17}$$

where the sum contemplates the possibility of several trajectories leading to a given n_f. Notice that, in line with our classical treatment, the different trajectories contribute probabilities – *not amplitudes* – and no phases are involved.

Moreover the possible values of n_f are not quantized and a continuous distribution of final actions is obtained. In the regular case a direct calculation using either expression is equivalent. Notice that there might be singularities at points where the derivative vanishes – an example of a *rainbow* – which occur at the frontier between classically allowed and forbidden values of n_f. In the irregular case, the second version of the formula is not directly applicable as the derivatives are not even defined at an infinite set of points but a discretized version of the first yields very good results. The initial variable is sampled uniformly and the value of n_f is binned to obtain the probability at a given scale of resolution. The results of such a calculation with $\approx$ 7000 points for the two cases considered above yield

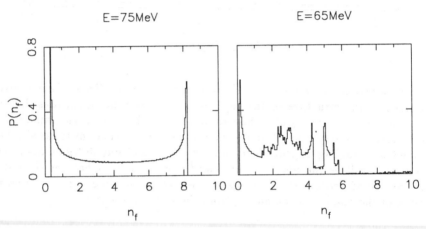

In the regular case the most prominent feature is the presence of the two rainbows at $n_f \approx 0$ and 8, occurring at the edges of the classically allowed regions. In the irregular case several secondary rainbows are visible, corresponding to the small regular regions in the reaction function, as well as a small component at $n_f \approx 9$. In the irregular case the shape of $P(n_f)$ depends on the size Δn_f of the bin, revealing more and more structure as it decreases (the size in the figure is $\Delta n_f = .1$). On the other hand, once a bin size is chosen the resulting distribution is independent of the number of sampling points, provided it is large enough.

The end procedure to compute these distributions is then like a Monte Carlo calculation in which initial values of the variables are sampled and the information contained in many trajectories is accumulated to construct cross sections and angular distributions. Such techniques have been commonly exploited in the analysis of heavy-ion reactions for both surface vibrations and rotations involving in some instances hundreds of degrees of freedom (cf. eg. refs. [19-21]).

6 Interference

The calculation of transition rates or cross-sections by simple addition of probabilities does not take into account the interference effects that occur in quantum mechanics when two (or several) classical trajectories lead to the same final values of the observables. The simple addition of probabilities must be replaced by the addition of *amplitudes* with the proper phases. The formal procedure to accomplish this is well known for scattering: Miller [18] has derived the approximate expression for the S-matrix valid in the semiclassical limit

$$\langle n_f | S | n_i \rangle_{sc} = \sum_{\gamma} p_{\gamma} \exp i\Phi_{\gamma}(n_f, n_i) , \qquad (18)$$

in which the sum is over all classical trajectories γ which satisfy the boundary conditions n_i and n_f for large values of r. Φ_{γ} is the phase with which they contribute

$$\Phi_{\gamma}(n_f, n_i) = -\nu_{\gamma} \frac{\pi}{2} - \int_{\gamma} \left(\frac{rdp}{\hbar} + \phi dn \right) \qquad (19)$$

and ν_{γ} is the additional Maslov phase that takes into account the points where the prefactor amplitude diverges. The prefactor p_{γ} is

$$p_{\gamma} = \left| \frac{1}{2\pi} \frac{\partial \phi_i}{\partial n_f} \right|^{\frac{1}{2}} \qquad (20)$$

and is just the square root of the classical contribution of each trajectory from (17). All quantities can be calculated for each scattering trajectory entirely in terms of classical mechanics: the action Φ by the numerical evaluation of the integral and the Maslov index by counting the number of times the trajectory crosses a caustic. These quantities, although not routinely calculated in classical reaction codes, are easily incorporated along with the integration of trajectories. Taking this semiclassical formula for the S-matrix we can compute the reaction probabilities as

$$P_{sc}(n_f, n_i) = |\langle n_f | S(E) | n_i \rangle_{sc}|^2 = \sum_{\gamma} p_{\gamma}^2 + \sum_{\gamma \neq \delta} p_{\gamma} p_{\delta} \exp i(\Phi_{\gamma} - \Phi_{\delta}) . \qquad (21)$$

The first term is just $P_{cl}(n_f)$ that we computed in (17). The second gives the desired interference in terms of the action differences among the contributing trajectories. This semiclassical result can in principle be evaluated for arbitrary real values of n_f but only for quantized integer values will it compare to the quantum calculation. The reaction probabilities can then be evaluated in three different ways: a purely classical one without taking into account interference, a semiclassical one with Miller's S-matrix and the exact coupled channel results. In the regular case at $E = 85$ MeV there are always two trajectories contributing to a given n_f and the resulting interference pattern is very simple

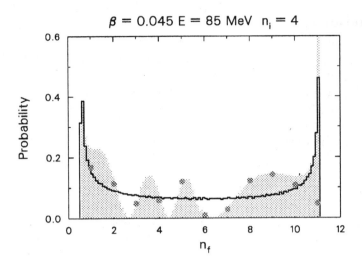

$\beta = 0.045 \ E = 85 \ \text{MeV} \ n_i = 4$

The shaded area is the semiclassical result computed as a continuous function of n_f to show more clearly the resulting interference pattern. At integer values it can be compared to the quantum results (heavy dots). The comparison is overall very good with significant deviations near the edges of the allowed regions where the presence of rainbows renders the semiclassical S-matrix invalid. There are well defined methods to deal with these effects that involve the consideration of complex trajectories [22] but we postpone for the moment their consideration.

It should be by now clear to the reader what is the additional difficulty that the presence of chaotic behaviour represents for this type of calculation. The number of classical trajectories that contribute to a given n_f is not well defined (it will increase as we increase the number of trajectories per interval of ϕ_i), but more crucially the sum over γ in (18) will have to take into account the interference of widely different trajectories with possibly different actions. Questions about the convergence and reliability of this procedure are very severe. It is certainly not very economical to have to compute and add an indefinite number of scattering trajectories to come up with a number that is at best an approximation to the quantum result. This state of affairs is quite similar to what happens for the computation of energy levels from periodic orbits [23] where an exponentially increasing number of trajectories have to be added to obtain a reasonable description of the spectrum of bound systems in terms of classical mechanics.

The results of this procedure in our case (with 3600 trajectories considered) is the following:

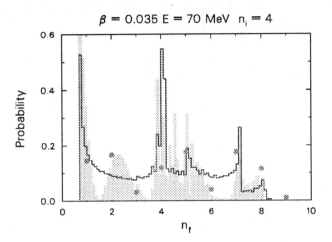

$\beta = 0.035\ E = 70\ \text{MeV}\ \ n_i = 4$

The agreement is still quite good in the regular regions and might be considered accidental in the chaotic regime around $n_f \simeq 5$. However preliminary, this result can be considered as a green light to proceed to more sophisticated techniques to perform the sum, but this we have not yet attempted.

7 Quantum reaction functions

There is another way to look at the coupled channel problem that displays more clearly the emerging chaotic behaviour. Remember that, classically, the most telling characteristic of chaotic behaviour was the irregular structure of the reaction function. Can we observe this behaviour – or some traces of it – in the *quantum* calculation?. We describe next a simple method, which is quite general, to construct quantum reaction functions from the knowledge of the S-matrix. First notice that the channel indices n_i, n_f are the quantized actions of the unperturbed oscillator. In these states the corresponding phase angle is totally undetermined. To allow some localization in angle we must allow some delocalization in action thus superposing several states $|n>$ with the proper phases. The best simultaneous specification of both n_i and ϕ_i in the initial state can be met, within the limitations of the uncertainty principle, by constructing the coherent oscillator state

$$|I\rangle = \sum_n \frac{I^n}{\sqrt{n!}} \exp -\frac{|I|^2}{2} \ |n\rangle \ , \tag{22}$$

where the complex number I is given by

$$I(n_i, \phi_i) = \sqrt{n_i + \tfrac{1}{2}}\left(\cos \phi_i + i \sin \phi_i \right). \tag{23}$$

An amplitude for populating the state n_f starting from the initial coherent state can then be constructed through

$$a_{n_i}(n_f, \phi_i) = \sum_n \langle n_f | S | n \rangle \, \langle n | I(n_i, \phi_i) \rangle \,, \tag{24}$$

and we can then call the probability distribution $W_{n_i}(n_f, \phi_i) = |a_{n_i}(n_f, \phi_i)|^2$ a *quantum reaction function*. The contours of this distribution have been plotted in the next figure for two situations in the regular (above) and irregular (below) regimes. Superimposed in the figure are the results of the classical calculation.

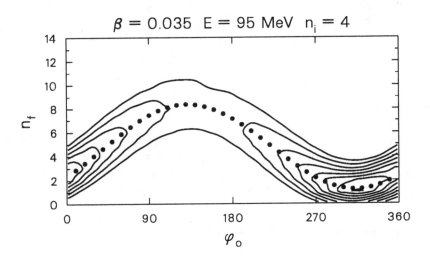

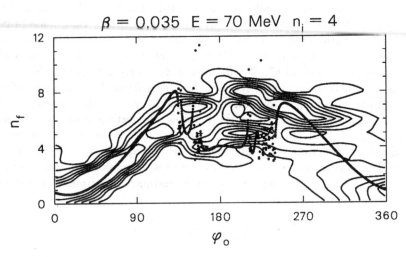

Even with the coarse angular resolution obtained with the interferences of a few quanta, the distribution shows a remarkable correspondence with the complex pattern that characterizes the fractal structure of the classical solutions. In this respect notice that the number of quanta that are significantly coupled gives the number of Fourier components present in the angle distribution. It is then clear that the classical pattern will only be an emergent feature of the quantum calculation showing more and more detail as the classical regime is reached and the barrier region is spanned by an infinity of channels. The quantum reaction function that we constructed display some properties of the S-matrix that are not probed in the inelastic probabilities $P(n_f)$. The relative phases of the different S-matrix components are important in the agreement that is obtained. Thus the quantum reaction functions complement the information contained in $P(n_f)$ and are closely correlated to the nature of the classical motion, therefore revealing quite clearly the presence of chaotic motion.

The procedure described above is very general. We have simply reexpressed the indeces of the S-matrix in the coherent state representation. For oscillator states this is a standard procedure which has also many analytical advantages. However the procedure is the same for any coupled channel calculation and for any S-matrix label. The essence of the procedure lies in the fact that we have constructed a localized wavepacket in action angle variables, the action being the discrete index n. The fact that this is an oscillator state is irrelevant and it could well have been an angular momentum, a particle number or any quantized torus action.

8 Perspectives and concluding remarks

In these lectures we have illustrated several aspects of chaotic motion in the context of heavy ion reactions by means of a simple model, which however has all the characteristics of more elaborate ones. Preliminary results [24] show that in fact it is possible to extract quantum reaction and deflection functions from realistic "state of the art" coupled channel codes like FRESCO [25], and that many features are closely correlated with classical trajectories, both regular and chaotic.

What have we learned?. It is clear that chaotic scattering will always be present – at the classical level – in near barrier phenomena in cases where the inner potential can be accessed. In this respect thinking in terms of chaotic motion should replace other mental pictures of reaction processes such as adiabaticity or integrability. Whether such a picture is useful or not will depend very much on two essential items yet to be investigated:

a) The description should be quantum mechanical. The relationship between classical trajectories and quantum effects in the chaotic region is complicated by the fact that contributing trajectories proliferate exponentially (cf. the construction of Miller's semiclassical S-matrix in Sec. 6). Even if summing the contributions proves to be feasible and accurate it is certainly not economical in realistic cases with many degrees of freedom. Hovever we have seen that the main features of the scattering matrix do have a basis in classical motion, even at the very low quantum numbers involved, and thus some role for the understanding in terms of few classical elements can be hoped for.

b) The role of absorption has to be addressed. The usual modeling by an optical potential of channels other than those taken explicitly into account at the dynamical level introduces new features that tend to blur the very fine structures of the chaotic regime. In this respect it should be noticed that a situation of relative transparency brings about a selective absorption of the more complicated classical trajectories. The possibility then arises for a convergent semiclassical calculation in terms of *few* trajectories which would be both economical and accurate. In this context absorption would play – in a very natural way – the same role as an imaginary part of the energy plays in bound-state calculations.

c) The task ahead is to look for experimentally verifiable consequences of these phenomena. For instance, calculations have indicated that excitation functions have unusual and pronounced structures in close correlation with the presence of chaotic motion. These are currently being searched for experimentally [26].

This work was supported in part by the grant DFG Bo 1109/1, the Heraeus Foundation, CONYCET Pid. 3233/92, the Fundación Antorchas, the Spanish CICYT project PB92-0663 and the EU Network contract CHRX-CT92-0075.

References

1. *Hamiltonian Systems: Chaos and Quantization*, A. M. Ozorio de Almeida, Cambridge University Press, Cambridge, 1988;

2. *Chaos in Classical and Quantum Mechanics*, M.C. Gutzwiller, Springer, New York, 1990

3. *The Transition to Chaos*, L.E. Reichl, Springer, New York, 1992

4. B. Eckhardt, Phys. Rep. **163** (1988) 205

5. *Stochastic Behavior in Classical and Quantum Hamiltonian Systems* (Proceedings, Como 1977) G. Casati and J. Ford eds., Lect. Notes in Physics **93**, Springer (Berlin, 1979)

6. *Chaotic Behaviour in Deterministic Systems* (Proceedings, Les Houches 1981), G. Iooss, R.G. Helleman and R. Stora eds., North Holland (Amsterdam, 1983)

7. *Chaos and Quantum Mechanics* (Proceedings, Les Houches 1989),M-J. Giannoni, A. Voros and J. Zinn-Justin eds., North-Holland (Amsterdam, 1991)

8. *Quantum Chaos* (Proceedings, International School "E. Fermi" Course CX-IX, Varenna, 1991), G. Casati, I. Guarneri and U. Smilansky eds., North Holland (Amsterdam, 1993)

9. Chaos **3** (1993) 417. (Focus issue on "Chaotic Scattering")

10. C.H. Dasso, M. Gallardo and M. Saraceno, Nucl. Phys. **A459** (1992) 265
11. C.H. Dasso, M. Gallardo and M. Saraceno, Braz. Jour. Phys. **24** (1994) 643
12. A. Rapisarda and M. Baldo, Phys. Rev. Lett. **66** (1991) 2581
13. M. Baldo, E.G. Lanza, A. Rapisarda, Chaos **3** (1993) 691
14. S. Landowne and A. Vitturi in *Treatise on Heavy-Ion Science*, Vol. 1, ed. D.A. Bromley, Plenum Publishing Co., 1984
15. M. Beckerman, Phys. Reports **129C** (1985) 145
16. *Proceedings of the International Conference on Fusion Reactions below the Coulomb Barrier*, MIT, June 13-15, 1984, Ed. S.G. Steadman, Lecture Notes in Physics **219**, Springer-Verlag, 1985
17. P.R. Christensen and A. Winther, Phys. Lett. **65B** (1978) 19
18. W.H. Miller, Adv. Chem. Phys. **25** (1974) 69
19. R.A. Broglia, C.H. Dasso and A. Winther, Proceedings of the International School of Physics "Enrico Fermi" on Nuclear Structure and Heavy Ion Collisions, ed. by R.A. Broglia, C.H. Dasso, and R. Ricci, North Holland, 1981, p. 327
20. C.H. Dasso and G. Pollarolo), Comp. Phys. Comm **50** (1988) 341
21. C.H. Dasso, T. Dossing, S. Landowne, R.A. Broglia and A. Winther, Nucl. Phys. **A389** (1982) 191; C.H. Dasso, in Proceedings of the XXI International Winter Meeting on Nuclear Physics, Bormio, Ric. Sci. Edu. **S30** (1983) 722
22. J. Knoll and R. Schaeffer, Ann. Phys. (N.Y.) **97** (1976) 307
23. CHAOS **2** (1992) (focus issue on periodic orbit theory)
24. G. Pollarolo, private communication
25. I.J. Thompson, Comput. Phys. Rep. **7** (1988) 167
26. A. Szanto de Toledo, private communication

Deterministic Chaos in Heavy–Ion Reactions[1]

Marcello Baldo, Edoardo G. Lanza and Andrea Rapisarda

Istituto Nazionale di Fisica Nucleare, Sezione di Catania
Dipartimento di Fisica, Università di Catania
Corso Italia 57, I-95129 Catania, Italy

Abstract: We review the phenomenon of chaotic scattering in heavy–ion reactions at energies around the Coulomb barrier. A model in two and three dimensions which takes into account rotational degrees of freedom is discussed both classically and quantum-mechanically. The typical chaotic features found in this description of heavy–ion collisions are connected with the anomalous behaviour of several experimental data.

1 Introduction

The discovery that dynamical systems governed by non-linear and non-integrable deterministic equations can exhibit a very irregular and unpredictable motion, i.e. *deterministic chaos* [1-3], represents a very important step forward for contemporary science. The transition form order to chaos is a novel important perspective which shades a new light in classical mechanics. Fluctuations which were previously rejected as spurious noise have found a sound theoretical and experimental explanation revealing simple and deterministic laws. On the other hand the assumption of many degrees of freedom is not any longer necessary for statistical mechanics to be applied.

These new and revolutionary ideas originate from the pioneering work of Poincarè at the beginning of the century. Unfortunately the exciting facets of relativistic theory on one hand and the puzzling aspects of quantum mechanics on the other one were already exhausting the mind of the scientists of the time. Hence it was only in the sixties with the fundamental work of the mathematicians Kolmogorov, Arnold and Moser (KAM theorem) that a new general interest on this field was raised. Then the recent availability of fast and powerful computers has given a strong boost to nonlinear science. Nowadays classical chaos with its precise and rigorous laws, is a well established concept (see Saraceno's lectures)

---

[1] talk given by E. G. Lanza

and is finding application in the most different fields: from physics to economics, from engineering to biology [3]. No doubt that this new paradigm will be one of the most important guidelines for research in the next decades.

If classical chaos is a sound concept, many questions rise when one goes to the quantum world. Is there "true" chaos or is there a "classical limit" which manifests in the quantum systems? Historically, the problem of whether there is chaos in quantum systems has been tackled by using the analogy with the classical world. Systems which show classical chaos have been quantized and studied to see whether they present characteristics which can be ascribed to the underlying classical behaviour. In fact a clear transition from order to chaos has been found also at the quantum level. Quantum chaos is an issue still at the frontierline of current research, but it seems to exhibit well defined features which do not exactly coincide with those found classically [4]. In general the definition of quantum chaos applies to the quantal behaviour of those systems which manifest classical chaos.

Probably the first example of quantum chaotic system was Niels Bohr's compound nucleus model [5], which was based on a statistical point of view. These ideas were quantified by the Wigner's suggestion to describe the neutron resonances in terms of a random matrix model. Then, the discovery of the so called Ericson's fluctuations gave an experimental evidence of this statistical description of the nucleus [6,7].

The most simple examples of fully chaotic classical systems are the billiards: the stadium and the Sinai's billiard. The quantization of such systems show that the fine properties of the discrete spectrum can be used to distinguish between a regular and a chaotic dynamics. In particular the fluctuations around the mean level spacing (nearest-neighbour spacing) depend very much on the characteristic of the system: for a chaotic quantum system they are described by the Wigner's Gaussian Orthogonal Ensemble (GOE), while for a regular one they follow the Poisson distribution.

In the nuclear case [7], the distribution of spacing of nearest-neighbour levels with the same spin and parity (1/2 s-wave neutron resonances) are well described by the GOE distribution. This is true also for other statistical correlations functions like the Δ_3 statistics. Random matrix theory has also been applied to cross section fluctuations both in the weak and strong overlapping resonance range. Ericson's fluctuations belong to the latter.

At neutron threshold, fluctuations of nuclear observables are consistent with the GOE predictions. At low energies, regular motion appear and is well described by the shell model and/or by the collective models. Very recently, it has been discovered [8] that low-lying levels in deformed nuclei behaves in a regular fashion: the fluctuations of the level spacing follows the Poisson distribution.

In this review we want to present a novel aspect of chaoticity in nuclear physics: the occurrence of chaotic scattering in nuclear reactions.

In the last years, the study of classical deterministic chaos has also been extended to the case of open systems. It has been found that scattering variables have an irregular behaviour as a function of the initial conditions when the interaction zone is chaotic [9-14]. Though scattering trajectories explore the real chaotic region only for a finite time, their behaviour can be so complicated that the final observables show strong and unpredictable fluctuations. These fluctuations are present on all scales of the initial conditions, revealing an infinite set of singularities with a Cantor-like fractal structure. Singularities are connected with those trajectories that remain trapped in the interaction region for very long times. In this sense the term chaotic has been extended also to scattering situations [9-12,15-19].

In the following we consider the reaction between a spherical and a deformed nucleus taking into account rotational degrees of freedom only. This is a simplified description of the way in which two nuclei can interact, but it can be considered very realistic for many heavy–ion reactions. We show that even a few degrees of freedom can produce a very complicated and unpredictable motion. This is a novel aspect of chaos in nuclear physics and we will show that it can be related to an anomalous behaviour of cross sections.

This subject has been already discussed in several published papers [13,15-17], in particular in the recent review of ref.[19]. Therefore we recall here only the main findings sending the interested reader to the original papers where further details can be found.

2 Classical scattering

In this section we introduce a three-dimensional model to describe the classical scattering between a spherical nucleus (1) and a deformed one (2). Using polar coordinates, the Hamiltonian depends on 5 degrees of freedoms, i.e. r, θ and ϕ to describe the motion of the spherical projectile and Θ, Φ for the deformed rotor, see fig.1. Thus the Hamiltonian can be written as

$$H = H(r, \theta, \phi, \Theta, \Phi) = T(r, \theta, \phi) + H_2(\Theta, \Phi) + V(r, \theta, \phi, \Theta, \Phi), \qquad (1)$$

where T is the kinetic term, H_2 the Hamiltonian of the deformed nucleus 2 and V is the ion-ion potential which contains the monopole and quadrupole term of the Coulomb interaction plus the nuclear part U_N

$$V = \frac{Z_1 Z_2 e^2}{r} + \frac{Z_1 Q_0 P_2(\cos\xi)}{2r^3} + U_N(r, \xi) , \qquad (2)$$

with

$$\cos\xi = \cos\Theta\cos\theta + \sin\Theta\sin\theta\cos(\Phi - \phi) , \qquad (3)$$

being ξ the angle between the rotor symmetry axis and the line joining the centers of the two nuclei. The symbol Q_o indicates the intrinsic quadrupole moment, while P_2 is the Legendre polynomial of order 2.

We have chosen as nuclear interaction the *proximity* potential [20,21]. This choice has nothing special and it has been considered only because this potential is one of the most commonly used for deformed nuclei.

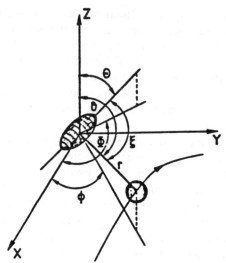

Fig. 1. Coordinate system used. Polar coordinates r, θ and ϕ specify the position of the spherical projectile nucleus, while Θ and Φ are the Euler angles of the intrinsic frame of the deformed target nucleus.

It should be noted that U_N represents the coupling between the relative motion and the internal (rotational) degrees of freedom. This coupling, breaking the central symmetry of the potential, is the one responsible of the onset of chaotic scattering as discussed later. Solving the equations of motion for the Hamiltonian (1) one can follow in time the evolution of the system. However these equations are very general and complicated, thus in order to show in a clear and simple way the typical features of chaotic motion, let us consider for the moment the scattering occurring on the x-y plane. In this case we have only 3 degrees of freedom, i.e. r , ξ and ϕ, while our Hamiltonian has two constants of motion, namely the total energy E and the total angular momentum L. Neglecting the ξ-dependence of the full ion-ion potential, the Hamiltonian is separable and thus integrable, because the internal angular momentum I and the orbital one ℓ are conserved separately. However the ξ-dependence of the ion-ion potential introduces a symmetry-breaking term leading to the conservation of L only and generating the onset of chaos. In reality the scattering problem is integrable asymptotically. It is the chaoticity of the interaction zone which makes the scattering become chaotic. The set of unstable phase space trajectories which are confined in the interaction region defines the so-called *repeller*. The latter has an unstable manifold which extends to asymptotic distances, thus scattering trajectories are trapped for long but finite times inside the phase space region. The erratic, though deterministic, motion of these trapped trajectories, which are those that come closest to the repeller, cause the unpredictability of the final scattering observables on all scales.

As a first example we take into account the reaction between the ^{28}Si nucleus considered spherical and the deformed ^{24}Mg. The potential $V(r,\xi)$ is shown in fig.2 as a function of r. The dependence on the angle, for the cases $\xi = 0^o$ and

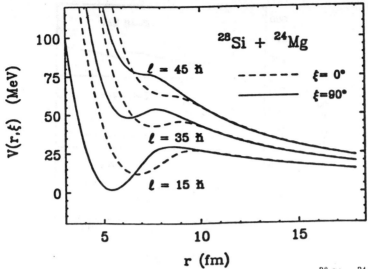

Fig. 2. The ion-ion potential adopted is plotted for the system $^{28}Si + ^{24}Mg$. Three value of orbital angular momentum $\ell = 15, 35, 45$ $\hbar$ are shown for two orientation angles, i.e. $\xi = 0°$ (dashed curve), and $\xi = 90°$ (full curve).

$90°$, is illustrated for three initial orbital angular momenta $\ell = 15, 35, 45\,\hbar$. The change of the orientation angle ξ from $90°$ to $0°$ lowers the height of the barrier and shifts the position of the minimum towards larger radii. Increasing ℓ the attractive pocket tends to disappear due to the enhancement of the centrifugal barrier. One should note that this is only a static picture. Actually, as the nuclei approach each other, due to the coupling between the relative motion and the internal degrees of freedom, the potential oscillates according to the variation of the orbital angular momentum ℓ and the angle ξ.

In fig.3 we show, for a fixed total angular momentum L, the final value of the scattering angle ϕ_f as a function of the initial rotor orientation Φ_i. The initial value of ϕ_i was always set equal to zero (then $\Phi_i = \xi_i$), while the rotor was considered always at rest $I_i = 0$ (then $\ell_i = L$). This choice has been kept through all the calculations presented here. The different trajectories were obtained varying the initial angle $\Phi_i = \xi_i$ from $0°$ to $180°$ and taking into account 1000 trajectories for each of the three different values of energy shown in fig.3. Below the Coulomb barrier - $V_B \sim 26.5 MeV$ - (bottom panel) we have a regular and smooth behaviour, while wild fluctuations show up as soon as the energy is increased (middle panel). These fluctuations tend to vanish and give again a regular motion with only a few singularities as the energy is further increased. The irregular fluctuations found are present at all scales of the initial conditions. Magnification plots show the persistence of regular and irregular islands with a well defined fractal structure[19]. This fact proves that for the reaction $^{28}Si + ^{24}Mg$ the scattering is chaotic just above the Coulomb barrier. Only above the barrier scattering trajectories can probe the chaoticity of the internal zone.

The system $^{28}Si + ^{24}Mg$ has no special characteristics and in fact we have seen that irregular scattering is rather typical for light heavy-ions[19]. Collisions

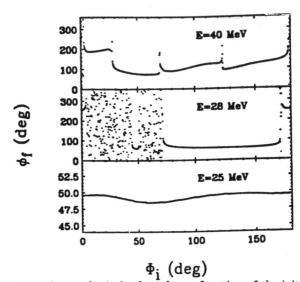

Fig. 3. The final scattering angle ϕ_f is plotted as a function of the initial rotor orientation Φ_i for the reaction $^{28}Si + ^{24}Mg$. For a fixed total angular momentum $L = 15\ \hbar$, three different energy values are considered. Planar scattering is considered in this case, see text.

between nuclei whose atomic mass number A lies in the range between A=4 and A=60, as for example the systems $^{4}He + ^{24}Mg$, $^{12}C + ^{24}Mg$, $^{86}Kr + ^{24}Mg$, show the same fluctuations found for $^{28}Si + ^{24}Mg$[19].

On the contrary, a reaction between two heavy ions, like for example $^{80}Kr + ^{152}Sm$, exhibit only a regular and smooth motion changing both E and L. This different behaviour has two main reasons. First, as the atomic number of the nuclei increases the enhanced Coulomb repulsion reduces the attractive nuclear pocket. Second, the greater are both the mass of the nuclei and the moment of inertia the slower is the motion of the barrier. The relative motion becomes faster than the one of the internal degrees of freedom, whose slow variation is not able to raise the barrier and trap the spherical nucleus. Therefore, in order to have chaotic motion the two characteristic time scales should be comparable.

Chaotic scattering is not peculiar of the simple 2-dimensional model. In fact taking into account the more general 3-dimensional Hamiltonian (1) the possibility for the scattering to be chaotic can even increase. All these features have been studied more quantitatively in ref.[19].

3 Quantum scattering

In the following we discuss the quantal analog of chaotic scattering between a spherical and a deformed nucleus in order to investigate the possible implications of classical chaos.

The quantum analog of classical scattering is given by the solution of the Schrödinger equation. In nuclear physics one usually introduces the deformation degrees of freedom as excited quantum states belonging to rotational bands. These states are coupled among each other and these couplings influence the quantum-mechanical evolution of the reaction. This method is often referred to as the coupled-channels approach.

We assume for simplicity and only for the moment that the reaction occurs on a plane, then the radial coupled–channels equations are

$$\left[\frac{d^2}{dr^2} + \frac{1}{r}\frac{d}{dr} - \frac{\ell^2}{r^2} + k_{L-\ell}^2(r)\right]\psi_\ell^L(r) - \frac{2m}{\hbar^2}\sum_{\ell'\neq\ell} V_{\ell'-\ell}(r)\,\psi_{\ell'}^L(r) = 0 \qquad (4)$$

with

$$k_{L-\ell}^2(r) = \frac{2m}{\hbar^2}(E - E_{rot} - V_o(r)) \qquad ; \qquad E_{rot} = \frac{I^2\hbar^2}{2\Im} \qquad (5)$$

$$V_{\ell'-\ell}(r) = \frac{1}{2\pi}\int_{-\pi}^{\pi} e^{i(\ell'-\ell)\xi}\, V(r,\xi)\,d\xi \qquad ; \qquad V(r,\xi) = V_{coul} + V_{nucl} . \qquad (6)$$

The moment of inertia $\Im$ as well as the ion–ion potential $V(r,\xi)$ are the same considered in the classical case. In (4) the coupling is taken to all orders and only between the nearest neighbours. For more details cfr. refs. [16,17,19].

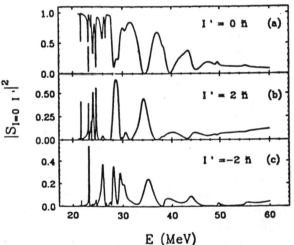

Fig. 4. Quantal elastic and inelastic transition probabilities as a function of incident energy. The calculations are the result of the 2D coupled channels approach described in the text. An energy step equal to 20 KeV is used.

Let us consider again as a typical example the reaction $^{28}Si +^{24} Mg$. Fig.4 shows the elastic ($I' = 0$) and two inelastic ($I' = \pm 2\,\hbar$) transition probabilities $\left|S_{I,I'}^L(E)\right|^2$, calculated at total angular momentum $L = 15\,\hbar$, as a function of incident energy. For an initial spin $I = 0\,\hbar$, 11 final channels were considered, $I = 0, \pm 2, \pm 4, \pm 6, \pm 8, \pm 10\,\hbar$. An energy step of 20 KeV was adopted for the calculations.

The S-matrix elements show rapid oscillations as a function of energy with a width ranging between 50 and 400 KeV, implying the occurrence of long-living intermediate states of the dinuclear system. Resonances exhibit larger widths, until their complete disappearance, as the energy is increased. A reduced energy step does not reveal any further structures. In ref. [15,16] it was shown that fluctuations manifest themselves only in the region of energy and angular momentum where classical chaos shows up, that is around the potential barrier. Thus one finds also at the quantum level a clear transition in correspondence of the classical order to chaos shift. In this sense we can say that this irregular behaviour is the manifestation of quantum chaos. However at variance with the classical case quantal fluctuations do not show singularities. The probability is irregular but smooth.

Another indication which gives support to this claim is the fact that the appearance of sharp and grouped structures depends in a sensitive way on the strength of the coupling term. In ref. [17] it was demonstrated that, by decreasing the strength, fluctuations rarefy and then they disappear completely. On the contrary, an increase of the coupling produces an enlargement of the energy region where fluctuations are present.

A quantitative study of the fluctuations shown in Fig.4 can be obtained by means of autocorrelation function analysis [10,11]. This can be done for both quantal and semiclassical cases [19]. Comparing then the quantal and the semiclassical values of the coherence length Γ, it turns out that, as in ref. [10,11] Γ_{cl} and Γ_{quan} are equal within a factor of two. This nice agreement links quantitatively classical chaos with its quantum counterpart.

4 Realistic calculations

The quantal approach we have used up to now is a simplified description of heavy ion scattering. A more realistic model should take into account: a) a three-dimensional description of the scattering; b) the effect of other degrees of freedom (like vibrations or nucleon transfer) by means of an absorption term in the potential; c) the calculation of cross sections, angular distributions and other observables directly comparable with experimental data.

The role of absorption was studied in ref. [15]. Adding an imaginary component to the interaction it was demonstrated that when the absorption at the barrier is strong enough the fluctuations in the transition probabilities can be completely washed out. As it is discussed in the next section, an important feature which has been found experimentally in the heavy ion reactions of the kind investigated here, is the superficial transparency of the potential. Thus in our case the assumption of a weak absorption is a very realistic approximation. Semiclassically this means that long lived trajectories give an appreciable contribution.

In the following, in order to consider a very realistic quantum description of the reaction between a deformed nucleus and a spherical one, we use the three-dimensional coupled channel code FRESCO [22]. The latter is a sophisticated code which can be considered the quantal analog of the three-dimensional classical picture described before. At the same time it gives us the possibility to take into account the three points discussed at the beginning of this section.

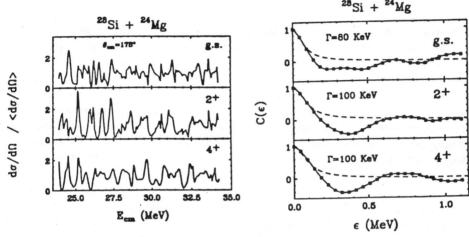

Fig. 5. Excitation functions for the system $^{28}Si +^{24} Mg$ at $\theta_{c.m.} = 178°$. The fluctuations are evidenced by dividing the cross section by its average local value obtained by considering an interval $\Delta E = 0.8$ MeV.

Fig. 6. Autocorrelation functions corresponding to the fluctuations displayed in Fig.8 (full squares). Lorentzian curves (dashed) are shown for comparison. The widths are also reported.

By using FRESCO we have calculated the excitation functions for the system $^{28}Si +^{24}Mg$. The real part of the nuclear ion-ion potential is the same used for the 2D calculations. A small absorption is considered. The parameters used are close to those adopted in ref. [23] for a similar system, see ref. [19]. We have considered only two rotational states, the 2^+ (at E=1.26 MeV) and the 4^+ (E=4.21 MeV) in the deformed nucleus plus the ground state 0^+. Only a coupling between the nearest neighbours is considered. The total number of exit channels is 9 [17,19]. The coupling factor is again given by Eq.(21). The energy step (in the center of mass frame) used is 0.046 MeV.

The excitation functions show very complicated fluctuations. In order to study them in a quantitative way we divide each point of the cross section by its local average value. That is we consider the quantity $X(E) = \frac{d\sigma/d\Omega}{<d\sigma/d\Omega>}$. This procedure eliminates the smooth behaviour of the cross section and at the same time enables one to investigate the fluctuations of a quantity which is dimensionless and vary over a few units [24]. In Fig.5 we plot $X(E)$ for the system $^{28}Si +^{24}Mg$ as function of the incident energy and for a fixed backward angle. The actual local average is done over an energy interval $\Delta E = 0.8$ MeV. Similar fluctuations are present also in the excitation function of the system [19] $^{12}C +^{24}Mg$.

One can now proceed in evaluating the autocorrelation functions. In this case, we adopt the standard formula used for cross sections [25,24]

$$\frac{< X(E)\ X(E+\epsilon) >}{< X(E) > < X(E+\epsilon) >} - 1 \ .$$

These autocorrelation functions are displayed in Fig.6 for the states considered in the calculations. The dashed curves are lorentzians whose widths (reported in the figure) are of the same order of the semiclassical ones.

Another quantity which can be determined experimentally is the angular distribution, i.e the differential cross section as a function of the detection angle for a fixed energy. By means of the code FRESCO we have calculated elastic and inelastic angular distributions as a function of the incident energy. It is found a strong oscillating behaviour as the energy is above the barrier and for backward angles. In general the angular distribution at large angles is dominated by the nuclear interaction, while the Coulomb one predominates at forward angles. Then the backward angles fluctuations are strictly connected to the internal part of the interaction which classically shows a chaotic dynamics.

In general no drastic change appear in the qualitative features of the scattering in passing from 2D to 3D. An irregular behaviour in cross section and angular distributions persist and can be connected to the underlying chaotic classical scattering.

5 Discussion and conclusion

Reactions between light nuclei, like $^{16}O + ^{28}Si$ and $^{12}C + ^{28}Si$, ^{32}S, $^{28}Si + ^{28}Si$, $^{24}Mg + ^{24}Mg$ and $^{24}Mg + ^{28}Si$ and many others, show strong fluctuations in the excitation functions [26]. In correspondence with these fluctuations, anomalous large and highly oscillating (at backward angles) angular distributions were observed. A dinuclear molecular model was proposed to explain these oscillations and fluctuations but the mechanism soon appeared much more complicated: systems leading to the same nuclear composite showed different structures and it was not always possible to understand the angular distributions in terms of only one single wave.

In the beautiful review paper of Braun-Munzinger and Barrette [26] it is clearly stated that the fluctuating fenomena in light heavy-ions seem to have a common nature: there are only quantitative, but not qualitative differences from system to system. All the advanced models failed - partly or completely - in reproducing the large set of existing data. The only model-independent consideration which comes out naturally from the experimental analysis is the unexpected presence of a very weak surface absorption. In other words, a relatively small number of channels is involved.

The connection between fluctuations in light heavy–ion reaction and a chaotic mechanism was stressed for the first time in the conclusions of ref. [24]. In the previous sections we have presented a model which exhibits chaotic scattering and is able to reproduce in a semiquantitative way the experimental phenomenology for light heavy–ion reaction discussed above. The puzzling irregularities observed experimentally find a natural explanation in the framework of chaotic scattering considering only a few degrees of freedom.

Having used rotational states, one could think that chaotic scattering is limited only to this kind of excitations. Actually, features very similar to those here discussed have been found both classically [14] and quantum-mechanically [18] for heavy–ion reactions when considering vibration modes too, see also M. Saraceno's lectures.

We can therefore conclude that irregular scattering has a well established theoretical and experimental foundation in light heavy-ion collisions.

In general the single fluctuations are not theoretically reproducible - quantum chaos seems to maintain a strong sensitivity on the input parameters - and one should compare instead autocorrelation functions, widths distributions or other statistical quantities. However absorption can help in increasing the theoretical predictive power smearing out the wildest fluctuations.

In our approach direct, semidirect and long lived reactions are taken explicitly into account. The widths of the resonances qre of the same order of the distance between resonances. This indicates a regime of chaoticity produced by a dynamical mechanism which differs from that of the Ericson's fluctuations. We want to stress that both regular and fully chaotic scattering are two extreme exceptional cases. In general one finds more often a mixed situation which lies in between. This situation is the most complicated to deal with, expecially in the quantum case where the chaos-to-order transition is more elusive. In this respect a lot of work has still to be done in order to characterize more precisely this transition.

In conclusion, it has been shown that chaotic scattering represents a real possibility in collisions between light heavy ions and that it can explain the irregular fluctuations observed experimentally. A few degrees of freedom can generate a very complicated and unpredictable motion expecially when semiclassical approximations are used. This is an important result both for nuclear physics and for more fundamental questions like the existence and the features of quantum chaos. These investigations allow to reinterpret standard approaches - although for the moment only in a generic way - in the new framework of the transition from order to chaos. The study of heavy–ion scattering is particularly interesting due to its privileged position between the classical and the quantum world.

References

1   E. Ott, *Chaos in dynamical systems*, (Cambridge University Press, 1993), and references therein.

2   M. Tabor, *Chaos and integrability in nonlinear dynamics*, (John Wiles and son, New York, 1989). , and references therein.

3   P. Civitanovic, *Universality in chaos*, (Adam Hilger, 1989) and references therein.

4   M.C. Gutzwiller, *Chaos in Classical and Quantum Mechanics*, (Springer-Verlag, 1990), and references therein; R.V. Jensen, Nature 355 (1992) 311. K. Nakamura *Quantum chaos - a new paradigm of nonlinear dynamics* , (Cambridge University Press, 1993) and references therein.

5   N. Bohr, Nature **137** (1936) 351.

6   T. Ericson and T. Mayer-Kuckuk, Ann. Rev. Nucl. Sci. **16** (1966) 183 and references therein.

7   O. Bohigas and H.A. Weidenmüller, Ann. Rev. Nucl. Part. Sci. **38** (1988) 421 and references therein.

8   J. D. Garrett and J. R. German, ORNL preprint, submitted to Phys. Rev. Lett.; J. D. Garrett, Proc. of Eighth Int. Symp. on "Capture Gamma-Ray Spectroscopy and Related Topics", ed. J. Kern (World Scientific, 1994, Singapore), in press; J. D. Garrett's Lectures at this school.

9   CHAOS **3** (1993), focus issue on chaotic scattering.

10   U. Smilansky, Proceedings of the International School on *Chaos and Quantum Physics*, Les Houches 1989, Eds. M.J. Giannoni, A. Voros and J. Zinn-Justin, (Elsevier Science Publishers B.V., 1991), and references therein.

11   R. Blümel and U. Smilansky, Phys. Rev. Lett. **60** (1988) 477.

12   P. Gaspard and S.A. Rice, J. Chem. Phys. **90** (1989) 2225, 2242, 2255.

13   A. Rapisarda and M. Baldo, Phys. Rev. Lett. **66** (1991) 2581.

14   C.H. Dasso, M. Gallardo and M. Saraceno, Nucl. Phys. **A549** (1992) 265.

15   M. Baldo and A. Rapisarda, Phys. Lett. **B279** (1992) 10.

16   M. Baldo and A. Rapisarda, Phys. Lett. **B284** (1992) 205.

17   M. Baldo, E.G. Lanza and A. Rapisarda , Nucl. Phys. **A545** (1992) 467c.

18   C.H. Dasso, M. Gallardo and M. Saraceno, NBI preprint 1993.

19   M. Baldo, E. G. Lanza and A. Rapisarda, Chaos **3** (1993) 691.

20   J.P. Blocki, J. Randrup, C. F. Tsang, and W. J. Swiatecki, Ann. of Phys. **105** (1977) 427.

21   R. A. Broglia and A. Winther, *Heavy Ion Reactions*, Lecture Notes, vol. I, (Benjamin, New York, 1981).

22   I.J. Thompson, Comp. Phys. Reps. **7** (1988) 167.

23   G. Pollarolo and R.A. Broglia, Nuov. Cim. **A278** (1984) 81. V.N. Bragin, G. Pollarolo and A. Winther, Nucl. Phys. **A456** (1986) 475.

24   A. Glaesner, W. Dünweber, M. Bantel, W. Hering, D. Konnerth, R. Ritzka, W. Trautmann, W. Trombik and W. Zipper, Nucl. Phys. **A509** (1990) 331.

25   G. Pappalardo, Nucl. Phys. **A488** (1988) 395c.

26   P. Braun-Munzinger and J. Barrette, Phys. Rep. **87** (1982) 209, and references therein.

Nuclear Level Repulsion, Order vs. Chaos and Conserved Quantum Numbers

J.D. Garrett [1], J.R.German [1] and J.M.Espino [2]

[1]Physics Division, Oak Ridge National Laboratory, Oak Ridge, TN
37831-6368, USA
[2]Departamento de Física Atómica, Molecular y Nuclear, Universidad de
Sevilla, Apdo. 1065, 41080 Sevilla, Spain

Abstract: A statistical analysis of the distribution of level spacings for states with the same angular momentum and parity is described in which the average spacing is calculated for the total ensemble. Though the resulting distribution of level spacings for states of deformed nuclei with Z = 62-75 and A = 155-185 is the closest to that of a Poisson distribution yet obtained for nuclear levels, significant deviations are observed for small level spacings. Many, but not all, of the very closely-spaced levels have K-values differing by several units.

1. Background Information

It has become increasingly fashionable to consider the information content of the spectrum of eigenstates of a quantal system in terms of statistical concepts [1]. The rich spectrum of nuclear states provides an excellent opportunity for such analyses. Indeed, an analysis [2] of the extensive, and "complete", spectroscopic data near the neutron and proton thresholds (typically at an excitation energy of about 8 MeV) provides evidence that the distribution of the spacing of levels, s, with the same angular momentum, I, and parity, π, can be reproduced by a parameter-free distribution derivable from the predictions of the Gaussian orthogonal ensemble (GOE) of random-matrix theory. Such level-spacing distributions (as shown in the upper portion of Fig. 1) have been taken as evidence [1,2] that the nuclear systems at low angular momentum and high excitation energy are "chaotic".

More recent level-spacing analyses extend to lower excitation energies [3-6] and attempt to extract information as a function of nuclear specie [3,4], for example, even-even versus odd-A, heavy versus light, or deformed versus spherical. Such analyses provide distributions intermediate to that expected for purely "ordered" (Poisson distribution) or purely "chaotic" (GOE or Wigner distribution)

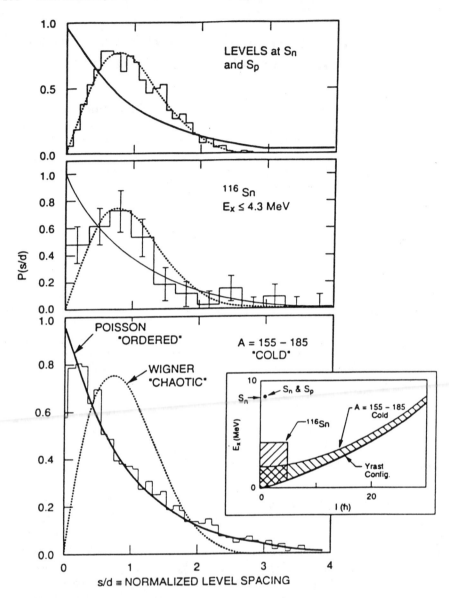

Fig. 1. Comparison of level-spacing distributions (histograms) for nuclear states with the same angular momentum and parity in three cases: (top) at the neutron and proton threshold [2]; (middle) levels of ^{116}Sn [6]; and (bottom) "cold" deformed rare-earth nuclei (present contribution). Curves corresponding to Poisson and Wigner distributions, often associated with "ordered" and "chaotic" behavior, respectively, are shown for comparison with each distribution of experimental levels. The inset indicates the ranges of excitation energy, E_x, and angular momentum, I, associated with each data set. Although the ensemble of nuclear levels considered for each distribution is very different, a progression from "ordered" to "chaotic" behavior seems to occur with increasing excitation energy.

systems (see e.g. the middle portion of Fig. 1) and perhaps a trend from GOE to Poisson proceeding from spherical regions to deformed regions [3,4]. To obtain sufficient data for such analyses it is necessary to sum over a large range of excitation energies and angular momenta in a single nucleus [5,6] and to really obtain meaningful statistics to sum over a variety of nuclei [3,4]. Since the level density increase with excitation energy is exponential, the sensitivity to the lowest-lying states is rapidly lost.

2. Near-Yrast Level-Spacing Distributions for Deformed Nuclei

In this contribution we present a statistical analysis of the distribution of level spacings for near-yrast states of the same angular momentum and parity in deformed nuclei with $Z = 62-75$ and $A = 155-185$. In contrast to earlier analyses [2-6], the average spacing, d, is calculated from the total data set, instead of for each individual angular momentum-parity group in a specific nucleus. Though the resulting distribution is the closest to that of a Poisson distribution yet obtained for nuclear levels, significant deviations are observed for small level spacings. Many, but not all, of the very closely-spaced levels have K-values differing by several units.

2.1 The Average Spacing

The modified prescription of obtaining d from the complete data set is partly justified by the fact that all levels considered are from deformed rare-earth nuclei, whose properties are known to be similar. (For example, the spread in the moment of inertia both for the yrast sequences of even-even nuclei at high angular momentum [7] and for a variety of decay sequences in neighboring odd- and even-A nuclei [8] in this mass region are significantly smaller than expected from simple models). Likewise, such a criterion circumvents the usual problem of obtaining sufficient data to establish an accurate average level spacing for each angular momentum-parity group in a specific nucleus. For example, in the previous attempts to obtain information for some of the best data sets at low excitation energy often there were as few as 3-4 levels per angular momentum-parity group in each nucleus [3]. The added statistics obtained by relaxing this requirement on the average spacing allows for the first time an analysis of level spacings for near-yrast states. Though the details of the level-spacing distribution shape are somewhat sensitive to the prescription for calculating the average spacing, the crux of the present contribution, the physics associated with the reduced number of very small level spacings obtained from this analysis, is not. A more complete discussion of this analysis and the data sets used is given in [9], as is an analysis as a function of angular momentum, parity and nuclear type (i.e. even-even, odd-even, etc.).

2.2 Level-Spacing Distributions

The distribution of 2522 level spacings obtained from deformed even-even and odd-A nuclei with $Z = 62\text{-}75$ and $A = 155\text{-}185$ is compared with calculated Poisson and Wigner distributions in the lower portion of Fig. 1. Except for the very smallest separations, the agreement of the experimental spacings with the Poisson distribution is exceptional. These and the other data shown in Fig. 1 also provide evidence for a systematic progression from a Poisson to a Wigner distribution with increasing intrinsic excitation energy. Such systematics illustrate the intimate relation between the mixing of nuclear states and "quantum chaos". When the nuclear states are well separated, as is the case for the majority of the low-lying states in deformed rare-earth nuclei ($d = 297$ keV), neighboring states have distinctively different wave functions, and they are termed "ordered". In the other extreme, illustrated by the closely-spaced levels at the particle threshold (upper portion of Fig. 1), the average spacing [10] is $d \approx 10$ eV, much less than the nuclear matrix element connecting neighboring states of the same angular momentum and parity, hence neighboring states have strongly-mixed wave functions. In this limit neighboring states contain only information about the average ensemble of levels, and they are termed "chaotic".

Though the level-spacing distributions for "cold" deformed nuclei are the closest to the Poisson distribution yet obtained for experimental nuclear levels, statistically-significant systematic deviations from Poisson are observed for small spacings, see Fig. 2. Such deviations are attributed to the level repulsion associated with the mixing of closely-spaced states with identical angular momentum and parity. These data indicate that this mixing apparently becomes increasingly important in deformed rare-earth nuclei for separations less than about 70 keV. Indeed, for separations less than 10 keV only 35 % of the spacings expected for a Poisson distribution remain. The excess of spacings for $70 < s < 110$ keV, see Fig. 2, also can be attributed to the repulsion of closely-spaced levels. Such systematics provide evidence for interaction matrix elements, V_{int}, distributed from as low as a few keV (spacings of the order of a few keV are observed, see Table 1) to about 35 keV (deviations from a Poisson distribution occur for $s < 70$ keV). These systematics also illustrate the intimate relation between the mixing of nuclear states and quantum chaos. At the lowest excitation energies, where the average spacing of nuclear states is large (297 keV for the ensemble of rare-earth states considered), the level spacing distributions show systematic deviations from a Poisson distribution for small spacings, implying deviations from a completely "ordered system" associated with small level spacings. Such deviations can be attributed to a significant mixing of many, but not all, of the closely-spaced levels.

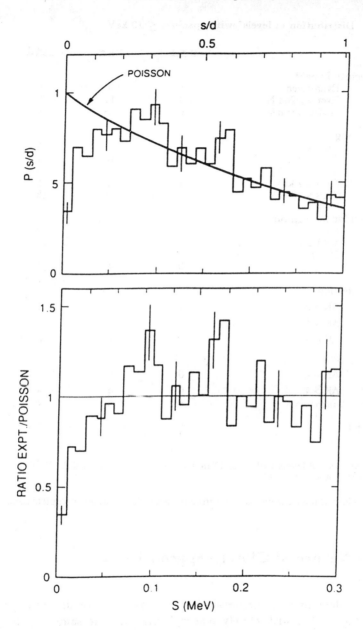

Fig. 2. The level-spacing distribution for small spacings of "cold" deformed rare-earth nuclei is shown in the upper portion. The same data are shown relative to a Poisson distribution in the lower portion. Note the precipitous drop relative to a Poisson distribution for very small spacings. For reference the complete level spacing distribution for these data is given in the lower portion of Fig. 1. Statistical uncertainties are indicated by the vertical error bars.

Table 1. Distribution of levels* with spacings $\leq$ 10 keV.

	$\pi = +$	$\pi = -$	Total
Nuclear Type**			
Even-Even	5	4	9
Even Z, Odd N	1	11	12
Odd Z, Even N	6	3	9
Spacing			
$s \leq 1$ keV	1	1	2
$1 < s \leq 3$ keV	1	0	1
$3 < s \leq 5$ keV	3	1	4
$5 < s \leq 7$ keV	1	4	5
$7 < s \leq 10$ keV	7	11	18
Angular Momentum			
$I \leq 5$	7	5	12
$5 < I \leq 10$	3	8	11
$10 < I \leq 15$	2	1	3
$15 < I \leq 20$	0	4	4
K Selection			
$\Delta K = 0$	1	1	2
$\Delta K = 1$	1	0	1
$\Delta K = 2$	5	1	6
$\Delta K = 3$	4	3	7
$\Delta K = 4$	0	6	6
$\Delta K > 4$	0	4	4
unknown K	1	3	4
Total	12	18	30

*Analysis of levels with same I^π in deformed nuclei with $62 \leq Z \leq 75$ and $155 \leq A < 185$.

**Odd-odd nuclei were excluded from the analysis due to less complete data.

3. The Nature of Closely-Spaced Levels

In order to demonstrate the physical basis of the very small interaction matrix elements associated with closely-spaced levels with the same angular momentum and parity, the distribution of the 30 level spacings in deformed rare-earth nuclei $\leq$ 10 keV (i.e. those corresponding to the lowest spacing interval shown in Fig. 2) is summarized in Table 1. Such closely-spaced, unperturbed levels are traditionally associated with additional conserved quantum numbers [1]. Indeed, even more esoteric nuclear structure effects, such as isobaric analog states and shape isomers [11], which also produce reduced interactions can be formulated in terms of conserved quantities [12].

Closely-spaced levels are observed for both parities, for even-even and for odd-A nuclei, and at low and high angular momentum (see Table 1) corresponding roughly with their occurrence in the data ensemble. Whereas the distribution as a function of spacing is strongly skewed toward the larger spacings, values of < 1 keV are observed, for example, between $I^\pi = 11/2^-$ states in ^{159}Er.

The observed asymmetry between positive- and negative-parity states with $s \leq 10$ in odd-A nuclei is the result of different densities of positive- and negative-parity single-particle (Nilsson) states near the Fermi level. For example, in the $N = 82$-126 neutron shell, which is being filled for the nuclei considered, there are 30 negative-parity neutron states (corresponding to the Nilsson components of the $h_{9/2}$, $f_{7/2}$, $f_{5/2}$, $p_{3/2}$ and $p_{1/2}$ shell-model states) and 14 positive-parity states (corresponding to the components of the $i_{13/2}$ intruder states) that span nearly the same excitation energy range. Hence an excess of closely-spaced negative-parity levels are expected for odd-N nuclei, as is observed (see Table 1). For protons, which fill the next lower shell (with 20 $g_{7/2}$, $d_{5/2}$, $d_{3/2}$ and $s_{1/2}$ positive-parity states and 12 $h_{11/2}$ negative-parity states) the opposite is true. The average level spacings of positive- and negative-parity levels for this mass region show asymmetries of a similar nature [9].

Probably the most striking feature of the closely-spaced levels is the correlation with the K quantum number (see Table 1). Sixty-five percent of the closely-spaced levels with known K-values correspond to states that differ in K by three or more units, and 88 % are associated with states differing in K by two or more units. Hence the majority, but perhaps even more interestingly not all, of the close spacing can be attributed to the conservation of the K quantum number in these deformed nuclei. Typical configurations for which close level spacings are observed are listed in Table 2. Several of the nontabulated closely-spaced levels differ in the number of excited quasiprotons and quasineutrons.

Table 2. Examples of levels with spacing ≤ 10 keV.

Configurations	ΔK^*	# Cases	Nuclei
ν 11/2⁻[505] & low-K	3 & 4	3	^{159}Dy, ^{159,161}Er
ν 7/2⁻[503] & ν 1/2⁻[510]	3	1	^{185}Os
π 9/2⁻[514] & π 1/2⁻[541]	4	3	^{171}Lu, ^{175}Ta, ^{181}Re
π 7/2⁺[404] & π 1/2⁺[411]	3	3	^{165,167}Tm
π 5/2⁺[402] & π 1/2⁺[411]	2	1	^{163}Tm
π 3/2⁺[411] & π 1/2⁺[411]	1	1	^{163}Ho
Very High-K Isomers	≥7	4	^{175}Hf, ^{180}Os
β & γ bands	2	1	^{170}Yb

*K is not a conserved quantum number in rotating systems. The strongest K admixture will occur for the $\Omega = 1/2$ configurations.

Though in the SU(3) symmetry limit the Interacting Boson Model predicts [13] the degeneracy of the β- and γ-vibrational states with the same angular momentum, only one such spacing, between 2^+ states in $^{170}\text{Yb}_{100}$, was observed to be < 10 keV. This spacing is 7.2 keV.

4. Discussion and Summary

A statistical analysis of level spacings is presented in which the average spacing is calculated for the total data set, which is restricted to nuclei with similar properties. The resulting distribution of level spacings for low-lying states of deformed nuclei with $Z = 62\text{-}75$ and $A = 155\text{-}185$ is the closest to that of a Poisson distribution yet obtained for nuclear levels. In contrast to most previous analyses [2-6], these data are concentrated on near-yrast states with small intrinsic excitation energies, see Fig. 1; thus, the average level spacing is large, $d = 297$ keV. Indeed, this value greatly exceeds that of the interaction matrix elements between these states (< 35 keV). Therefore, neighboring states with the same quantum numbers are expected to retain individual character. However, even for the low-lying states of deformed rare-earth nuclei, significant deviations are observed for spacings less than about 70 keV (see Fig. 2). Such deviations are attributed to the level repulsion associated with the mixing of closely-spaced states with the same quantum numbers. Indeed, many, but not all, of the observed levels with spacing of ≤ 10 keV have K-values that differ by several units, indicating that K can be considered as a good quantum number for classifying low-lying states of deformed rare-earth nuclei. Though the details of the level-spacing distributions are expected to be somewhat sensitive to the prescription for calculating the average spacing, the crux of the present analysis, a sharp deviation from a Poisson distribution for small spacings, would not be expected. Neither would such an artifact be expected to be correlated with K values.

Therefore, a coherent picture emerges. When nuclear states are well separated, such as all but the most closely-spaced states considered in the present analysis of rare-earth data, each state contains distinctly different nuclear structure information from that of its neighbors with the same I^π, the level-spacing distribution is Poisson, and the situation is described as "ordered". The other extreme is illustrated by closely-spaced nuclear levels at the particle thresholds where $d \ll V_{\text{int}}$. In this limit neighboring states have strongly-mixed wave functions, the level-spacing distribution is Wigner, and the situation is described as "chaotic". Thus the wave functions of the individual states contain only general information on the average ensemble of nuclear states in a specific region of excitation energy. When the level spacing is $\approx V_{\text{int}}$, an intermediate situation emerges. Two contrasting examples of this are demonstrated in the present contribution: (i) when $d \approx 2V_{\text{int}}$ (the case of ^{116}Sn, shown in Fig. 1, with $d \approx 110$ keV), the level-spacing distribution is intermediate between that of a Poisson and a Wigner distribution; and (ii) when $d \gg V_{\text{int}}$ (the case of the present analysis of "cold" deformed rare-earth nuclei shown in Figs. 1 and 2), significant deviations remain for spacings $s < 2V_{\text{int}}$.

Acknowledgements

The authors acknowledge informative discussions with S. Åberg, C. Baktash, O. Bohigas, I.Y. Lee, B.R. Mottelson, W. Nazarewicz, S. Raman, F. Sakata, J.F. Shriner, T. von Egidy and H.A. Weidenmüller. Oak Ridge National Laboratory is managed by Martin Marietta Energy Systems, Inc. for the U.S. Department of Energy under Contract No. DE-AC05-84OR21400.

References

1.   O. Bohigas and H.A. Weidenmüller, Ann. Rev. Nucl. Part. Sci. **38** 421 (1988).
2.   R. Haq, A. Pandey and O. Bohigas, Phys. Rev. Lett. **48** 1986 (1982).
3.   T. von Egidy, A.N. Behkami and H.H. Schmidt, Nucl. Phys. **A454** 109 (1986) and Nucl. Phys. **A481** 189 (1988).
4.   J.F. Shriner, Jr., G.E. Mitchell and T. von Egidy, Z. Phys. **A338** 309 (1991).
5.   G.E. Mitchell, E.G. Bilpuch, P.M. Endt and J.F. Shriner, Jr., Phys. Rev. Lett. **61** 1473 (1988) and Z. Phys. **A335** 393 (1990).
6.   S. Raman, T.A. Walkiewicz, S. Kahane, E.T. Jurney, J. Sa, Z. Gacsi, J.L. Weil, K. Allaart, G. Bonsignori and J.F. Shriner, Jr., Phys. Rev. **C43** 521 (1991).
7.   J.M. Espino and J.D. Garrett, Nucl. Phys. **A492** 205 (1989); and D.F. Winchell, D.O. Ludwigsen and J.D. Garrett, Phys. Lett. **289B** 267 (1992).
8.   C. Baktash, J.D. Garrett, D.F. Winchell and A. Smith, Phys. Rev. Lett. **69** 1500 (1992) and Nucl. Phys. **A557** 145c (1993).
9.   J.D. Garrett, J.R. German, L. Courtney and J.M. Espino, in Future Directions in Nuclear Physics with 4π Gamma Detection Systems of the New Generation, ed. J. Dudek and B. Haas, AIP Conference Proceedings 259 (AIP, 1992, New York) pg. 345.
10.   S.F. Mughabghab and D.I. Garber, Neutron Cross Sections, BNL 325, 3rd ed. (National Technical Information Service, 1973, Springfield, Va.) vol. 1.
11.   See e.g. J.D. Garrett, in Microscopic Models in Nuclear Structure Physics, ed. M. Guidry, et al. (World Scientific, 1989, Singapore) pg. 254.
12.   W. Nazarewicz, Nucl. Phys. **A557** 489c (1993).
13.   See e.g. R.F. Casten and D.D. Warner, Rev. Mod. Phys. **60** 389 (1988).

Nuclear Physics and Nuclear Astrophysics with Radioactive Nuclear Beams

J. Vervier

Institut de Physique Nuclèaire, Universitè Catholique de Louvain
B-1348 Louvain-la-Neuve, Belgium

Abstract: The various methods to produce radioactive nuclear beams, i.e. the fragmentation and ISOL methods, are reviewed and compared. The applications of these beams to Nuclear Astrophysics are investigated in some details, both at low and high energies. Informations on Nuclear Physics which could be obtained with these beams are shortly summarized.

1 Introduction

In this contribution, we give a general view of the present status in the field of nuclear physics and nuclear astrophysics as studied with Radioactive Nuclear Beams. Three topics will be covered : the methods used to produce Radioactive Nuclear Beams ; their uses in Nuclear Astrophysics ; their uses in Nuclear Physics. The substance of the first topics heavily relies on a serie of lectures presented by P. Leleux at the "Ecole Joliot-Curie de physique nuclèaire 1993" entitled "Techniques de production de faisceaux exotiques", which will appear in the Proceedings of this School [1]. For the second topics, we somewhat extend the subjects covered by the invited paper we have presented during the "Fifth International Conference on Nucleus-Nucleus Collisions" at Taormina, Italy, on May 30 - June 4, 1994, which will appear in the Proceedings of this Conference [2]. The third topics is covered with somewhat less details than the first two ones, since it represents a wide and quickly-changing domain ; we summarize the problems in Nuclear Physics which can be tackled with Radioactive Nuclear Beams, and we present a few specific and recent examples of such studies. We thereby try to fulfil the purpose of the Fifth La Rabida International Summer School, which should be "devoted to topics along which nuclear physics will develop in the next few years".

2 Production of Radioactive Nuclear Beams

2.1 Introduction

Two general methods can be used to produce Radioactive Nuclear Beams (RNB), the fragmentation method and the Isotope Separator On Line (ISOL) method. In the first method, a high- energy (several tens to hundreds of MeV/A) primary heavy-ion stable beam is sent on a thin (primary) target, where it is fragmented into many nuclei, some of which are radioactive ; these are mostly emitted in the forward direction with about the same velocity as the primary beam. The desired RNB, with energies comparable to the one of the primary beam, are separated from the latter and from the other fragments by a suitable method. This secondary RNB is then sent on a (secondary) target to study nuclear reactions, or is collected by a suitable catcher to study its spectroscopic properties. For the ISOL method, large quantities of radioactive nuclei are produced by bombarding a thick (primary) target with high-intensity primary stable beams, protons or heavy ions, produced by a first accelerator, or by irradiating a fissile element with high-flux thermal neutrons in a reactor. Many different kinds of radioactive nuclei are thereby produced, depending on the type and energy of the projectiles, which are stopped in the target. These nuclei are extracted from the target as atoms or molecules, transformed into ions by a suitable ion source, mass-separated by an Isotope Separator On Line and accelerated by a second accelerator. The radioactive secondary beams thereby obtained are sent on a (secondary) target or collected by a catcher, as in the fragmentation method. Details on these two methods and a comparison between their respective performances are given in the following Subsections.

2.2 The fragmentation method

As the reaction mechanism for the first method is the fragmentation of a projectile by the target, nuclei not very different from the projectile are produced, either on the proton-rich or neutron- rich side with respect to the projectile. The cross section for the production of a given fragment weakly depends on the primary beam energy, provided it is high enough ($\geq$ 50 MeV/A), and on the projectile- target combination [3,4]. It is advantageous to choose a neutron-(proton)-rich projectile to produce neutron-(proton)-rich fragments, as shown by experiments with ^{48}Ca beams, which yielded neutron-rich Cl, Ar and K isotopes [5] , and with ^{112}Sn and ^{124}Xe beams, which recently allowed the discovery of the very neutron-deficient ^{100}Sn [6,7]. Empirical formulae exist which predict reasonably well these production cross sections [8].

Concerning the kinematics of the produced fragments, the momentum vector $\vec{p}$ of a given fragment can be decomposed into $p_{//}$ and $p_{\perp}$ which are parallel and perpendicular to the incident beam, respectively. The distribution of $p_{//}$ is almost gaussian, with an average value $< p_{//} >$ which is slightly smaller than the beam momentum [9] ; its width $\sigma_{p_{//}}$ only depends on the projectile and fragment masses M_P and M_F and on a constant σ_0, which represents the Fermi momentum of the nucleons removed from the projectile (about 90 MeV/c), through [10] :

$$\sigma_{p_{//}} = \sigma_0[M_F(M_P - M_F)/(M_P - 1)]^{1/2}$$

The values of $< p_{//} >$ and $\sigma_{p_{//}}$ are almost independent on the target mass and on the beam energy. The distribution of $p_\perp$ agrees with an isotropic emission of the fragments in the projectile reference frame ; the width $\sigma_{p_\perp}$ is close to $\sigma_{p_{//}}$, and the resulting angular dispersion of the fragments in the laboratory frame amounts to a few degrees. The fragment velocity is very close to the incident beam velocity.

The target has a weak influence on the fragment production rate. The fragmentation cross section is roughly proportional to the surface of the target nucleus, i.e. to $M_T^{2/3}$ where M_T is the target mass. However, the number of target nuclei per unit length of the projectile path in gr/cm^2 is proportional to M_T^{-1}. As a consequence, the yield of the fragment is proportional to $M_T^{-1/3}$, i.e. weakly depends on M_T. Multiple scattering of the incident beam and of the emitted fragments in the target somewhat broadens the $p_{//}$ and $p_\perp$ distributions. The energy losses of the beam and of the fragments in the target are not the same ; as a consequence, a fragment emitted at the end of the target will have a lower $p_{//}$ than at the beginning. In any case, the widening of the $p_{//}$ and $p_\perp$ distributions with respect to the purely kinematical contributions, due to the target, are generally not very large, except for thick (gr/cm^2) targets [11].

The fragments emitted at angles different from $0°$ with respect to the beam are polarized, sometimes to a high (20 %) degree. This fact has been used in some experiments [12].

The methods used to separate the desired fragments from the primary beam and from the other fragments are based on the following two remarks. First, the velocity of the various fragments is almost the same for all of them, and very close to the beam velocity ; consequently, the momentum of the fragments is proportional to their mass A. Second, at the energies considered, the fragments are almost completely stripped from all their electrons ; accordingly, their charge is proportional to their atomic number Z. In a magnetic field B, the radius ρ of the circular trajectory of a charged particle is proportional to its momentum and inversely proportional to its charge and to B ; as a consequence, the product $B\rho$ is proportional to A/Z, and a magnetic analysis yields an A/Z selection. To remove the ambiguity between the various fragments with the same A/Z ratio, the fragments are sent through an energy degrader, where the energy loss they experience has another dependance on A and Z then A/Z, being mainly a function of Z. In practice, a first magnetic analysis in A/Z yields a dispersion in momentum (due to the momentum distribution of the fragments, see above). A wedged-shape energy degrader compensates for this dispersion, and a properly-designed second magnetic analyser completes an achromatic system. As a result, a large fraction of the fragments resulting from the primary beam interaction with the target, with given A and Z, are concentrated at the focal plane of the second magnetic analyser [13]. Velocity filters, consisting of crossed electric and magnetic fields, are sometimes used, after the device just described, to improve the beam purity [14]. A superconducting lens, located in front of the target, improves the acceptance of the selecting device and hence its efficiency. Various fragment separators of this kind have been built, at GANIL, MSU, RIKEN, GSI, ... They are generally very long (20 to 100 m), have $\Delta p/p$ acceptances of a few percents and solid angle acceptances of a few msr.

The characteristics of the RNB obtained by the fragmentation method are the followings. The lifetimes of the radioactive nuclei can be very short, being only limited by the time of flight in the analysing device, i.e. a few hundreds of ns. The energies E of the RNB are high, since the fragmentation mechanism only works above several tens of MeV/A. They can be decelerated, but then have to be cooled in order to preserve reasonable emittances ; cooler and storage rings have been constructed for that, with a further limitation in the reachable lifetimes due to the fact that cooling needs time. The energy dispersion ΔE of the RNB is rather large, since $\Delta p/p$ is of the order of a few percents, $\Delta E/E = 2 \Delta p/p$ and E is large ; it is accordingly difficult to perform with them reaction studies which need a good energy resolution, or to measure accurately spectroscopic properties of the fragments such as their masses, except if time-of-flight methods can be used in addition to the selection techniques described above.

It should finally be mentioned that lower-energy RNB can be obtained by exploiting the kinematics of other reactions than fragmentation to separate the desired RNB from a primary beam. As an example, at the University of Notre Dame, ^{8}Li beams of a few MeV/A have been obtained by a transfer reaction of a ^{7}Li primary beam on a suitable target (D or Be) ; the ^{8}Li nuclei , emitted at an angle close but different from 0° , are separated from the scattered ^{7}Li nuclei using a solenoidal magnetic field, while the ^{7}Li primary beam is stopped at 0° in a beam dump. This method has also allowed to get ^{18}F^m ($J = 5^+$, $\tau = 160$ ns) isomeric beams. The RNB obtained by this method have low energies (a few MeV/A) and low intensities (below 10^7 particles per sec) [15].

2.3 The ISOL method

The production of radioactive nuclei in the ISOL method differs, in the various schemes using it, by the choice of the primary beam projectiles (protons, heavy ions or neutrons) and their energies (tens of MeV, GeV, tens of MeV/A, thermal energies). The yield of these radioactive nuclei is given by the general schematic formula :

$$N = I_0 \int_{E_0}^0 \sigma(E).n_T(E).dE \qquad (1)$$

where : I_0 is the projectile intensity ; $\sigma(E)$ is the cross section for the production reaction at an energy E of the projectiles ; $n_T(E)$ is the number of target nuclei at energy E per cm^2 of target and per unit energy interval dE (which is related to the inverse of the stopping power of the projectiles in the target) ; and E_0 is the initial projectile energy.

When low-energy (≤ 100 MeV) protons are used as projectiles, the primary beam intensity can be very high (approaching 1 mA), and they have short ranges in the target (a few cm) ; as a consequence, very large power densities have to be dissipated in the target, which does not have to be heated externally (in order to release the produced radioactive nuclei) but rather cooled. As an example, a 30-MeV proton beam with an intensity of 500 μA dissipates a 15-kW power in a volume which has some 3 cm diameter and 1 cm length. The radioactive nuclei produced are not very far from the line of stability, since the (p,n) reaction dominates at these energies with some contributions from the (p,2n), (p,2p) and (p,αn) reactions, and they are proton rich. Enriched targets often have to be used. There are generally no strong safety problems due to the radioactivity induced by the beam in the target, since few radioactive nuclei, generally with short halflifes, are produced. The yields may be quite high, reaching 1-2 10^{-3} radioactive nuclei per incident proton for (p,n) products [16].

High-energy (hundreds of MeV to GeV) proton beams have lower intensities (up to a few μA), and their ranges in the target are longer (several tens of cm) ; their power densities are accordingly lower, and the target has to be externally heated. For instance, a 1 GeV, 1 μA proton beam dissipates its 1-kW power in a target which has some 60 cm length. Radioactive nuclei very far from the line of stability can be produced by evaporating many nucleons and alphas from the target nucleus through a spallation-type reaction, and they are either neutron- or proton - rich. A few well-chosen natural targets can generally be used. Strong safety problems have to be faced, since many different highly radioactive nuclei, sometimes with long halflifes, are produced in these reactions, requiring remote handling of the irradiated targets. High yields can be reached for radioactive nuclei near the target nucleus, which quickly decreases when many nucleons have to be evaporated [17].

The "light" heavy-ion beams (for example ^{12}C) with energies below 100 MeV/A also have lower intensities, up to 1-2 particle μA, and their ranges in the target are comparable to those of low- energy protons ; their power densities are thus comparable to those of the latter projectiles. They also produce nuclei very far from the line of stability, neutron- or proton - rich, by various mechanisms, fragmentation, fission ,spallation, ... Natural targets can often be used and the safety problems are comparable to those of high-energy protons. The cross sections for producing the most exotic nuclei are often larger than for the latter projectiles, an advantage which is somewhat compensated by the higher stopping powers of the heavy ions with respect to the protons, which reduce the effective number of target nuclei along the paths of the former projectiles in equation (1).

Thermal neutrons are available with very high fluxes (10^{14} neutrons per cm^2 and per sec) in high-power nuclear reactors. When sent on fissile element targets, mainly ^{235}U, they can produce very neutron-rich exotic nuclei by fission, with very large cross sections (hundreds of barns instead of barns for charged projectiles). The yields for these nuclei are thus much larger than those for charged projectiles, high-energy protons or heavy ions [18]. The practical problems to extract these nuclei from a place close to the center of the reactor, and to ionize them with a reasonable efficiency, still have to be solved.

The problems related to the design of the targets used with charged projectiles are : to allow a fast and efficient dissipation of the beam power ; to extract the produced radioactive atoms with a high- efficiency and a high speed (to limit decay losses) ; (if possible) to select the desired elements and reject (i.e. not extract) the unwanted ones. In these respects, each nucleus, except the noble gases, presents specific problems, and few general solutions to them have so far been developed. A small (a few standard cm^3 per hour) gas flow is often used to increase the extraction efficiency for gaseous elements from the target. The solutions to these problems often depends on the type of ion source used in conjunction with the target, so that both the target and the ion source, and their connection, have to be designed as a single unit [19,20].

Various kinds of ion sources can be used to ionize the radioactive atoms (or molecules) extracted from the target [19,21]. In surface-ionization sources, alcalines (halogenes) elements can loose (gain) one electron when striking a hot surface, for example rhenium at 3000 K [22]. For hot- cathode arc sources of the FEBIAD type, electrons emitted by a heated cathode and guided by a magnetic field ionize the gas extracted from the target [23]. Electron Cyclotron Resonance (ECR) ion sources include an axial magnetic "bottle", produced by axial coils and multipolar permanent magnets, which (more or less) confines a plasma ; the latter is created by electrons which (multiply) strike the atoms extracted from the target and which receive their energy from a microwave generator through the electron cyclotron resonance in the magnetic field. Multiply-charged ions are thereby obtained with high yields, which are extracted axially by a puller electrode. The yield of the ECR sources strongly depends on the pressure inside the source, especially for high charge states, so that the unwanted gases extracted from the target by the beam impact have to be eliminated [24]. In laser ion sources, the gas emanating from the target is irradiated by a very high power (pulsed) laser ; resonant absorption of the laser light by specific atoms results in a high element (i.e. Z) selectivity. The duty cycle of these sources, and hence their efficiency, is (so far) rather low. Electron Beam Ion Sources (EBIS) are capable of ionizing, with high efficiencies, already ionized (for example 1^+) atoms into higher charge states.

The separation of the various ions thereby obtained according to the mass A and charge Z of the radioactive nuclei is generally performed with an Isotope Separator On Line, in which a magnetic field deviates them according to their mass over (ionic) charge ratio as outlined above. In the focal plane of such a device, the distribution of the ions with a given charge and a given separation mass M is a gaussian with a width ΔM. The separation power $M/\Delta M$ of this ISOL has to be high, of the order of 10.000 or more, since a radioactive nucleus has (almost) always a stable isobar with the same A value and with a mass different from its mass by about 10^{-4} in relative value [25]. The transmission efficiency of this ISOL has also to be high, a condition not always easy to fulfill given the emittances of the ion sources used.

The (post)acceleration of the resulting ions can be performed with tandems, cyclotrons or linacs. Tandems have very good energy definition and variability, and very good beam qualities (emittances and dimensions). They require negative ions, so that the positive ions produced by the ion source have to be converted into negative ions, with a corresponding loss of intensity. They produce DC beams, which may be an advantage for some coincidence experiments and a limitation for time-of-flight measurements. The energies of the RNB thereby produced is limited to a few MeV/A. Cyclotrons do not have good energy definition and easy energy variability, and their beam qualities are worse than for tandems. To be efficient, they (generally) need multiply charged ions, so that they are often used in conjunction with ECR sources. Their beams are pulsed (with a time interval between the pulses in the 50-100 ns range), which may be an advantage for time-of-flight measurements. Their main advantage is that they are (very) good mass analysers, so that an ISOL is not required in conjunction with them. They can accelerate ions from a few tenths to a few tens of MeV/A, depending on the ion charge state. Linacs have average energy definition and variability, and average beam qualities. They can accelerate low-charge state ions (for example 1^+), so that they can be used with sources producing such ions ; in that case, they require a preaccelerator, generally a Radio Frequency Quadrupole (RFQ) accelerator. Their beams are pulsed (time interval in the ms range), and they do not by themselves separate the isobars.

The characteristics of the RNB obtained by the ISOL method are the followings. The lifetimes of the radioactive nuclei are longer than for the fragmentation method (above a few tenths of a sec), since they are limited by the diffusion time of the atoms or molecules outside the targets, the transfer time between the target and the ion source and the sticking time in the wall of the ion source. Their energies depend on the accelerator used, and range from the low (tenths of MeV/A) through the medium (MeV/A) to the high (tens of MeV/A) regions. The beam qualities (energy resolution and emittance) depends on the accelerator as outlined above, but are in any case better than for the fragmentation method, allowing reaction experiments to be performed with good resolution.

2.4 Comparison between the 2 methods

The advantages of the fragmentation method are : the possibility of producing RNB with short lifetimes ; the weak radioactivity of the target ; the high collection efficiency of the fragments produced ; the use of a single accelerator. Those of the ISOL method are : high yields for the RNB ; good beam qualities ; the accessibility to a wide energy range.

3 Nuclear Astrophysics with Radioactive Nuclear Beams

3.1 Why Are Radioactive Nuclear Beams Useful for Nuclear Astrophysics ?

It is well known [26] that the production of energy and the synthesis of elements in stars occur through nuclear reactions. In "quiet" star as our sun, which are in equilibrium conditions, the rate of nuclear reactions is much lower than the radioactive decay rates, so that only nuclear reactions involving stable nuclei are important. In "violent" stellar events, the inverse situation prevails : the nuclear reaction rates are comparable or higher than the decay rates, so that radioactive nuclei which are sometimes formed in these reactions do not have time to decay before being involved in other nuclear reactions. As a consequence, nuclear reactions between radioactive nuclei and (very often) protons or alpha particles have to be taken into account [27]. Examples of such events are [26] : the rapid-neutron capture process (r-process), which is responsible for the nucleosynthesis of the heaviest elements known in Nature, the Th and U isotopes ; the primordial nucleosynthesis, which occured during the first three minutes of the Universe and led to the formation of a few light elements ; type-II supernovae, which represent the last few seconds of the life of a massive star. Other examples imply binary systems (about 50% of the stars in the sky belong to such systems) in which one of the two stars is either a white dwarf, i.e. essentially a giant atom, or a neutron star, i.e. a giant nucleus. Matter ejected from the companion "normal" star, mostly hydrogen, is attracted by the white dwarf or the neutron star and, after some spiraling around it, is "accreted" on it, leading to the explosion of its surface if it is a white dwarf - this is the novae phenomenon -, or to the (more or less) periodic emission of X-rays - this is the X-ray bursts phenomenon [26]. In these cases, a new nucleosynthetic process may occur, the so-called rapid-proton capture process (rp-process) [28], in which many successive proton captures and β-decays, starting from elements in the Ne region, lead to the nucleosynthesis of nuclei up to A = 65 to 75 and maybe higher. To understand these "violent" events, to confirm - or contradict - the scenarios proposed by the astrophysicists to describe them, it is necessary to know the cross sections for nuclear reactions involving radioactive nuclei. Up to about 1990, these cross sections were derived from theoretical calculations. Since then, it has become possible to mesure them, either directly with low- energy ($\leq$ 1 MeV/A) RNB, or indirectly with high-energy (several tens to hundreds of MeV/A) RNB.

The "paths" of the r- or rp-processes lead to very exotic nuclei, either neutron- or proton-rich, which are sometimes close to the neutron or proton drip lines. To calculate these "paths", for example the so-called "waiting points" of the r-process, it is thus necessary to know many spectroscopic properties of these nuclei : lifetimes, masses, decay modes, level densities, fission barriers, ... These nuclei can be prepared and separated, and their properties can be measured, with the help of RNB.

It thus appears that the RNB needed for Nuclear Astrophysics (NA) must have, either low energies ($\leq$ 1 MeV/A) or high energies (several tens to hundreds of MeV/A), to allow direct or indirect determinations of the cross sections for nuclear reactions involving radioactive nuclei and to study the spectroscopic properties of very exotic nuclei.

3.2 Low Energy Radioactive Nuclear Beams for Nuclear Astrophysics

The first direct measurement of the cross section for an astrophysically important nuclear reaction has dealt with the $^{13}N(p,\gamma)^{14}O$ reaction, which has to do with the so-called CNO cycles [26,27]. The cold CNO cycle corresponds to the following reaction and decay sequence :

$$^{12}C(p,\gamma)^{13}N(\beta^+)^{13}C(p,\gamma)^{14}O(\beta^+)^{14}N(p,\gamma)^{15}O(\beta^+)^{15}N(p,\alpha)^{12}C \qquad (2)$$

It leads to the transformation of 4 protons into 1 alpha-particle using ^{12}C as catalyst, and represents the principal source of energy in main-sequence stars somewhat heavier than the sun. At high stellar temperatures and/or densities, the hot CNO cycle can occur, with the following reaction and decay sequence :

$$^{12}C(p,\gamma)^{13}N(p,\gamma)^{14}O(\beta^+)^{14}N(p,\gamma)^{15}O(\beta^+)^{15}N(p,\alpha)^{12}C \qquad (3)$$

It is believed that the latter cycle plays a major role in the novae phenomenon. The conditions under which either of these 2 cycles dominates are determined by the competition between the β^+ decay rate of ^{13}N with a ten minutes halflife, and the rate for the $^{13}N(p,\gamma)^{14}O$ reaction. In the astrophysically interesting region, the cross section for this reaction is dominated by a resonance at 0.545 MeV in the $^{13}N + p$ center-of-mass, corresponding to the 1^- first excited level of ^{14}O at 5.17 MeV. Three experiments have been carried out at Louvain-la-Neuve on this reaction using ^{13}N beams [29] : (i) the direct measurement of the capture cross section in the resonance region, through the detection of the capture γ-rays ; (ii) the study of the resonance in the $^{13}N + p$ elastic scattering, using a novel method ; (iii) the measurement of the cross section for the $^{13}N(d,n)^{14}O_{g.s.}$ reaction, through the radioactivity of ^{14}O. These experiments have yielded the following quantities : (i) the radiative width Γ_γ of the resonance ; (ii) the energy E_R and total width Γ of the resonance ; (iii) the spectroscopic factor S for the $^{13}N(d,n)^{14}O_{g.s.}$ reaction. These results have allowed to calculate the cross section for the $^{13}N(p,\gamma)^{14}O$ reaction over the whole energy region interesting for nuclear astrophysics and the rate of this reaction in stars : the resonance contribution to the cross section is directly proportionnal to Γ_γ ; the resonance energy enters into the reaction rate in a factor $\exp(-E_R/kT)$ where k is Boltzmann's constant and T the absolute temperature, factor which is very sensitive to E_R if kT is not too large ; the spectroscopic factor S allows to calculate the direct capture contribution to the cross section and its interference with the resonance contribution. These data allow to calculate the so-called astrophysical S(E)-factor, and the regions of the "phase" diagramme of the proton density <u>versus</u> the temperature wherein the cold and hot CNO cycles dominate. The range of density and temperature which prevail in novae is clearly in the hot CNO region. These results have been used to confirm the scenario proposed by the astrophysicists to describe the novae phenomenon and to calculate some properties of the novae which could be observed experimentally [30].

A second astrophysically important nuclear reaction is ^{19}Ne(p,γ)^{20}Na. An "escape" from the CNO cycles could occur through the reaction and decay sequence [28] :

$$^{15}O(\alpha,\gamma)^{19}Ne(p,\gamma)^{20}Na(\beta^+)^{20}Ne \tag{4}$$

The competition between the β^+ decay of ^{19}Ne to ^{19}F and the ^{19}Ne(p,γ)^{20}Na reaction is important, since ^{19}F would lead back to the CNO elements through the ^{19}F(p,α)^{16}O reaction, whereas ^{20}Na and/or ^{20}Ne would be the starting point of the rp-process. Four resonances could contribute to the ^{19}Ne(p,γ)^{20}Na reaction above the ^{19}Ne + p threshold, at 0.447, 0.658, 0.797 and 0.887 MeV center-of-mass energies, the lower one being probably the most important. Three kinds of experiments have been carried out at Louvain-la-Neuve on this reaction using ^{19}Ne beams [31] : (i) the direct measurement of the capture cross section in the region of the four resonances, through the radioactivity of ^{20}Na as observed through its β^+ and α decays ; (ii) the study of these resonances in the ^{19}Ne + p elastic scattering ; (iii) the measurement of the cross section for the ^{19}Ne(d,n)^{20}Na reaction leading to the bound states of ^{20}Na, through the ^{20}Na radioactivity. These experiments are being analysed. Partial results of them are as follows : the 0.447-MeV resonance has a low γ-width Γ_γ, smaller than 20 meV ; the other resonances have Γ_γ values in the 100 meV region; the 0.447-MeV resonance has a total width Γ smaller than about 0.5 keV ; the upper two resonances have energies E_R, spins and parities J^π, and total widths Γ which are 0.797 MeV, 1^+, 20 keV, and 0.887 MeV, 0^+, 36 keV, respectively. The detailed analysis of the results, and their astrophysical significance for the rp-process, still have to be worked out.

Work has been done on the ^{8}Li(α,n)^{11}B reaction, which is important for the primordial nucleosynthesis in non-homogeneous big-bang scenarios [32].

Important spectroscopic informations for nuclear astrophysics have been obtained using (very) low-energy RNB (50-100 keV) as produced by Isotope Separators On Line. They have dealt with "waiting points" in the r-process around N = 50 and 82, more specifically $^{79}_{29}$Cu$_{50}$ and $^{130}_{48}$Cd$_{82}$, and have given indications on the r-process scenario [33]. The β^- decay of ^{16}N to levels of ^{16}O which are unbound for their decay into ^{12}C + ^{4}He has recently been studied; these experiments have given interesting informations on the cross section for the ^{12}C(α,γ)^{16}O reaction at very low energies, which is important in the nucleosynthesis of ^{12}C and ^{16}O in red giants [34,35].

3.3 High-Energy RNB for NA

Indirect determinations of the cross section for astrophysically important nuclear reactions can be carried out by Coulomb dissociation experiments [36]. These are based on the principle of detailed balance which states that the cross sections for inverse reactions such as A(b,γ)C and C(γ,b)A are simply related (for the same quantum states of A, b and C), due to the time-reversal invariance of the strong interaction. The source of photons for the $C(\gamma,b)A$ reaction is the virtual photon spectrum due to the Coulomb interaction of a high-energy ($\geq$ 50 MeV/A) RNB beam of the type C with a high-Z target, for example ^{208}Pb. The nucleus C thereby dissociates into the fragments A and b in the Coulomb field of the target. The advantages of this method over the direct determination of the $A(b,\gamma)C$ cross section are : higher cross sections (mb versus μb), and hence higher counting rates or need of lower intensities RNB ; the possibility of studying lower-energy resonances in the A + b system, which should be unaccessible to the direct method due to their very low cross sections. The limitations of the Coulomb dissociation method are the following. The reaction mechanism has sometimes been questioned, and the roles of the nuclear interaction and of postacceleration effects have to be taken into account. Only the transition between the ground states of A and C can be measured, and not those to the excited bound levels of C in the $A(b,\gamma)C$ reaction. A high resolution has to be achieved in the A + b relative energy spectrum, in order to resolve the $A_{g.s.}$ to $C_{g.s.}$ transition proceeding through the resonance of C under study, from the other resonances in C and from the transition to the bound excited levels of A.

The first Coulomb dissociation experiment of this kind has been performed on the ^{13}N(p,γ)^{14}O reaction, which has been discussed in Subsection 3.2, through the ^{14}O(γ,p)^{13}N reaction as studied with a ^{14}O beam. The above-mentioned limitations of the method can be overcome, since the ^{14}O first excited level has a long lifetime and is rather isolated from the other levels of ^{14}O, and since ^{13}N has no bound excited level. The results of two experiments, performed at RIKEN [37] and GANIL [38] , have yielded values of Γ_γ for the 0.545-MeV resonance in ^{13}N + p mentioned in Subsection 3.2 which are in very good agreement with the results of the direct measurement of the ^{13}N(p,γ)^{14}O reaction cross section.

A second example is the ^{11}C(p,γ)^{12}N reaction which has to do with the so-called hot p-p chain :

$$H(H,\beta^+)D(p,\gamma)^3He(\alpha,\gamma)^7Be \tag{5}$$

$$^7Be(\alpha,\gamma)^{11}C(p,\gamma)^{12}N(\beta^+)^{12}C \tag{6}$$

The first reaction and decay sequence above occurs in the sun, whereas the second could occur in hot and dense environments and would then lead to the nucleosynthesis of ^{12}C. The experiment on the ^{12}N(γ,p)^{11}C reaction has been performed at GANIL, with a 70.9 MeV/A ^{12}N beam of about 10^6 particles per sec sent on a 120 mg/cm^2 ^{208}Pb target [39]. The protons were recorded in 12 CsI(Tl) detectors, and the ^{12}C fragments, by a magnetic spectrometer. The results include the radiative widths Γ_γ of the 1.190 MeV level in ^{12}N (6 $\pm$ 4 meV) and of the 1.800 MeV level (70 $\pm$ 20 meV). Their astrophysical consequences still have to be worked out in detail, but point towards a weak contribution of the above-mentioned reaction sequence for the nucleosynthesis of ^{12}C as compared to the "traditional" triple-alpha process in red giants [26].

A third example is the ^{7}Be(p,γ)^{8}B reaction which is related to the solar neutrino problem. The nucleus ^{7}Be, produced as outlined above (5), can, in the sun, capture a proton and lead to ^{8}B. The decay of the latter nucleus yields most of the solar neutrinos which are observed in the "chlorine" experiments based on the $^{37}Cl(\nu_e,e^-)^{37}Ar$ reaction. A lowering of the $^7Be(p,\gamma)^8B$ reaction cross section would help resolving the observed deficit of these solar neutrinos. A recent experiment at RIKEN, using a ^8B radioactive beam, has yielded a cross section which is somewhat lower than the ones measured by previous direct measurements which did not always agree with each other [40]. The astrophysical consequences of this result are being worked out.

Spectroscopic informations related to nuclear astrophysics and obtained with high-energy RNB are very diverse. The lifetime of ^{65}As, as measured at MSU with a 75 MeV/A ^{78}Kr beam, shows that this nucleus is not the end point of the rp-process, which can accordingly proceed beyond A = 65 [41]. Some neutron-rich Cl, Ar and K isotopes closed to N = 28 have been investigated, lifetimes and decay modes, using a 60 MeV/A ^{48}Ca beam at GANIL ; their properties help to explain the observed low $^{46}Ca/^{48}Ca$ solar abundance ratio [5]. The decay scheme of ^{20}Mg to proton-unbound levels of ^{20}Na, studied at RIKEN [42], MSU [43] and GANIL [44] with high-energy RNB, has yielded information on the ^{20}Na levels above the $^{19}Ne + p$ threshold which are important for the $^{19}Ne(p,\gamma)^{20}Na$ reaction discussed in Subsection 3.2. The recent discovery, at the GSI-Darmstadt, of the bound-state β – decay [45], has confirmed the existence of this new decay mode of the nucleus, which has been postulated in some astrophysical scenarios.

3.4 Conclusions and perspectives

The results discussed in the present Section 3 give a flavour on the various types of nuclear astrophysical questions which can be tackled with radioactive nuclear beams ; they deal with primordial nucleosynthesis and type-II supernovae, the r-process and the rp-process, the cold and hot CNO cycles, novae and X-ray bursts, the nucleosynthesis of ^{12}C, ^{16}O, $^{46,48}Ca$, ... Many more astrophysical problems can be attacked with RNB. Their study will be made easier by the development of new RNB facilities with improved performances. At low energies, the Louvain-la-Neuve ARENAS[3] project will include a new accelerating cyclotron, specifically designed for the astrophysically interesting region (0.2 - 0.8 MeV/A) and with a higher accelerating efficiency [46] ; new facilities will come into operation, at the INS (Japan) [47] and ORNL (USA) [48] laboratories. Most high-energy RNB facilities, at MSU, RIKEN, GANIL, GSI, ... have programmes for increasing their intensities. The field of NA as studied with RNB is thus likely to grow during the next few years to come.

4 Nuclear Physics with Radioactive Nuclear Beams

One of the general topics in Nuclear Physics which can be studied with RNB is the investigation of the isospin degrees of freedom in nuclei. We recall that the isospin has been introduced in Nuclear Physics on the basis of the approximate charge independence of nuclear forces. This fact leads to the assignment of an isospin $t = 1/2$ to the nucleon, with a third component $t_z = +1/2$ for the neutron and $t_z = -1/2$ for the proton (i.e. the inverse convention as in Elementary Particle Physics). Nuclei with Z protons and N neutrons have isospins $T \geq T_z = \frac{N-Z}{2}$. If the internucleon forces were exactly charge independent, the same quantum states with the same isospin T in different isobaric nuclei with different T_z would be identical. One clear deviation from this invariance arises from the Coulomb force, which is obviously charge dependent and whose influence in nuclei cannot be neglected. Nuclear forces, on the other hand, are charge symmetric to a high degree of accuracy (i.e. are the same for the proton-proton and neutron-neutron cases) ; their charge independence is however less well satisfied (i.e. the proton-neutron interaction somewhat differs from the proton- proton and neutron-neutron ones in the same quantum states), due to a small difference (about 3 MeV) in the constituent-quark masses of the up and down quarks [49]. To study the effects, in nuclei, of these violations of the isospin symmetry in the nucleon-nucleon case, one has to compare the properties of nuclear levels in isobaric nuclei with the same T but different T_z, correct them for the "trivial" effects of the Coulomb force, and look for "residual" effects due to the isospin asymmetry in nuclear forces just outlined.

Stable beams do not represent ideal tools for such studies. For a given A, there is only one stable beam which can be produced if A is odd, and at most two (three for A = 40, 50, 26, 124, 130, 138, 176, 180) with quite different T_z if A is even. The availibility of RNB raises this limitation, and allows the acceleration of isobaric beams with a wide range of T_z. Examples are : mirror beams with T = 1/2, $T_z = \pm 1/2$ ($^{13}_{6}C_7$ and $^{13}_{7}N_6$, $^{19}_{10}Ne_9$ and $^{19}_{9}F_{10}$, ...) ; isobaric beams with T = 1, $T_z = \pm 1$ ($^{18}_{10}Ne_8$ and $^{18}_{8}O_{10}$, $^{30}_{14}Si_{16}$ and $^{30}_{16}S_{14}$ ...) ... Such beams will allow an extended study of the isospin degrees of freedom in nuclei.

RNB have been - and will probably still be - used for the study of nuclear structure [50]. Many new exotic nuclei have been discovered by the fragmentation method described in Subsection 2.2. As a consequence, the proton and neutron drip lines have been reached up to about A = 40, and many nuclei at the proton drip line are known up to about A = 65. Ground-state properties of these nuclei, halflifes, masses, decay modes, ... have been determined, mainly by the fragmentation method, but also by the ISOL method at low energy (near Isotope Separators On Line such as ISOLDE at CERN). Investigation of the excited states of these exotic nuclei, through Coulomb excitation, inelastic scattering, nucleon transfer reactions ... has just started using the fragmentation method, and will probably also be carried out with the ISOL method at medium energies. The high-spin properties of these nuclei, for example their isospin dependence and the search for predicted superdeformed and hyperdeformed bands, will be carried out by the ISOL method, using fusion-evaporation types of reactions. Finally, the production of the long-predicted superheavy nuclei above $Z = 110$ may become possible using neutron-rich RNB.

In the field of nuclear reactions [50] , RNB have allowed to discover and study a new and very interesting phenomenon in Nuclear Physics, the occurence of halo nuclei [51]. These are nuclei in which a core with "normal" size (i.e. whose radius is given by the usual $A^{1/3}$ law) is surrounded by a "cloud" of nucleon(s) with a rms radius much larger than the one of the core. One well-studied example is ^{11}Li, whose ^{9}Li core has a radius of about 2.5 fm, and is surrounded by the last two neutrons at distances of the order of 10 fm. Within the experimental evidences for these halo structures, one finds : large total interaction cross sections of halo nuclei with various targets as compared to neighbouring non-halo nuclei ; narrow distributions for the transverse momentum of the core, which imply large extensions of the halo nucleons through the uncertainty principle. The present situation in this field displays the identification of many neutron-halo light nuclei (^{8}He, ^{11}Li, ^{11}Be, ...) but very few, if any, proton-halo nuclei. RNB will also allow to study the isospin degrees of freedom in various nuclear reactions, i.e. : elastic scattering and the optical potential ; subbarrier fusion and the shape of the barrier ; deep inelastic scattering ...

Recent examples of nuclear structure studies with RNB are the followings. More than 50 new neutron-rich elements have been identified at the GSI-Darmstadt through an investigation of the fission of 750 MeV/A ^{238}U nuclei by a lead target, in a reverse kinematics scheme [52]. The proton-rich $^{100}_{50}$Sn$_{50}$ nucleus has been discovered at the GSI-Darmstadt [7] and GANIL Caen [0] by fragmentation of ^{124}Xe and ^{112}Sn beams, respectively. The reduced E2 transition probability B(E2) for the first 2^+ level in $^{56}_{28}$Ni$_{28}$ has been measured at the GSI-Darmstadt by proton inelastic scattering [53].

The role of isospin in nuclear reactions has been recently studied at Louvain-la-Neuve by comparing the ^{13}C + ^{12}C and ^{13}N + ^{12}C, ^{13}C + ^{13}C and ^{13}N + ^{13}C elastic scatterings at various energies [54].

5 Conclusions

The study of Nuclear Physics and Nuclear Astrophysics with Radioactive Nuclear Beams is a field of Science which is in a fast growing stage since about 5 years. The problems in these two disciplines which can be tackled with RNB are very numerous and very diverse. Several RNB facilities will come into work during the next few years, in Europe, Japan and the United States. Most existing ones have improvement plans to increase the RNB intensities presently available. One can thus foresee a continuous development of the whole field for the next decade.

References

1   P. Leleux, Compte-rendu de l'Ecole Joliot-Curie de Physique Nuclèaire 1993, IN2P3 (Paris).

2   J. Vervier, Proc. Fifth International Conference on Nucleus-Nucleus Collisions, Taormina, 1994 ; Nucl. Phys. (to be published).

3   G.D. Westfall et al., Phys. Rev. Lett. 43 (1979) 1859 ; C. Dètraz et al., Phys. Rev. C19 (1979) 164.

4   M. Langevin et al., Phys. Lett. 150B (1985) 71 ; D. Guerreau et al., Phys. Lett. 131B (1983) 293.

5   A. Sorlin et al., Phys. Rev. C47 (1993) 2941.

6   D. Guillemaud-Mueller et al., ref. 2.

7   G. Munzenberg et al., ref. 2.

8   K. Sümmerer, Proc. Int. Workshop Physics and Techniques of Secondary Nuclear Beams, Dourdan, 1992, J.F. Bruandet, B. Fernandez, M. Bex eds. (Frontières, Gif-sur-Yvette), p. 273.

9   D.E. Greiner et al., Phys. Rev. Lett. 35 (1975) 152.

10 A.S. Goldhaber and H.H. Heckmann, Ann. Rev. Nucl. Sc. 28 (1978) 161.

11 I. Tanihata, Hyperf. Int. 21 (1985) 251.

12 K. Asahi et al., Phys. Lett. 251B (1990) 488.

13 B.M. Sherrill, Proc. Sec. Int. Conf. Radioactive Nuclear Beams, Louvain-la-Neuve, 1991, Th. Delbar ed. (Adam Hilger, Bristol) p. 3.

14 A.C. Mueller and R. Anne, Nucl. Instr. Meth. Phys. Res. B56/57 (1991) 559.

15 F.D. Becchetti et al., Nucl. Instr. Meth. Phys. Res. B56/57 (1991) 554.

16 Sindano wa Kitwanga et al., Phys. Rev. C42 (1990) 748.

17 G. Rudstam, Zeits. Naturf. 21A (1966) 1027.

18 J.L. Belmont et al., ref. 8, p. 407.

19 H.L. Ravn and B.W. Allardyce, Treatise on Heavy Ion Science, vol. 8, D. Allan Bromley ed., p. 363.

20 D. Darquennes et al., Nucl. Instr. Meth. Phys. Res. B47 (1990) 311 ; P. Decrock et al., ibid B70 (1992) 182.

21 P. Van Duppen et al., ref. 8, p. 289.

22 T. Bjornstad et al., Phys. Scripta 34 (1986) 578.

23 R. Kirchner and E. Roeckl, Nucl. Instr. Meth. 133 (1976) 187.

24 P. Decrock et al., Nucl. Instr. Meth. Phys. Res. B58 (1991) 252.

25 K.S. Sharma et al., Nucl. Instr. Meth. B26 (1987) 362.

26 C.E. Rolfs and W.S. Rodney, Cauldrons in the Cosmos, The University of Chicago Press, 1988.

27 W. Fowler, Rev. Mod. Phys. 56 (1984) 149.

28 A.E. Champagne and M. Wiescher, Ann. Rev. Nucl. Part. Sci. 42 (1992) 58.

29 P. Decrock et al., Phys. Lett. B304 (1993) 50.

30 M. Arnould et al., Astron. Astrophys. 254 (1992) L9.

31 R. Page et al., Proc. Third Int. Conf. Radioactive Nuclear Beams, East Lansing, 1993, D.J. Morrissey ed. (Frontières, Gif-sur-Yvette), p. 489 ; R. Coszach et al., ibid, p. 323.

32 R.N. Boyd, ref. 31, p. 445.

33 K.L. Kratz et al., Z. Phys. A325 (1986) 489, A430 (1991) 419.

34 J.D. King et al., ref. 31, p. 483.

35 Z. Zhao et al., ref. 31, p. 513.

36   G. Baur et al., Nucl. Phys. A458 (1986) 188.
37   T. Motobayashi et al., Phys. Lett. B264 (1991) 258.
38   J. Kiener et al., Nucl. Phys. A552 (1993) 66.
39   A. Lefèbvre et al., ref. 31, p. 477.
40   T. Motobayashi et al., to be published.
41   J. Winger et al., ref. 31, p. 597.
42   S. Kubono et al., Phys. Rev. C46 (1992) 361.
43   J. Goerres et al., Phys. Rev. C46 (1992) R833.
44   A. Piechaczek et al., ref. 31, p. 495.
45   M. Jung et al., ref. 31, p. 291.
46   M. Loiselet et al., ref. 31, p. 179.
47   I. Katayama et al., ref. 31, p. 87.
48   J.D. Garrett, ref. 8, p. 311.
49   G.A. Miller et al., Phys. Rep. 194 (1990) 1.
50   European Radioactive Beam Facilities, NuPECC report (1993).
51   B. Jonsson, ref. 2.
52   P. Armbuster et al., GSI Scientific Report 1993, p.70.
53   G. Kraus et al., ref. 31, p. 365.
54   P. Liènard et al., ref. 2.

List of Participants

Prof. Dr. Sven Åberg
Department of Mathematical Physics
Lund Institute of Technology
S-220 07 Lund. Sweden
e-mail: sven.aberg@matfys.lth.se

Dr. Clara E. Alonso
Dept. Fís. Atóm., Mol. y Nuclear
Facultad de Física
Aptdo. 1065
41080 Sevilla. Spain
e-mail: alonso@cica.es

Dr. José E. Amaro Soriano
Departamento de Física Moderna
Universidad de Granada
18071 Granada. Spain.
e-mail: amaro@ugr.es

Dr. María V. Andrés
Dept. Fís. Atóm., Mol. y Nuclear
Facultad de Física
Aptdo. 1065
41080 Sevilla. Spain
e-mail: andres@cica.es

Dr. F. Mikael Björnberg
University of Helsinki
Deparment of Physics
Siltavuorenpenger 20 M
SF- 00014 Helsinki. Finland
e-mail: bjornberg@phcu.helsinki.fi

Dr. Fiorella Burgio
INFN Sezione di Catania
57 Corso Italia
I-95129 Catania. Italy
e-mail: burgio@catania.infn.it

Dr. Juan L. Aguado
Dept. Física Aplicada
Escuela Politécnica
Universidad de Huelva
Huelva. Spain

Dr. Luis Álvarez
Dept. Fís. Atóm., Mol. y Nuclear
Facultad de Física
Bloque C, 2a. planta
Burjassot. Valencia. Spain.

Dr. Luís Amoreira
Dept. de Física
Universidade da Beira Interior
P-6200 Covilhá. Portugal
e-mail: m_amoreira@alpha1.ubi.pt

Dr. José M. Arias
Dept. Fís. Atóm., Mol. y Nuclear
Facultad de Física
Aptdo. 1065
41080 Sevilla. Spain
e-mail: ariasc@cica.es

Dr. Lucilia P. Brito
Departamento de Física
Universidade de Coimbra
P-3000 Coimbra. Portugal
e-mail: lucilia@fteor1.uc.pt

Dr. Juan A. Caballero
Instituto de Estructura de la Materia
CSIC. Serrano 123
28006 Madrid. Spain.
e-mail: emjuanan@iem.csic.es

Prof. Dr. Xavier Campi
I.P.N. Division Physique Theorique
B.P. 1
91406 Orsay Cedex. France.
e-mail: campi@ipncls.in2p3.fr

Dr. José Caro
Departamento de Física Moderna
Facultad de Ciencias
Universidad de Granada
E-18071 Granada. Spain.
e-mail: jcaro@ugr.es

Dr. Sara Cruz-Barrios
Dept. Física Aplicada
Facultad de Física. Aptdo. 1065
41080 Sevilla. Spain
e-mail: sara@cica.es

Dr. Sasa Cvijetić
St. dom "Cvjetno naselje" 208/6
Odranska 8
HR-41128 Zagreb. Croatia
e-mail: sac@magnix.kfunigraz.ac.at

Prof. Dr. Carlos Dasso
Niels Bohr Institute
17 Blegdamsvej
DK-2100 Kobenhavn Ø
Denmark
e-mail: dasso@nbivax.nbi.dk

Dr. Brahim Elattari
Institut de Physique Nucleaire
91406 Orsay Cedex
France
e-mail: elattari@ipncls.in2p3.fr

Dr. Carlos Esebbag
Departamento de Matemáticas
Universidad de Alcalá de Henares
Aptdo. 20
28871 Alcalá de Henares, Madrid
Spain
e-mail: mtesebbag@alcala.es

Dr. José M. Espino
Dept. Fís. Atóm., Mol. y Nuclear
Facultad de Física
Aptdo. 1065
41080 Sevilla. Spain
e-mail: espino@cica.es

Prof. Dr. Bernard Frois
DAPNIA/SPHN
C.E. Saclay
91191 Gif-sur-Yvette
France
e-mail: frois@cernvm.cern.ch

Dr. M. Isabel Gallardo
Dept. Fís. Atóm., Mol. y Nuclear
Facultad de Física
Aptdo. 1065
41080 Sevilla. Spain
e-mail: gallardo@cica.es

Prof. Dr. J.D. Garrett
Oak Ridge National Lab.
Bldg 6000 MS 6371
P.O. Box 2008
Oak Ridge
TN 37831. U.S.A.
e-mail:
garrett@orph01.phy.ornl.gov

Dr. Wouter Geurts
Dept. of Physics and Astronomy
Free University
De Boelelaan 1081
1081 HV Amsterdam
The Neetherlands
e-mail: geurts@nat.vu.nl

Dr. Joaquín Gómez-Camacho
Dept. Fís. Atóm., Mol. y Nuclear
Facultad de Física
Aptdo. 1065
41080 Sevilla. Spain
e-mail: gomez@cica.es

Dr. Edoardo G. Lanza
INFN Sezione di Catania
57 Corso Italia
I-95129 Catania
Italy
e-mail: lanza@catania.infn.it

Prof. Dr. Manuel Lozano
Dept. Fís. Atóm., Mol. y Nuclear
Facultad de Física
Aptdo. 1065
41080 Sevilla. Spain
e-mail: lozano@cica.es

Dr. Ismael Martel
Dept. Fís. Atóm., Mol. y Nuclear
Facultad de Física
Aptdo. 1065
41080 Sevilla. Spain
e-mail: ismael@cica.es

Prof. Dr. Eulogio Oset
Dept. Fís. Atóm., Mol. y Nuclear
Facultad de Física
Bloque C, 2a. planta
Burjassot. Valencia. Spain.
e-mail: oset@evalvx.ific.uv.es

Prof. Dr. Alfredo Poves
Dept. de Física Teórica C-XI
Universidad Autónoma de Madrid
Cantoblanco
E-28049 Madrid. Spain.
e-mail: poves@vm1.sdi.uam.es

Dr. Enrique Gutiérrez de San Miguel
Dept. Física Aplicada
Escuela Politécnica
Universidad de Huelva
Huelva. Spain

Prof. Dr. Roberto Liotta
Manne Siegbahninstitutet of Physics
Frescativagen 24
S-10405 Stockholm
Sweden
e-mail: liotta@msi.sunet.se

Dr. Piotr Magierski
Institute of Physics
Warsaw University of Technology
ul. Koszykowa 75
PL-00-662 Warsaw. Poland
e-mail: magiersk@if.pw.edu.pl

Dr. Servan Misicu
Institute for Atomic Physics
Center fot Earth Physics
Bucharest-Magurele, P.O. BoX MG-6
Bucharest. Romania
e-mail: misicu@ifa.ro

Dr. Francisco Pérez-Bernal
Dept. Fís. Atóm., Mol. y Nuclear
Facultad de Física
Aptdo. 1065
41080 Sevilla. Spain
e-mail: fperez@cica.es

Dr. Weronika A. Ptóciennik
Institute of Physics
Warsaw University of Technology
ul. Koszykowa 75
PL-00-662 Warsaw. Poland
e-mail: plocien@if.pw.edu.pl

Dr. José M. Quesada
Dept. Fís. Atóm., Mol. y Nuclear
Facultad de Física
Aptdo. 1065
41080 Sevilla. Spain
e-mail: quesada@cica.es

Prof. Dr. Peter Ring
Physikdepartment der Technischen
Universität München
D-85747 Garching
Germany
e-mail:
ring@physik.tu-muenchen.de

Dr. José Rodríguez
Dept. Fís. Atóm., Mol. y Nuclear
Facultad de Física
Aptdo. 1065
41080 Sevilla. Spain
e-mail: jquinter@cica.es

Dr. Nicolae Sándulescu
Royal Institute of Technology
Physics Department Frescati
Frescativägen 24
S-10405 Stockholm
Sweden
e-mail: sandulescu@msi.sunet.se

Dr. Vladimir Spevak
Department of Nuclear Physics
School of Physics
Tel Aviv University
Tel Aviv 69978. Israel
e-mail: spevak@tauphy.tau.ac.il

Dr. Jordi Ventura
Departamento ECM
Facultat de Física
Universitat de Barcelona
Avda. Diagonal 647
08028 Barcelona. Spain.
e-mail: jordi@dirac.ecm.ub.es

Prof. Dr. Achim Richter
Institut für Kernphysik
Technische Hochschule Darmstadt
Schoβgartenstr. 9
64289 Darmstadt. Germany
e-mail:
richter@linac.ikp.physik.th-darmstadt.de

Dr. Ramón Rodrigo
Dept. Fís. Atóm., Mol. y Nuclear
Facultad de Física
Bloque C, 2a. planta
Burjassot. Valencia. Spain.
e-mail: bartual@evalux.uv.es

Dr. Francisco Salzedas
Centro de Fisica Teorica
Universidade de Coimbra
P-3000 Coimbra
Portugal
e-mail: salzedas@fteor.uc.pt

Prof. Dr. Marcos Saraceno
Departamento de Física Nuclear
Comisión Nacional de Energía Atómica
Av. del Libertador 8250
Buenos Aires
Argentina
e-mail: saraceno@tandar.edu.ar

Dr. Federico Vaca
Dept. Física Aplicada
Escuela Politécnica
Universidad de Huelva
Huelva. Spain

Prof. Dr. Jean Vervier
Institut de Physique Nuclèaire
Université Catholique de Louvain
B-1348 Louvain la Neuve
Belgium
e-mail: baras@fynu.ucl.ac.be

Dr. Xavier Viñas
Departamento ECM
Facultat de Fisica
Universitat de Barcelona
Avda. Diagonal 647
08028 Barcelona. Spain.
e-mail: xavier@ecm.ub.es

Dr. John N. Wilson
Department of Physics
Oliver Lodge Laboratory
University of Liverpool
P.O. BoX 147
Liverpool L69 3BX
U.K.
e-mail: jnw@ns.ph.liv.ac.uk

Prof. Dr. John D. Walecka
CEBAF Theory Group
12000 Jefferson Ave.
Newport News
Virginia 23606
U.S.A.
e-mail: walecka@cebafvax.bitnet

Lecture Notes in Physics

For information about Vols. 1–403
please contact your bookseller or Springer-Verlag

Vol. 404: R. Schmidt, H. O. Lutz, R. Dreizler (Eds.), Nuclear Physics Concepts in the Study of Atomic Cluster Physics. Proceedings, 1991. XVIII, 363 pages. 1992.

Vol. 405: W. Hollik, R. Rückl, J. Wess (Eds.), Phenomenological Aspects of Supersymmetry. VII, 329 pages. 1992.

Vol. 406: R. Kayser, T. Schramm, L. Nieser (Eds.), Gravitational Lenses. Proceedings, 1991. XXII, 399 pages. 1992.

Vol. 407: P. L. Smith, W. L. Wiese (Eds.), Atomic and Molecular Data for Space Astronomy. VII, 158 pages. 1992.

Vol. 408: V. J. Martínez, M. Portilla, D. Sàez (Eds.), New Insights into the Universe. Proceedings, 1991. XI, 298 pages. 1992.

Vol. 409: H. Gausterer, C. B. Lang (Eds.), Computational Methods in Field Theory. Proceedings, 1992. XII, 274 pages. 1992.

Vol. 410: J. Ehlers, G. Schäfer (Eds.), Relativistic Gravity Research. Proceedings, VIII, 409 pages. 1992.

Vol. 411: W. Dieter Heiss (Ed.), Chaos and Quantum Chaos. Proceedings, XIV, 330 pages. 1992.

Vol. 412: A. W. Clegg, G. E. Nedoluha (Eds.), Astrophysical Masers. Proceedings, 1992. XX, 480 pages. 1993.

Vol. 413: Aa. Sandqvist, T. P. Ray (Eds.); Central Activity in Galaxies. From Observational Data to Astrophysical Diagnostics. XIII, 235 pages. 1993.

Vol. 414: M. Napolitano, F. Sabetta (Eds.), Thirteenth International Conference on Numerical Methods in Fluid Dynamics. Proceedings, 1992. XIV, 541 pages. 1993.

Vol. 415: L. Garrido (Ed.), Complex Fluids. Proceedings, 1992. XIII, 413 pages. 1993.

Vol. 416: B. Baschek, G. Klare, J. Lequeux (Eds.), New Aspects of Magellanic Cloud Research. Proceedings, 1992. XIII, 494 pages. 1993.

Vol. 417: K. Goeke P. Kroll, H.-R. Petry (Eds.), Quark Cluster Dynamics. Proceedings, 1992. XI, 297 pages. 1993.

Vol. 418: J. van Paradijs, H. M. Maitzen (Eds.), Galactic High-Energy Astrophysics. XIII, 293 pages. 1993.

Vol. 419: K. H. Ploog, L. Tapfer (Eds.), Physics and Technology of Semiconductor Quantum Devices. Proceedings, 1992. VIII, 212 pages. 1993.

Vol. 420: F. Ehlotzky (Ed.), Fundamentals of Quantum Optics III. Proceedings, 1993. XII, 346 pages. 1993.

Vol. 421: H.-J. Röser, K. Meisenheimer (Eds.), Jets in Extragalactic Radio Sources. XX, 301 pages. 1993.

Vol. 422: L. Päivärinta, E. Somersalo (Eds.), Inverse Problems in Mathematical Physics. Proceedings, 1992. XVIII, 256 pages. 1993.

Vol. 423: F. J. Chinea, L. M. González-Romero (Eds.), Rotating Objects and Relativistic Physics. Proceedings, 1992. XII, 304 pages. 1993.

Vol. 424: G. F. Helminck (Ed.), Geometric and Quantum Aspects of Integrable Systems. Proceedings, 1992. IX, 224 pages. 1993.

Vol. 425: M. Dienes, M. Month, B. Strasser, S. Turner (Eds.), Frontiers of Particle Beams: Factories with $e^+ e^-$ Rings. Proceedings, 1992. IX, 414 pages. 1994.

Vol. 426: L. Mathelitsch, W. Plessas (Eds.), Substructures of Matter as Revealed with Electroweak Probes. Proceedings, 1993. XIV, 441 pages. 1994

Vol. 427: H. V. von Geramb (Ed.), Quantum Inversion Theory and Applications. Proceedings, 1993. VIII, 481 pages. 1994.

Vol. 428: U. G. Jørgensen (Ed.), Molecules in the Stellar Environment. Proceedings, 1993. VIII, 440 pages. 1994.

Vol. 429: J. L. Sanz, E. Martínez-González, L. Cayón (Eds.), Present and Future of the Cosmic Microwave Background. Proceedings, 1993. VIII, 233 pages. 1994.

Vol. 430: V. G. Gurzadyan, D. Pfenniger (Eds.), Ergodic Concepts in Stellar Dynamics. Proceedings, 1993. XVI, 302 pages. 1994.

Vol. 431: T. P. Ray, S. Beckwith (Eds.), Star Formation and Techniques in Infrared and mm-Wave Astronomy. Proceedings, 1992. XIV, 314 pages. 1994.

Vol. 432: G. Belvedere, M. Rodonò, G. M. Simnett (Eds.), Advances in Solar Physics. Proceedings, 1993. XVII, 335 pages. 1994.

Vol. 433: G. Contopoulos, N. Spyrou, L. Vlahos (Eds.), Galactic Dynamics and N-Body Simulations. Proceedings, 1993. XIV, 417 pages. 1994.

Vol. 434: J. Ehlers, H. Friedrich (Eds.), Canonical Gravity: From Classical to Quantum. Proceedings, 1993. X, 267 pages. 1994.

Vol. 435: E. Maruyama, H. Watanabe (Eds.), Physics and Industry. Proceedings, 1993. VII, 108 pages. 1994.

Vol. 436: A. Alekseev, A. Hietamäki, K. Huitu, A. Morozov, A. Niemi (Eds.), Integrable Models and Strings. Proceedings, 1993. VII, 280 pages. 1994.

Vol. 437: K. K. Bardhan, B. K. Chakrabarti, A. Hansen (Eds.), Non-Linearity and Breakdown in Soft Condensed Matter. Proceedings, 1993. XI, 340 pages. 1994.

Vol. 438: A. Pękalski (Ed.), Diffusion Processes: Experiment, Theory, Simulations. Proceedings, 1994. VIII, 312 pages. 1994.

Vol. 439: T. L. Wilson, K. J. Johnston (Eds.), The Structure and Content of Molecular Clouds. 25 Years of Molecular Radioastronomy. Proceedings, 1993. XIII, 308 pages. 1994.

Vol. 440: H. Latal, W. Schweiger (Eds.), Matter Under Extreme Conditions. Proceedings, 1994. IX, 243 pages. 1994.

Vol. 441: J. M. Arias, M. I. Gallardo, M. Lozano (Eds.), Response of the Nuclear System to External Forces. Proceedings, 1994, VIII, 293 pages. 1995.

New Series m: Monographs

Vol. m 1: H. Hora, Plasmas at High Temperature and Density. VIII, 442 pages. 1991.

Vol. m 2: P. Busch, P. J. Lahti, P. Mittelstaedt, The Quantum Theory of Measurement. XIII, 165 pages. 1991.

Vol. m 3: A. Heck, J. M. Perdang (Eds.), Applying Fractals in Astronomy. IX, 210 pages. 1991.

Vol. m 4: R. K. Zeytounian, Mécanique des fluides fondamentale. XV, 615 pages, 1991.

Vol. m 5: R. K. Zeytounian, Meteorological Fluid Dynamics. XI, 346 pages. 1991.

Vol. m 6: N. M. J. Woodhouse, Special Relativity. VIII, 86 pages. 1992.

Vol. m 7: G. Morandi, The Role of Topology in Classical and Quantum Physics. XIII, 239 pages. 1992.

Vol. m 8: D. Funaro, Polynomial Approximation of Differential Equations. X, 305 pages. 1992.

Vol. m 9: M. Namiki, Stochastic Quantization. X, 217 pages. 1992.

Vol. m 10: J. Hoppe, Lectures on Integrable Systems. VII, 111 pages. 1992.

Vol. m 11: A. D. Yaghjian, Relativistic Dynamics of a Charged Sphere. XII, 115 pages. 1992.

Vol. m 12: G. Esposito, Quantum Gravity, Quantum Cosmology and Lorentzian Geometries. Second Corrected and Enlarged Edition. XVIII, 349 pages. 1994.

Vol. m 13: M. Klein, A. Knauf, Classical Planar Scattering by Coulombic Potentials. V, 142 pages. 1992.

Vol. m 14: A. Lerda, Anyons. XI, 138 pages. 1992.

Vol. m 15: N. Peters, B. Rogg (Eds.), Reduced Kinetic Mechanisms for Applications in Combustion Systems. X, 360 pages. 1993.

Vol. m 16: P. Christe, M. Henkel, Introduction to Conformal Invariance and Its Applications to Critical Phenomena. XV, 260 pages. 1993.

Vol. m 17: M. Schoen, Computer Simulation of Condensed Phases in Complex Geometries. X, 136 pages. 1993.

Vol. m 18: H. Carmichael, An Open Systems Approach to Quantum Optics. X, 179 pages. 1993.

Vol. m 19: S. D. Bogan, M. K. Hinders, Interface Effects in Elastic Wave Scattering. XII, 182 pages. 1994.

Vol. m 20: E. Abdalla, M. C. B. Abdalla, D. Dalmazi, A. Zadra, 2D-Gravity in Non-Critical Strings. IX, 319 pages. 1994.

Vol. m 21: G. P. Berman, E. N. Bulgakov, D. D. Holm, Crossover-Time in Quantum Boson and Spin Systems. XI, 268 pages. 1994.

Vol. m 22: M.-O. Hongler, Chaotic and Stochastic Behaviour in Automatic Production Lines. V, 85 pages. 1994.

Vol. m 23: V. S. Viswanath, G. Müller, The Recursion Method. X, 259 pages. 1994.

Vol. m 24: A. Ern, V. Giovangigli, Multicomponent Transport Algorithms. XIV, 427 pages. 1994.

Vol. m 25: A. V. Bogdanov, G. V. Dubrovskiy, M. P. Krutikov, D. V. Kulginov, V. M. Strelchenya, Interaction of Gases with Surfaces. XIV, 132 pages. 1995.

Vol. m 26: M. Dineykhan, G. V. Efimov, G. Ganbold, S. N. Nedelko, Oscillator Representation in Quantum Physics. IX, 279 pages. 1995.

Vol. m 27: J. T. Ottesen, Infinite Dimensional Groups and Algebras in Quantum Physics. IX, 218 pages. 1995.